MAPPING TIME

E. G. Richards was formerly a Senior Lecturer in the Department of Biophysics at King's College, University of London. His interest in the calendar was sparked when he wrote and published computer programs for converting dates from one calendar to another. An historical note on the various calendars included in the exercise was intended to accompany the programs but as the author's appetite for knowledge about calendars grew, so did the note. It eventually became, after many years of research, this book.

Dr Richards and his wife live in London.

MAPPING TIME

The Calendar and its History

E. G. RICHARDS

OXFORD
UNIVERSITY PRESS

*This book has been printed digitally and produced in a standard specification
in order to ensure its continuing availability*

OXFORD
UNIVERSITY PRESS

Great Clarendon Street, Oxford OX2 6DP

Oxford University Press is a department of the University of Oxford.
It furthers the University's objective of excellence in research, scholarship,
and education by publishing worldwide in

Oxford New York

Auckland Cape Town Dar es Salaam Hong Kong Karachi
Kuala Lumpur Madrid Melbourne Mexico City Nairobi
New Delhi Shanghai Taipei Toronto
With offices in
Argentina Austria Brazil Chile Czech Republic France Greece
Guatemala Hungary Italy Japan South Korea Poland Portugal
Singapore Switzerland Thailand Turkey Ukraine Vietnam

Oxford is a registered trade mark of Oxford University Press
in the UK and in certain other countries

Published in the United States
by Oxford University Press Inc., New York

ISBN 978-0-19-286205-1

Preface

🝢

SEVERAL years ago, I had occasion to be writing computer programs for converting dates from one calendar to another. It seemed appropriate to include, with my programs, a note on the history of the calendars. This simple exercise whetted my appetite for more knowledge about calendars and my historical note grew longer and longer. Eventually it became, after years of research, this book.

There are several people without whose encouragement and help this book would either not have seen the light of day or would have been a poorer object; so, my thanks are due to my friends Emily Wilkinson, Walter Gratzer, Colin Liebenrood, and Denise Swallow, and to my wife, Dallas, for encouragement, for reading the manuscript, and for making such helpful comments. I am particularly indebted to Walter for searching the indexes of *Nature* covering nearly a century and providing me with photocopies of articles about the calendar.

I am grateful to the staff of the Cambridge University Library, the Science Museum Library in London, the London Library, the Libraries of University College, the School of Oriental and African Studies of London University, the Warburg Institute, and my local public library for helping me to locate so many dusty and out of print books.

Thanks are also due to Dr Graham Parlett of the Victoria and Albert Museum for help with some of the Indian philology; to Geoffrey Wilson for information about the French Revolutionary calendar; to Dr Yangxi Wang and Dr Francisco Real for providing me with the names of the days of the week in Chinese and Basque; to Maurice Lee for obtaining information from the Library of Congress; to Dr Skelton for help and advice about the Indian calendar; to Professor Thurston for information about early measurement of the length of the year and to Dr Anna Adinolfi and the Superintendent of Archaeology at Pompeii for tracking down FIG. 21.1.

Finally, thanks also to Michael Rodgers of Oxford University Press. Without his encouragement, this book might well have foundered.

This book is dedicated to my cats Henry and Freddie who so often sat beside me and purred soothingly as I wrote.

London   E. G. RICHARDS
June 1998

Contents

𝕀

List of tables ix
List of algorithms xii
List of illustrations xiii
Introduction xix

PART I

THE CALENDAR IN THEORY

1 The calendar 3
2 The astronomical background 17
3 Time and the clock 42
4 Writing and libraries 64
5 Numbers and arithmetic 72
6 The variety of calendars 89
7 The reform of the calendar 110

PART II

THE CALENDARS OF THE WORLD

8 Introduction 127
9 Prehistoric calendars 130
10 The calendars of Babylon and the Near East 145
11 The Egyptian calendar 150
12 The calendars of China and East Asia 161
13 The calendars of India 172
14 The Mayan and Aztec calendars 186
15 Four European calendars 196

16 The Roman and Julian calendars 206

17 The Jewish calendar 220

18 The Islamic and Bahá'i calendars 231

19 The Gregorian calendar 239

20 The French Republican calendar 257

21 The week 265

PART III

CALENDAR CONVERSIONS

22 Calendar conversions 287

23 Mathematical notes 292

24 To calculate the day of the week 299

25 The conversion of regular calendars 310

26 The Jewish calendar 326

27 The Mayan calendar 335

PART IV

EASTER

28 A short history of Easter 345

29 The date of Easter Sunday 354

30 A Book of Hours 379

APPENDICES

I Astronomical constants 387

II The names of the days of the week 391

III The names of the days of the year in the French Republican calendar 398

Glossary 401

Further reading 411

Index 429

Tables

🏛

2.1 The signs of the zodiac 27

2.2 The houses of the sun at the equinoxes and solstices at different periods 28

2.3 Determinations of the mean length of the tropical year as expressed in mean solar days 33

2.4 Determinations of the mean length of the synodic period of the moon expressed in mean solar days 35

2.5 The five planets known to the ancients 37

3.1 Divisions of time 45

6.1 The length in days of the lunation and astronomical year 93

6.2 The refinement of the lunisolar and solar year 98

6.3 Chronological cycles of years and days 105

6.4 Eras 107

7.1 The dedications of the months in the calendar of Auguste Comte 114

7.2 Some details of proposals for a new calendar 120

8.1 The stages of man's early history 127

8.2 Landmarks in the history of calendars 129

9.1 The divisions of the hypothetical Bronze Age year 143

10.1 The names of the Babylonian months 148

11.1 Periods and dynasties in the history of Egypt 151

11.2 The names of the months in calendars based on the Egyptian calendar 154

11.3 Nations which adopted the Egyptian or Alexandrian calendar 158

12.1 The ten heavenly stems 164

12.2 The twelve earthly branches (Zhi) 164

12.3 The 24 'fortnightly' periods 166

12.4 The length of the tropical year 168

13.1 The Sanskrit names of the signs of the zodiac 175

13.2 The lengths of the sidereal years assumed in the siddhântas 176

13.3 The seasons of the solar year 177

13.4 The names of the lunar months 178

13.5 Some eras used in India 184

14.1 The chronology of Mesoamerica 187

14.2 The Mayan and Aztec names of the 20-day periods 190

14.3 The Mayan and Aztec names of the veintena 191

14.4 The units used in specifying the long count 192

15.1 The months of the Celtic calendar 201

15.2 The principal festivals of the Celtic calendar 202

15.3 Quarter days in England and Scotland 202

16.1 The days in the months of three Roman calendars 212

16.2 The Roman Julian calendar 213

17.1 The names of the Babylonian and post-captivity Jewish months 222

17.2 Relevant factors in estimating the year of the creation from biblical data 225

17.3 The distribution of the days of the year among the months 229

18.1 The months of the Islamic calendar 232

18.2 The names of the early years of the era of the Hegira 234

18.3 The Bahá'i months and days 237

19.1 Dates on which some countries adopted the Gregorian calendar 248

19.2 Estimates of the length of the year available in the late sixteenth century 250

20.1 The names of the Republican months translated into other languages 262

21.1 The assignment of planets to the hours of the day 271

21.2 Planets and gods and their equivalents 280

21.3 The names of the planets in Sanskrit 283

22.1 Regular calendars 290

23.1 Days in the months 295

23.2 Embolismic years in the Metonic cycle 297

24.1 The numbers and letters of the days of the week 300

24.2 Relation between the position of a year in the solar cycle and its dominical letter 303

24.3 Regulars and concurrents 306

25.1 Parameters for the conversion of dates in regular calendars 311

25.2 Definition of terms 313

25.3 The starts of the first cycles 317

25.4 Parameters for calculating the Gregorian correction 320

25.5 The Roman months 323

26.1 Characters of years 332

26.2 The aggregate number of days preceding each month 333

29.1 The Julian lunar almanac 357

29.2 Distribution of full and hollow lunations among the 19-year cycle. 359

29.3 The Dionysian paschal table 362

29.4 The dates of the notional paschal new and full moons corresponding to various values of the epact 373

30.1 The number of days in the four seasons as indicated in *Les très riches heures* (TRH) and in modern ephemerides 382

1.1 The mean synodic period of the moon 389

1.2 The length of the mean tropical year 390

Algorithms

A   To calculate the day of the week for a date in the Julian calendar 307

B   To calculate the day of the week for a date in the Gregorian calendar 308

C   To calculate the day of the week for a date in the Julian calendar—shorter algorithm 309

D   To calculate the day of the week for a date in the Gregorian calendar—shorter algorithm 309

E   To calculate the day number of a date in a regular calendar 323

F   To calculate the date in a regular calendar of a day number 324

G   To calculate the day number of the first day of Tishri for any year in the Jewish calendar 330

H   To calculate the year in the Jewish calendar in which any day falls 331

I   To calculate the date in the Jewish calendar of any day number 334

J   To calculate the day number of any date in the Jewish calendar 334

K   To calculate the Mayan calendar round date of any day number 340

L   To calculate the day number of the calendar round date 341

M   To calculate the date of Easter Sunday according to the Dionysian canon 364

N   To calculate the date of Easter Sunday according to the Dionysian canon—shorter algorithm 364

O   To calculate the date of Easter Sunday according to the Gregorian canon 375

P   To calculate the date of Easter Sunday according to the Gregorian canon—shorter algorithm 376

Illustrations

1.1   A string calendar from Sumatra. A tally of the days of a month is made by threading a string through the 30 holes in turn. 4

1.2   Part of a Scandinavian calendar. The rule is marked off in days, seven to a week. Other symbols represent festivals. 4

1.3   Nostradamus (1505–66). 13

2.1   The observer and his horizon: A Altitude; B Azimuth; C Zenith distance. 18

2.2   A long-time exposure directed to the north celestial pole. 19

2.3   The celestial sphere: V Vernal equinox; first point of Aries; S Summer solstice; A Autumnal equinox; first point of Libra; W Winter solstice. 20

2.4   The movement of the stars in the course of a night is different at different latitudes. 21

2.5   Declination and right ascension: A First point of Aries; L First point of Libra; D Declination; R Right ascension; N North polar distance. 23

2.6   Orbits of the earth and moon. 36

2.7   The surviving part of the meridian arc of the quadrant of Ulugh Beg (1394–1449), at Samarkand, which was about 40 metres radius, set in the meridian. 39

2.8   The gnomon of Guo Shoujing (1231–1316) at Dengfeng. The position of the shadow of the crossbar—which was about 9 metres above the horizontal scale which was in the meridian—could be read with fair accuracy and used to determine the day when the shadow was shortest. 40

3.1   An Egyptian shadow clock. The shadow of the bar reached mark A two 'hours' after sunrise, and marks B, C, and D at hourly intervals thereafter. At noon when the sun was overhead, the clock was reversed to point to the west. 47

3.2   Telling the time in Borneo. Many people could estimate the time of day from the position and length of the shadow of a vertical stick used as a gnomon. 49

3.3   A Roman sundial from Utica, Tunisia. 50

3.4   A hand dial. A stick held in the open hand and orientated towards the pole acts as a simple sundial and casts a shadow on the fingers which can be used to tell the time. 51

3.5   A portable sundial described by Chaucer. The hole on the circumference casts a shadow on the scale inside the ring, which indicates the time of day. It could be adjusted both for the season and also the latitude of the user. 52

3.6   Sixteenth-century nocturnal. With the nocturnal held vertically at arm's length, the pole star is viewed through the central hole. The radial bar is then turned till it is aligned on the pointer stars of the Great Bear (or Dipper). The time of night can then be read from the scale after adjusting it for the date. 53

3.7   The development of the clepsydra. The chronological scale runs from 500 BC to 100 BC. 54

3.8   A seventeenth-century hourglass, attached to the pulpit of a church and used to time sermons. 54

3.9   A Chinese incense seal carved from wood in about AD 1329. The single continuous groove was filled with incense of different aromas; lit at one end, the incense burnt for maybe 12 hours. The time could be determined from the odour of the incense or the position of the area of burning. 55

3.10   The Tower of the Winds in the Agora at Athens, built by Andronicus of Kyrrhos in about 50 BC. Originally there were sundials on the eight walls and a wind vane on top. 56

3.11   Su Sung's water clock, built in AD 1090. The rotation of the water-wheel was controlled by a mechanism which released the wheel when a bucket was full. The time was indicated by the puppets, and the armillary sphere at the top moved so that a star could be kept in view as the earth rotated—as in modern astronomical telescopes. 57

3.12   The accuracy of clocks improved in western Europe from about 15 minutes error a day in the fourteenth century to a minute fraction of a second in the twentieth century. 59

4.1   The Rosetta stone was discovered by Pierre Bouchard, an officer in Napoleon's expedition to Egypt in 1798. It contains a decree by Ptolemy V of 196 BC which is written in Egyptian heiroglyphs, a demotic script, and Greek. It enabled Champollion to decipher ancient Egyptian writing. 68

4.2   Jean-François Champollion (1790–1832). He first deciphered Egyptian heiroglyphs. 69

5.1   A medieval scheme for communicating numbers with the fingers. 73

5.2   An Inca guipu used as a tally of various objects. Clusters of knots denoted the digits of numbers. Different strings, maybe of different colours, tallied different objects. 74

5.3   Some numbers as written by the Maya: a dot stands for one, a bar for five, and a lozenge for zero. Numbers are written vertically in a vigesimal system (base 20). 77

5.4   A Babylonian mathematical text with cuneiform numbers. 78

6.1   Most things were in short supply in Britain in 1944. 103

7.1   Auguste Comte (1798–1857). 115

8.1   The first civilizations of the Western world. Civilizations arose in Mesoamerica, but much later. 128

9.1   Bone plaque from Abri Blanchard, Dordogne. This late palaeolithic piece of bone was inscribed some 30 000 years ago. Marshack supposes that the marks correspond to the days of two or more lunations and are to be read in the order indicated. 131

9.2   Marshack's interpretation of the Blanchard plaque. 133

9.3   The Winnebago calendar stick. This stick, of square cross-section, about 132 cm long, was made in 1825 by Winnebago chief Tshi-zun-hau-kau. Each side of the stick bears marks representing the days in six months; each month contains 30 or 29 days. The days of invisibility and the last crescent are marked for each month, which is divided generally into three 10-day periods. The top photograph shows the ninth month of the first year; the lower diagram clarifies the salient features. 134

9.4   Rock carving from Monterrey, Mexico, from about 2500 BC. It is believed that the marks represent days which are divided into periods which match the lunar phases. 135

9.5   Stonehenge from the air. 137

9.6   Alignments at Stonehenge. Various stones (which are numbered in the diagram) are supposed to be aligned on various solar and lunar events. 139

9.7   New Grange. An elevation showing the path of the rays of the sun at sunrise on the summer solstice. They pass through the roof box on to the stone at the rear of the passage. 141

9.8 The war memorial at Melbourne. The rays of the sun shine through a hole in the roof on to the remembrance stone at the 11th hour of the 11th day of the 11th month. 11 November is the day dedicated to the remembrance of those who fell in the first World War. 142

9.9 The position of sunrise at a Hopi village. The village elders could tell for which days to organize festivals by noting the position of the setting sun on the mountainous horizon. 144

13.1 (a) The intercalation of a lunar month in the lunar calendar. Note how both Vaishaknas start after Mesha, so Adhika-Vaishakha has been intercalated. (b) The extracalation of a lunar month in the lunar calendar. Note how no lunation starts between Dhanus and Makara, so that Paulus, which normally comes before magha, has been extracalated. 180

14.1 Glyphs representing the 18 uinals and uayeb of the Mayan calendar. 189

14.2 A Mayan stela from Guatemala. It bears the long count date 9.16.15.0.0 (or 18 February AD 766). 194

15.1 The Coligny calendar. A reconstruction of the 153 surviving fragments of the bronze tablet. It is possible to discern columns of small holes into which pegs may have been inserted to mark the passage of the days. 200

16.1 Part of a calendar from the Roman republic; reconstruction of the shattered stone tablet. 209

16.2 Julius Caesar (107–44 BC). Courtesy of the Vatican Museum. 214

16.3 'Thirty days hath Tacitus . . .' what our children might have had to learn had the renaming of the Roman months by later emperors stayed with us. 217

19.1 Pope Gregory XIII (Ugo Boncampagni), 1572–1585. St Peter's Basilica, Rome. 242

19.2 Aluise Baldassar Lilio (1510–76), originator of the Gregorian calendar. 243

19.3 Christopher Clavius (1537–1612), who wrote the definitive book on the Gregorian calendar. 245

19.4 The papal bull; the title page published on 24 February 1582. 246

19.5 A medal commemorating the new calendar, struck by Pope Gregory XIII. 247

19.6 Philip Stanhope, the Earl of Chesterfield (1694–1773). He guided the bill to reform the English calendar through Parliament. 254

19.7 'Humours of an election entertainment' published by Hogarth in February 1755. The banner on the floor carries the inscription 'Give us back our eleven days'. 255

20.1 Pierre Sylvain Maréchal (1750–1803). 258

20.2 Philip François Nazaire Fabre d'Eglantine (1750–94), author of the names of the months and the days of the year in the Republican calendar. 259

20.3 A decimal watch (1795) manufactured in Condom. Separate hands show the time according to the 10-hour republican day and the 24-hour traditional day. 260

20.4 A five-franc piece struck by Napoleon in 13 ER. 262

21.1 Roman graffiti from Pompeii. This shows, on the left, abbreviations for the names of the planets in the order of the days of the week. The inscriptions suggest that they were part of a school lesson on the Roman calendar. 270

21.2 Baron Samedi, head spirit of the Gédé and counterpart of St Expedit. Painting by Edward Duval Carrié. 281

22.1 The identification of days. 289

29.1 The Easter canon of Hyppolytus. This was discovered in 1551 and is now in the Vatican Library. 374

30.1 September from *Les très riches heures*. 381

Introduction

THE calendar is an abstract thing and our knowledge of ancient calendars is partial and incomplete, limited as it is to the written traces that survive. These consist, in part, of dates incised on tombs and memorials or written on parchment, the bark of trees, paper, and the like. They also exist in the works of history—sometimes ambiguous or confused—written in the ancient world itself. On occasions classical writers quote earlier authorities, and it may be suspected that some of their statements are based on legends confused by 'chinese whispers'.

From the Renaissance onwards, we have an increasing number of works about the calendar, written by scholars who decipher, describe, and analyse calendars—ancient and not so ancient. Such scholars are now from several disciplines: classical scholarship, archaeology, ethnography, philology, history, astronomy and the history of astronomy, mathematics, and no doubt others. There was a flood of such scholarly works in the late nineteenth century, many by erudite German scholars.

Today, the number of manuscripts, books, and articles in learned journals that bear on the calendar is immense and to read them all would require several lifetimes and a working knowledge of the major European languages as well as Latin, Greek, Hebrew, and Sanskrit—to say nothing of Babylonian cuneiform scripts, Egyptian hieroglyphs, Chinese and other Oriental languages.

I have been forced to restrict my reading to a tiny selection written in the languages I comprehend. Thus, the book is in no sense a work of original scholarship; it is a wholly derivative work intended for the general reader, and there are no footnotes, references to learned journals, or quotations in other languages.

It would be naive to take everything that has been written about the calendar at face value. Aristotle is renowned for repeating far-fetched travellers' tales. In more recent times, a certain type of historian (if that is what we should call him) has built fanciful histories in which conjecture is piled upon conjecture. Descriptions of calendars, that vary from the soberly documented to the wildly fantastic, can be found in books. It is not always easy to untangle the truth but I have tried to eliminate from my consideration some of the 'wilder' authors, to cull information only from reputable sources, and to avoid 'moon beams from the larger lunacy'. Sometimes, however, I have mentioned theories or opinions that may be inaccurate or even downright wrong but which are interesting, widely held, or worth repeating. I have tried to qualify these with suitable remarks.

The book is in four parts. Part I (Chapters 1 to 7) deals with general aspects of the calendar in the abstract. Chapter 1 introduces calendars and their uses. Chapter 2 provides a background to the astronomy which underlies all calendars; the calendar is based on attempts to reckon time by the cycles displayed in the behaviour of the sun and the moon. It is therefore appropriate to describe, in this chapter, the basic astronomical facts as they appear to people ignorant of our sophisticated modern astronomical explanations. In Chapter 3, I digress to describe briefly the history of the measurement of time, particularly as this impinges on the calendar. The existence of a calendar requires some ability to count and to write; I therefore digress further in Chapters 4 and 5 to present a brief history of writing and of numbers and arithmetic. In Chapter 6, I return to the calendar and discuss some of the principles involved in a variety of calendars and how the years are counted. Finally, in Chapter 7, I discuss some reforms that have taken place or been proposed.

In Part II, Chapter 8 provides a brief introduction to the calendars described. In Chapters 9 to 20 I describe several calendars, both ancient and modern, that have been or are used throughout the world, while Chapter 21 deals with the week—that artificial but ubiquitous period of time. This second part could have been expanded to several times its current length and I have selected examples of calendars that are known in some detail or are otherwise important.

In Part III, I delve into the mathematics of calendars and present a number of algorithms for calculating the day of the week and for converting dates in one calendar to those in another. These can be used with a pocket calculator, or they can be programmed into a computer in whatever language you prefer. Alternatively, of course, you can use them with a pencil and paper and a little more effort. I have attempted to explain them as clearly as I know how, and the mathematics required does not go beyond elementary arithmetic and algebra. Any difficulty that may arise most probably emanates, not from the mathematics, but in the logic of converting a verbal description of how a calendar works into a simple formula.

Part IV is devoted to Easter. After a discussion of the history of Easter and the extraordinary difficulties experienced by the Christian Church in agreeing on a method of deciding when this festival should be held, I go on to describe algorithms for calculating the date of Easter Sunday. Finally, I discuss briefly *Les très riches heures*—a medieval Book of Hours containing interesting calendrical information including Easter.

There are three appendices at the end of the book, a glossary, and a list of sources and books where further information can be found. I have thought to provide brief descriptions of what you will find in the books and learned papers mentioned. Be warned though, it represents only a tiny fraction of what has been written. Calendars have their own specialized vocabulary, as does astronomy. Many of the relevant terms used in the book are defined in the glossary. If you come across a term in the body of the text which is unfamiliar to you, try the glossary, it may be there.

There is one matter that I have deliberately avoided. Undoubtedly, one of the most important reasons for having a calendar was, at any rate for ancient peoples, to establish the dates for various religious practices—rites, feasts, fasts, and sacrifices. Only very rarely do I discuss what went on during these activities or why they were held to be important. It would be impossible to describe them in any but the most perfunctory way, except in a work many times the size of this one.

My intention was to write a book to entertain and possibly instruct the general reader. I do this partly by remembering that the calendar is one of the works of man so, wherever it is possible, I have tried, however inadequately, to bring the protagonists, or at least the societies in which they lived, to some sort of life. I have thus included anecdotes and verses as appropriate. Most learned papers on the calendar are written by scholars for scholars in a terse style which assumes many years of study of the subject. I have tried to be more discursive and, I hope, readable. Only you can say if I have been successful.

Much of the material in this book is historical in nature. I am reminded of the remark by A. W. Crosby that 'writing a history book without making mistakes is like living a life without sin; worth the effort but impossible'. Any historical mistakes are my responsibility, as are the simplifications. I regret the former and hope that professional historians will forgive the latter.

Several of the conversions described in Part III, and others besides, have been implemented in a computer program. Versions for the PC and for the Atari ST(E) and similar computers are available for a small charge. Further details and a list of any misprints found in this book may be obtained from the author's web site: http://www.users.zetnet.co.uk/egrichards/

PART I

THE CALENDAR IN THEORY

CHAPTER I

The Calendar

God made the days and nights but man made the Calendar
ANON.

⚉ Calendars and clocks

Some have likened the calendar to a clock; this is, of course, a mistake. A calendar hangs by my desk but there is no way I can tell the date just by looking at it. The calendar is more abstract; it is a systematic way of naming the days by allocating each to a year and a month and maybe a week. It enables us to label the days in the past and in the future and to arrange them all in order. If we know the date of two events we know which was the earlier and which the later; the calendar enables us to make or impose unambiguous commitments for the future; without a calendar, a diary would be a muddle. In another meaning of the word, a calendar is an almanac, a program of future events or a record of past events, each assigned to a day or a year.

The calendar is thus a human invention. If I want to know the date I may refer to a newspaper or the radio or look at my calendar; I could ask a friend. If I were on my own however like Robinson Crusoe, I would have to remember the date on which I was marooned and note the days as they passed, maybe marking them with notches on a stick. If I were to lose count, I would have extreme difficulty in recovering the date. It could only be done if I remembered the year and knew of the precise date of some future astronomical event such as an equinox or an eclipse that I might be able to observe. If, like Rip Van Winkle, I had slept for many years and lost track of the year, there would be no way I could ever recover the precise date; I would be cast adrift in time with no land in sight.

All calendars are based on the succession of days and nights punctuated by the waxing and waning of the moon or the rhythm of the seasons and the movement of the sun. In the beginning, our remote ancestors had no traditional knowledge of the regularity of the motions of the sun and the moon to guide them. Only after they had learnt to count and do simple arithmetic, and

FIG. 1.1 A string calendar from Sumatra. A tally of the days of a month is made by threading a string through the 30 holes in turn.

FIG. 1.2 Part of a Scandinavian calendar. The rule is marked off in days, seven to week. Other symbols represent festivals.
© Museum of the History of Science, Oxford.

after many nights of careful observation of the heavens, did the calendar begin to take shape. Before that, they had no calendar to help them plan ahead or organize their experience of time—just a succession of days and moons and seasons.

Any picture attached to the calendar is there purely for decoration, but reference to the printed form tells us the day of the week of a particular day or how many days are left till Christmas. The calendar in the abstract has many other significant functions.

𝕏 The uses of the calendar

Our ancestors were already recognizably like ourselves some 10 000 generations back. At first they lived by searching out edible berries, roots, nuts—whatever was edible; by scavenging and by eating such game as they could catch. They hunted and they gathered. They spread across the globe, and as they travelled they had to learn when and where to find the local food; they had to follow the game and maybe migrate themselves to avoid the winter cold. Any signs that they could discern of the timetable of natural events was crucial to their survival. They may well have watched the sun, the moon, and the stars for clues.

Some 10 000 years ago, after the glaciers had receded for the last time, people learnt to grow crops and to domesticate animals. As food became more plentiful, they became even more dependent on the seasons, for they had to know when to sow and when to reap their crops.

About 4000 years later, food had become sufficiently plentiful to support cities in the great fertile river valleys of the world. Standing armies were needed if only for protection from marauding barbarians, and great civil engineering projects were carried out. To feed the soldiers and the workers, the peasants were taxed, and to do this properly, dated records had to be kept; bureaucrats, especially, need a calendar. The rich cities attracted conquering barbarians who in turn adopted the way of life they found. Empires were created and trade started between the centres of the ancient world. Money was lent and bargains agreed; records of these agreements needed to be signed and dated. Civilization was well on its way and with it came a pressing need for a calendar.

Emperors and kings kept their power and privileges by ensuring that there was enough food and that marauders from beyond the city walls were kept at bay. One way of doing this was to intercede with the 'unseen powers behind the scenes'—the gods. These had to be placated with sacrifices and rituals and there were proper days for these activities. The calendar came to be organized by the priests.

The intimate links between the behaviour of the heavenly bodies, the seasons, and the calendar maybe suggested that the sun and the moon and the planets were themselves associated with gods, perhaps even were gods. Thus the uneasy alliance between religion and astrology came about. It became necessary to watch the heavens for omens and only undertake important activities on days when these were good. An adequate calendar is essential for these prognostications. In time astrology gave rise to astronomy as we understand it today.

Today each of the major religions has its own calendar which is used to programme its religious ceremonies, and it is almost as true to say that each calendar has its religion. The Christians, the Moslems, the Jews, the Buddhists, the Jains, the Hindus, the Zoroastrians, and, more recently, the adherents of Bahá'í, all have their calendars. There are several calendars used in Christian communities: the Gregorian calendar of Western christianity; the Julian calendar of the Eastern church; the calendars of the ancient Coptic and Ethiopian churches. More recently calendars have been invented for anti-religious purposes, and others in honour of Mammon.

Sometimes two calendars were used simultaneously, one for religious purposes and another with more secular uses, for example, for agriculture, taxation, and trade. The Egyptians operated a solar calendar side by side with a religious lunar calendar. The Christian countries still do this by using the Gregorian calendar for secular purposes and the lunar ecclesiastical calendar to fix the date of Easter. Jews and Moslems similarly use both the Gregorian calendar and their own religious calendar. At one stage the Ottoman Empire employed a financial calendar, based on the Julian calendar, as well as the Islamic calendar. Today international trade would be almost impossible without the near universal use of some calendar; this happens to be the Gregorian calendar.

Today, if there were no calendar, then examinations, concerts, races, parties, football matches, to mention just a few activities, could not be arranged for more than a few days ahead with any expectation that the participants would all turn up on the designated day. The compilers of airline and shipping schedules and railway timetables would have real difficulties. The lawyer can argue that if the suspect was dead on the fourteenth day, he could not have committed the crime on the fifteenth; the astronomer must know not only where to point his telescope but when; the farmer must know by which date he must plant his potatoes and on which day the market is held to sell them; the surgeon and patient must meet on the same day; to the historian, the calendar is all important for ordering and making sense of the past. The list goes on and on.

All manner of things from the ancient and modern world have the date written on them. Among the most durable are coins and clay tablets and engravings on stones used as gravestones, cenotaphs, dedications, and commemorations. The less durable items, which have nevertheless survived when protected from the elements, include papyrus, strips of bamboo, parchment, and paper. Even computer files on magnetic tape and disk have dates, though how durable these will turn out to be remains to be seen.

The first coins to be dated were minted in Syria from about 312 BC; these were stamped with the year of the Seleucid era when they were minted. Later coins from a variety of places were dated with the year of the reign of the monarch whose head they bear. Nearly all Islamic coins carry dates, but the practice was slow to be adopted in the Christian world; the first English coin to be dated was the gold half sovereign of Edward VI minted in 1548. Edward had reformed the coinage which had been debased by Henry VIII whose treasury was exhausted.

𝕀𝕀 Days, months, and years

The fundamental unit of time used in nearly all calendars is the day—the period of rotation of the earth about its axis or, from our human viewpoint, the time taken by the sun to complete its journey and return to the same point in the heavens. The passage of the days is marked, except in the polar regions, by the alternation of night and day; these are the ticks of the celestial clock. The day is an interval of time deeply rooted in our biology and that of other creatures, we need to sleep every day, and other organisms have similar diurnal rhythms manifest in their physiology and behaviour.

We may count the days as they pass but as the numbers mount we might well lose count; it is possible to make and keep an appointment for three days' time, but the chances of two people meeting in 537 days' time without the aid of a calendar are slim. Larger units of time are required. The calendar is a scheme for grouping the days into longer units, the months, and grouping the months into years. Sometimes groupings smaller than a month, such as our week, are used.

Months are based on the moon, the pale lantern of the night, which waxes and wanes with convenient regularity. The period of a lunation, the cycle of the moon as it moves in its orbit about the earth, is close to the period of the human menstrual cycle. Evocations of rebirth and fecundity, and the provision of light invest the moon with a deep emotional significance. Lovers delight in it, poets muse on it, dogs howl at it, and lunatics madly celebrate when it is full. Full moon is the time to make journeys at night or hold long celebrations in its light.

The cycle of the seasons gives us the year. Before the astronomical signs were properly understood, people learnt to foretell the coming of the seasons by a myriad of natural signs: the buds on the trees in spring, swallows in the summer, the yellow falling leaves of autumn, the ice of winter. The Greek poet, Hesiod, recorded these and many other natural events in his poetic *Works and days*, some 700 years before Christ. In Egypt there was the annual flooding of the Nile to tell the farmers when to sow their crops. It may be that the passage of the seasons is embedded in our instincts: we revel in the joys of spring and feel the gloom of winter when (at any rate until recently) disease and cold carries off our loved ones; even the sun seems to be dying.

These natural signs can be unreliable—misread them and a famine might ensue. It was therefore important to discover better methods of determining the season of the year. The year represents the time taken by the earth to revolve around the sun in its orbit, or as the ancients saw it, for the sun to move in the full circle of the zodiac against the backdrop of stars; but there was no clear start to the year and no obvious way to determine its length until people learnt to recognize the solstices and the equinoxes.

Some societies went to prodigious lengths to establish the length of the year. The building of Stonehenge began in about 3100 BC, possibly earlier. Some believe that one purpose behind this effort was to provide a means of establishing the day of the summer solstice. On this day, the sun rises at a particular point on the horizon as viewed from the centre of the stone circle. At that point the builders may have started counting the days of the year. There are many other megalithic structures in Western Europe which may have been erected for similar purposes. Others learnt to observe the heliacal risings of the stars which occur when they first show above the horizon at dawn, just ahead of the sun; the early Egyptians once began their years when the star Sirius made its heliacal rising. Some noted the day on which the noonday shadows were shortest; and others, when the sun was directly overhead.

It was now possible to reset the count of lunations each time a new year started so that, just as the days cycled in the months, the months cycled in the year. There are few natural cycles longer than a year that were apparent to the ancients, and various schemes were invented for numbering the years themselves. Some nations started the numbering anew when a new king was enthroned; others invented artificial cycles of years, some of great length. Yet others, like ourselves, count the years from a notable event such as the birth of Christ.

Meanwhile the great problem of calendar design became apparent: the counting of days in a lunation sometimes yielded 29 and sometimes 30; the count of days in the year gave, once the start was adequately defined, sometimes 365 and sometimes 366; and there were 12 new moons or sometimes 13 in each year. Twelve lunations accounted for some 354 days with a disconcerting 11 days or so left over. It was apparent that there was not a whole number of days in a lunation; neither was there a whole number of days or lunations in a year—the days, lunations, and years are incommensurate.

⌛ Anniversaries

The need to celebrate anniversaries seems to be universal. Silver jubilees are celebrated with parades, fireworks, and special postage stamps. Relatives come from distant places to celebrate golden weddings. But why make a twenty-fifth or fiftieth anniversary an occasion, but not a twenty-seventh or forty-first? No rational reason can be given; it is a matter of convention and a preference for round numbers.

Every child looks forward to presents on its birthday every year, but we may ask what is special about an anniversary; why not some multiple of, 297 days, rather than $365\frac{1}{4}$ or so (and that only on average)? A contrived and naive explanation might be that after a year, the earth—and the child—has returned to the same point in space. But of course that is specious if not meaningless; the solar system drifts through space relative to the stars and in that odd quarter of a day's uncertainty in the time, the earth moves some 400 000 miles in its orbit. The only sensible answer is that anniversaries betray a predilection for the completion of a calendrical cycle, the year, however many days that might be. In the Islamic world anniversaries occur more frequently since the year is shorter. Jewish anniversaries occur irregularly, and even Christian ones are not observed after constant intervals of time.

Most societies have celebrated rites of passage at the onset of puberty and of adulthood. These lingered on in the Christian world and at the age of 21, men, at any rate, were given the 'key of the door' and eventually the vote. The number 21 is possibly related to the mystic properties of the number seven.

As well as yearly celebrations, there are weekly rites. The Jews, and later the Christians and the Moslems, worship regularly, every week. This period may have its origins as an approximation, but not a very good one, to a quarter of a lunation; it is most probably pure convention, for there is nothing in nature that lasts just seven days. Although the period of these yearly or weekly celebrations is conventional, do not dismiss them as mere superstition or the like. They give pleasure and satisfaction, and what more can one ask of a practice than that?

⌛ The millennium

I am aware that soon the world will celebrate the start of a new millennium (from the Latin *mille annum* or 'a thousand years')—or at any rate the Christian world will; people using other calendars will presumably pass it by. The only question is when will the celebrations take place? Surely most will answer on 1 January of the year AD 2000, maybe forgetting that this date is only 1999 years from the start of the Christian era which began on 1 January of the year AD 1—there being no year 0. However, 1999 is hardly a round number. Maybe the proper answer should be 1 January 2001; this is exactly 2000 years from the start—but 2001 is definitely not round. No doubt the world will celebrate on 1 January 2000, itself a nice round date, and ignore these quibbles.

Beware of basing your answer to this conundrum on the notion that we are about to celebrate the two thousandth anniversary of the birth of Christ. To begin with, the available evidence suggests he was born in 4 BC, or even earlier, so this anniversary passed off in 1997 or before. Next we must note that the exact date is problematic; even if we take 25 December as his official birthday, we need to consider whether this is to be reckoned according to the Julian or Gregorian calendar. Two thousand years on, according to Julian reckoning, from 25 December 4 BC works out as 7 January 1998 in our Gregorian calendar.

Similar controversies have dogged the celebration of the turn of the century, at least from 1699/1700. Stephan Jay Gould has noted that the establishment (whom we might call the 'pedants') has nearly always preferred the later date for such celebrations, whereas the populace (the 'vulgar') has preferred the earlier. This introduction of the class struggle into the debate is perhaps irrelevant to those looking forward to a good party.

Whenever people decide to celebrate this coming millennium, problems must be anticipated on 1 January 2000. Many computer programs prefer to store only the last two digits of the year of dates; the designers of the programs argue that it is pointless to store the two extra digits or ask the operators to type them in when they are always the same. As the first day of the new millennium dawns—first in New Zealand, or maybe in Pacific Islands further East—many computers will come to believe that the date is not Saturday, 1 January 2000, but Monday, 1 January 1900. A wave of computer mishaps will follow the rising sun as it sweeps round the world. The computer programmers probably told themselves that they would be enjoying their retirement by the time the millennium came round.

⌛ Millenarianism

Predictions of the end of the world, the Apocalypse, the triumph of good over evil, the Battle of Armageddon, and the like, have been with us at least since the start of the Christian era. Many of the more heretical or nonconformist sects, both ancient and modern, have set a date for such an event. Two modern examples must suffice, though there are plenty more: William Miller, founder of the Millerite sect in America, opted for 22 October 1844, and Taze Russell, founder of the Jehovah's Witnesses chose 1914. The appeal of such prognostications lies, at any rate to the poor and downtrodden, in the understanding that the old, oppressive order will be done away with on the fateful day. Such is millenarianism.

Frequently the prophet is unlettered, sometimes mad. But even some who are associated with modern science, such as Isaac Newton, Robert Hooke, and Jacob Bernoulli, have believed that the end, or at least some world-shattering event was nigh. Newton, after study of the Book of Revelations set the date, not of the end of the world, but of the end of the 'reign of the beast' at 1867. One can understand the gratification of the prophet when he succeeds in per-suading his following that the end is near; less comprehensible is his willing-ness to risk looking sheepish when it fails to happen. Several methods have been used to avoid the latter: the prophet might revise his prediction to a later date, just before the matter would be put to the test; he could specify a date after his likely demise or even be deliberately vague about the date; or, more tragically, urge his followers to mass suicide. Despite all this I do not mind betting that the world will end someday—but not for a while. Such apocalyptic predictions are curiously similar to modern warnings of global warming, the spread of terrible diseases from darkest Africa, or of thermonu-clear warfare—although these seem to be based on sounder reasoning.

Sometime in the sixteenth century more orthodox thinkers developed the-ories of the millennium founded, it seems, on traditional prophesies such as those in the Book of Revelation. In Revelation 20: 1–15, for example, the history of the world is mapped into thousand-year periods culminating in the Second Coming and the millennium, a period of peace and prosperity. Other ingredients of these murky intellectual exercises were, a statement by St Peter, recorded in his second Epistle (2 Peter 3: 8), that: 'one day is with the Lord, as a thousand years, and a thousand years as one day', and the belief that God made the world in six days, resting on the seventh.

One problem was to ascertain when God's great construction project was started. Archbishop Ussher (1581–1656) calculated that it began at noon on 23 October 4004 BC (see also Chapter 17). The implication is that the final, seventh stage of the history of the world, marked by the Second Coming, started in 1997 (or 1998 according to our reckoning).

Modern scholarship has cast doubt on traditional reports that the turn of the previous millennium, at the end of the tenth century, was marked by dire predictions and riots. Modern man is unable to take all this seriously, but has taken on the idea that the year 2000 (or 2001), which marks the start of the new millennium, is some sort of watershed, or at least an excuse for fire-works and a party. In doing this, he has shifted the meaning of the word 'mil-lennium' from something originally resembling a golden age, to something more ephemeral and pessimistic.

Other nations with different beliefs have held that history was circular and that every so often the world would be destroyed and recreated. This notion was popular with, among others, the Maya of Central America, the Chinese, and the Indians. It is curiously similar to the modern cosmological idea that the universe, born at the 'big bang' will eventually disappear into the 'big crunch'—a sequence of events that may, according to some, be repeated endlessly. The starts of the cycles of the circular cosmologies are sometimes supposed to be marked by some remarkable astronomical event such as a simultaneous conjunction in which all the planets are lined up, or, more mundanely (if that is the right word), an eclipse.

It may be that we are in for another wave of millenarianism (or 'chiliasm' to give it its Greek name) as we approach the 2000 mark. Certainly there are several long-standing predictions of doom still outstanding. The Maya priests proposed the world would end on the Sunday before Christmas in 2012 and Michele de Nostradame (Nostradamus, 1503–66) wrote:

> L'an mil neuf cens nonate neuf sept mois
> Du ciel viendra grand Roy deffraieur
> Rescusciter le grand Roy d'Angolmois
> Avant apres Mars regner par bonheur.

> (The year 1999 seven month
> Will come from the sky the great king of terror
> To bring back to life the great king of the Mongols
> Before and after Mars to rule luckily.)
>
> QUATRIAN X-72

I am not too sure what to expect in the fateful July, but the date is clear enough ('Anglemois' is, I am mysteriously informed, an anagram of 'Mongolois'). Maybe these prophecies are only a little inaccurate and really refer to the computerized apocalypse predicted above. Maybe Nostradamus knew of the eclipse of the sun scheduled for 29 July 1999 (in his Julian calendar, or 11 August in ours); this will be visible, weather permitting, as a total eclipse in parts of Southern England and Northern France.

⏳ The science of chronology

Chronology is concerned with the dating of the past and establishing a common temporal framework spanning different cultures and empires. This it does by relating dates in one calendar with those in another. The modern science of chronology started with the publication in 1583 of *De emendatione temporum* by Joseph Scaliger (1540–1609)—though several classical historians

FIG. 1.3 Nostradamus (1505–66). © Getty Images.

had considered the matter earlier. The standard calendar to which all other dates are usually reduced is conventionally taken to be the Julian calendar which was used throughout Christendom till 1582.

The means at the disposal of the chronologist are first and foremost inscriptions and documents. With luck these may provide a date in terms of two or more calendars, and thus a link between the two. They may also give lists of kings and the lengths of their reigns; these may serve to decipher regnal dates. Lists of kings and their dates are now available for many ancient dynasties, and the Archons of Athens and the Consuls of Rome are known over a span of several centuries. Coins are frequently inscribed with the year in which they are minted and often the profile and name of the king. These can usually be dated by archaeological means, and provide a correlation between regnal and our current year numbers. Sometimes the date of an eclipse is mentioned in a document. The dates of such, even those in the remote past, can be calculated with great exactitude. These may provide anchor dates for aligning a whole dynasty with known dates or they may serve to confirm other deductions. Likewise other astronomical events—conjunctions and oppositions of the planets, for instance—can be calculated and may correlate with the dates recorded for ancient observations. Attempts to identify the star which the Wise Men of the East followed to the infant Christ are, alas, too equivocal for us to be able to date the nativity in this way.

It is thus possible to assign Julian dates to much of human history back to the eighth century BC, with only a few gaps. Prior to that the margin of error increases, and eventually we reach a time before dates were committed to writing, perhaps about 4000 BC. There may be artefacts which have survived from before this time and which can be dated by the methods developed by archaeologists or by technical methods such as carbon 14 dating—but the history remains unknown.

Radiocarbon dating is capable of dating organic material—fragments of charcoal, wood, bone, hair, shells, cloth—back 40 000 years or more, but has only been accurately calibrated to about 4500 BC by means of tree ring dating or dendrochronology. In the South-West of the United States there lives the bristlecone pine, a tree that grows for immense periods of time. The pattern of rings observed in its logs reflects the weather as it was when the tree was laying them down. Logs of different ages can provide matching rings that overlap and extend the scale into the remote past. These have been used to calibrate carbon 14 dating. Other series of tree rings have been used in other parts of the world.

▣ The naming of parts of the calendar

Most calendars group the days into months and the months into years. The specification of a day then requires three numbers representing the day, month, and year, with perhaps a fourth designating the day of the week. An alternative is to refer to these components by names.

The names of the days of the week in several languages, both European and Oriental, are given in Appendix II and discussed in Chapter 21; here we shall merely mention the fact that there are three basic naming schemes; the first, as in Portuguese or the names used in the Eastern churches, simply provides a number (one to seven); the second refers each day to a planet or wandering star (for example, Monday = day of the moon) and is of astrological origin; while the third takes the Roman god associated with the planet and names the day after the corresponding god in a local pantheon. (Tuesday = Tiw's day = *Dies*

Martis). There are exceptions in various languages but these are the main patterns.

The names given to the months of the year are sometimes of extreme antiquity. The modern names of the months, used in many European languages, are derived from the Latin. Several are based on Latin numbers (September, October, November, and December are from the Latin for seven, eight, nine, and ten); others are based on the names of Roman calendar reformers (July, August); while others still are of a more obscure origin, but probably refer to gods (*Juno, Janus*) or have a religious, ceremonial origin (February). The names of the months in other calendars have sometimes an uncertain source, but in the Islamic calendar there are vestiges of a counting system: Jomada I and Jomada II. In others, some of the names derive from the names of seasons or are otherwise connected to meteorological phenomena or agricultural practices (as in the French Republican calendar); such seasonal names cannot have worldwide appeal.

The naming of the days of the month is rare; perhaps because months tend to be variable in length and there would be too many names to remember. One exception is in the Baha'i calendar which names the nineteen months of its years, and gives the same names to the nineteen days contained in each month. The Romans numbered the days of the month using an elaborate system of counting backwards from three salient days—the *Kalends*, *Nones*, and *Ides*—as explained in Chapter 16. By the eleventh century, this system was giving way to the simpler one we know today, which originated in Egypt and Syria, where the days are given numbers from 1 to 31. In the eighth and ninth centuries, a curious halfway stage was in use in Italy in which the days of the first half of the month were numbered forwards, while in the second half they were numbered backwards. Yet another popular system was to number the days from some saint's day; for instance, the day before St Bartholomew's day, 24th August, was the eve of St Bartholomew, and the 26th was the second day of St Bartholomew. The Christmas song tells us what your true love was up to on the *N*th day of Christmas.

Several sets of names have been given to the full quota of 365 or so days of the year. The Roman Church allocated one or more saints to each day of the year and have sometimes required that a child be baptized with the name of the saint allocated to their birthday. The French revolutionaries concocted a full set of names for every day of the year; these were chosen from those of animals, plants, minerals, or artefacts in general use in eighteenth-century France. Several other sets were constructed from lists of warriors, benefactors of humanity, and so forth, by other would-be reformers of the calendar in eighteenth- and nineteenth-century France.

Years are rarely named, perhaps because they are not often referred to in cycles. One exception is the cycle of 12 Chinese years, with names such as the Year of the Dragon. Another is the first 10 years of the Islamic era which were named after important events in the ministration of the Prophet.

Today, the full name of a day, such as 'the second day of March in the year of Our Lord, 1995' is rarely spelt out in full. Various abbreviations include shortening the name of the month to three letters ('Mar' for March) or replacing it with its equivalent number ('3' for March). The shortest form might be 2/3/95, but different nationalities place the day, month, and year in different orders: Americans use month/day/year; most Europeans use day/month/year; and some Eastern Europeans, year/month/day. This last convention has much to recommend it: the largest unit comes first and the smallest last, as in our way of writing numbers. This has the bonus that if dates are written without spaces and allowing two digits for both month and the day (for example, 2 March 1995 = 19950302), they can be sorted into temporal order just by arranging them in numerical order. This is perhaps of use mainly with computers. Another custom, fast disappearing, is the use, in business letters, of inst. and ult. to refer to this month and last month.

Computers habitually store dates in an abbreviated form. For instance, MS-DOS utilizes 2-byte numbers. The least significant five bits represent the day (1–31); the next 4 bits represent the month (1–12); and the high-order 7 bits represent the year after 1980 has been subtracted. After the year 2107, this system will become unusable. Other computers store the last two digits of the date and so will work happily till the millennium, when the problems already noted may arise. Such problems are not mere conjecture. In 1992, Mary Bandar of Winona, Minnesota reached the age of 104 and received a letter inviting her to join a kindergarten. The social services computer, unable to cope with ages over 100, had concluded she was four.

I will end this section on a lighter note. Anthony Burgess, in *Joysprick*, explains the word play to be found in the writings of James Joyce. He describes a game in which the participants, in parody of Joyce, 'punbaptise the names of the months from the point of view of a confirmed drunkard'. He suggests: 'Ginyouvery, Pubyoumerry, Parch, Grapeswill, Tray, Juinp, Droolie, Sawdust, Siptumbler, Actsober, Newwinebar, Descendbeer'.

CHAPTER 2

The Astronomical Background

⚏ Introduction

Our interpretation of what we see of the heavens is very different from that of the ancients. We believe, following Copernicus, that the earth revolves around the sun and rotates about its own axis. The ancients believed, following Ptolemy and common sense, that the earth was stationary and that the stars and sun and moon moved in cycles above it. If we were to observe the solar system looking down on it from the north, we would see the earth spinning on its axis in an anticlockwise direction and moving in its orbit in the same sense. Ptolemy, again looking south, would say that the sun and the stars rotated above the earth in a clockwise direction from east to west.

All the relevant astronomical measurements on which the calendar depends had been made by the time of Copernicus in the sixteenth century. Thus to understand the history of the calendar we must learn to see the heavens as the ancients saw them. It may be helpful for us in trying to appreciate the disposition and behaviour of the heavenly bodies to take a Copernican standpoint, but to understand the calendar and the problems posed and overcome in its development we must remain Ptolemaic.

To avoid repetition I will describe the heavens as they appear to a person north of the equator. Denizens of 'down under' must make a mental adjustment to the text.

The motions of the heavenly bodies in our solar system are at first sight straightforward, but it is as well to recognize that the earth, the sun, the moon, and the other planets interact with one another in a most complicated manner. This has the consequence that the obvious motions are slightly perturbed and what we observe is not quite as simple as we might suppose. These effects are often small, so they were not recognized by the ancients.

⚏ The night sky and the celestial sphere

Stand outside on a clear night, far from the lights of a town—as was easy a century or more ago—and look around you. Surrounding you is the circle of

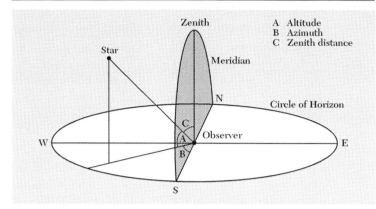

FIG. 2.1 The observer and his horizon: A Altitude; B Azimuth; C Zenith distance.

your horizon (FIG. 2.1); it defines a plane and you are standing at its centre. Directly above you is your zenith in a direction perpendicular to the plane of your horizon. The opposite direction is your nadir—in the direction in which a plumb line points.

Imagine a vertical semicircular disc with yourself at its centre; it passes directly above you through your zenith and it is orientated in a north–south direction (see FIG. 2.1); this is your meridian. The points where your meridian meets your horizon define your north–south line; at right angles to this is your east–west line. People at the same longitude on the earth's surface share the same meridian; those at different longitudes have different meridians. The prime meridian at longitude 0° passes through Greenwich in London, England. There, you can see a line drawn on the ground to mark it exactly.

Above you is half of the celestial sphere; with a great deal of patience you might be able to discern in it about 2000 stars with your naked eye. Watch for a few hours; you will see some stars sink below the horizon in the west and new ones rise up in the east; others, the circumpolar stars, will neither rise nor set but move in circular arcs. Point your camera at the sky and leave the shutter open; the developed film will show the movement of the stars, extended into arcs, streaking across the sky (FIG. 2.2). The east is the Orient (from the Latin for 'to grow, be born, or rise'); the west is the Occident, from 'to fall down, die, or set'). The stars die in the west; objects that have died have 'gone west'; as the film ends, the hero walks off towards the sunset.

FIG. 2.2 A long-time exposure directed to the north celestial pole.

The centre of the circular paths of the circumpolar stars lies in your meridian, due north of where you are standing; close by it is a star named Polaris, the Pole star. It is as if the whole celestial sphere were rotating about an axis, the pole, pointing towards Polaris. If you insist on a Copernican standpoint, the earth rotates relative to the heavens about this axis.

A circle whose diameter passes through the centre of the celestial sphere is a great circle; it is as large as a circle drawn on the sphere can be. Conversely, the diameter of a small circle misses the centre. The great circle which is perpendicular to the pole is the celestial equator (FIG. 2.3); the earth's equator lies in the plane of the celestial equator which passes through a well-defined series of constellations and stars, so that with practice you should be able to locate it easily in the sky.

If you were standing on the equator at latitude 0°, none of the stars would be circumpolar, for the axis of rotation would be horizontal and the pole star would be close to the horizon (FIG. 2.4). As you move from the equator to higher and higher latitudes, the pole is seen to mount higher and higher in the heavens and more and more stars become circumpolar. Finally when you reach the north pole at latitude 90°, the pole star is directly overhead, at your zenith, and all of the stars are circumpolar about it.

As the stars arc through the heavens they may pass through your meridian;

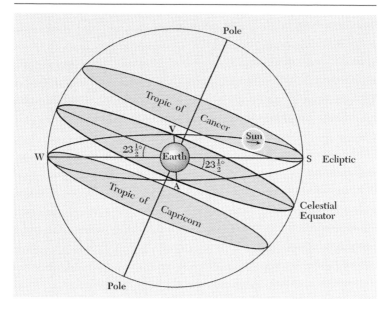

FIG. 2.3 The celestial sphere: V Vernal equinox; first point of Aries; S Summer
solstice; A Autumnal equinox; first point of Libra; W Winter solstice.
© DigitalVision.

they are said to transit when they do this. The circumpolar stars may pass
through the meridian while above the pole (an upper transit) or below it (a
lower transit). The lower transit of other stars takes place below the horizon
and cannot be seen. At its upper transit, the star is at its highest point in the arc
of its path and is then said to culminate.

Keep watching night after night and become acquainted with the stars. In
time you will invent patterns from their positions which help you to recognize
them. You will have invented the idea of the constellation. If you wish to tell
others about them you will agree on names for the constellations as the ancient
Babylonians did. Ptolemy described 48 constellations. Astronomers today
have mapped the entire celestial sphere into 88 named constellations, though
these differ considerably in the area of the sky that they cover; when told that a
comet is in Ursa Minor, they all know where to look.

The names that different societies have given to the constellations have
been usually quite arbitrary and fanciful. Some are named after animals (Ursa
Major or the Great Bear), some after household objects (Libra, the Scales),

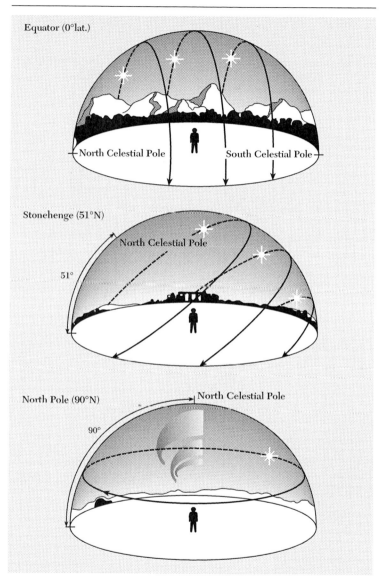

FIG. 2.4 The movement of the stars in the course of a night is different at different latitudes.

and some after mythological heroes (Orion). Many of the names we use today were inherited from the Babylonians and Greeks. The stars themselves were sometimes given proper names too (such as Aldebaran, of Arabic origin), but others are named according to their apparent brightness and the constellation in which they lie—thus the brightest star in Ursa Minor is Alpha Ursae Minoris, the next brightest is Beta Ursae Minoris, and so on. Some stars have aliases, so that, for example, Polaris is also called α-Ursae Minoris.

Most stars do not move relative to one another by a discernible amount, so that the constellations appear today almost as they did when the pyramids were built. Any small change in the position of a star relative to its neighbours is called its proper motion. A very faint star called Barnard's Star shows the greatest proper motion, but even this moves through less than a third of a degree in a century. Raise your forefinger at arm's length; its width subtends about $1°$ to your eye; the moon subtends an angle of about $\frac{1}{2}°$.

Although the true relative positions of the stars are unchanging (apart from their proper motion), their apparent positions when seen by us are subject to several small variations. Changes which vary periodically over the course of a year include aberrations of up to $20''$ of arc caused by an effect originating in the finite velocity of light, and parallax due to changes in the angle at which the nearer stars are observed as the earth moves round its orbit. Other variations include the effect of refraction by the earth's atmosphere (up to half a degree), and other, even smaller effects.

As you continue to watch the sky throughout the night you will eventually see the sun rise in the east, and as its rays flash out, the stars will fade from view. Another day has started. The sun, our nearest star, seemingly anchored on the celestial sphere will ascend in the sky only to sink again and eventually disappear below the horizon in the west as the day progresses from dawn to dusk. After dusk, the stars reappear and another night has started.

⚁ The position of a star in the celestial sphere

The actual distance between one star and another is vast and rarely of interest. Instead we specify the angular distance between them—that is the angle between two lines drawn from them to the Earth.

Several co-ordinate systems have been devised for specifying the position of a star in the sky; we shall mention three of them. Each is analogous to the system of latitude and longitude used to specify the position of a place on the surface of the earth. (There are mathematical rules for converting co-ordinates in one system to those in another, but these need not detain us.)

The altitude and azimuth of a star are its latitude and longitude on the celestial sphere as it appears to you (see FIG. 2.5). The equivalent of the equator is your horizon, and azimuths are measured from your meridian. The altitude of the pole star is (almost) equal to your terrestrial latitude. The altitude and azimuth of a star both depend on your position and the time, as the celestial sphere carrying the star rotates.

Alternatively, the declination and right ascension of the star are its fixed latitude and longitude in the celestial sphere (FIG. 2.5); the equivalent of the equator is now the celestial equator, and the right ascension is measured from a point on the sphere called the first point of Aries (see page 28). The complement of the declination is called the north polar distance of the star and is the angle between the pole and the star. The declination

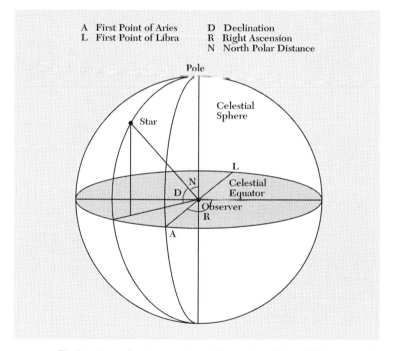

FIG. 2.5 Declination and right ascension: A First point of Aries; L First point of Libra; D Declination; R Right ascension; N North polar distance.

and right ascension of a fixed star do not change (except for the effects already noted and those due to the precession of the equinoxes (see page 29), but those of the sun, moon, and planets do change, for these move in the celestial sphere.

The ancient and medieval astronomers used a third system based upon a sphere whose equator was the plane of the ecliptic (see page 26). Thus celestial latitude is measured from the plane of the ecliptic; and celestial longitude is measured eastwards from the first point of Aries.

If the north polar distance of a star is less than your latitude, it is circumpolar and never sets; other stars spend some of their time above and some below the horizon. Stars with a positive (that is, north) declination spend the greater part of their time above the horizon; they rise and set at points north of your east—west line. Stars whose declination is zero divide their time equally between being above and below the horizon; they rise due east of you on your east—west line and set due west. Stars with negative declinations are south of the celestial equator; they spend most of their time below the horizon and rise and set south of your east—west line. Stars with north polar distances greater than 180° minus your latitude are never visible.

⧗ The day

Each day at noon the sun makes its transit through the meridian; it is then always due south and highest in the sky. The shadow of a vertical post or gnomon then points north and is shorter in length than at any other time. The time interval between two successive transits of the sun defines the apparent solar day.

During the night you can observe the transit of any star and the interval between two successive transits of a star is the sidereal day. As we shall see, it is shorter than the mean solar day by about four minutes.

Without accurate clocks, the discrepancy between the solar and sidereal day was not easily determined by the ancients, though it was known to Hipparchus of Nicaea in 130 BC. We in our privileged historical position can easily establish the difference and explain it from a Copernican standpoint. The earth moves round the sun, but at the same time it rotates on its axis. The sidereal day measures the time it takes to rotate once on its axis relative to the stars. During this time it has moved a little further on in its orbit about the sun, so that an observer who was directly facing the sun at noon of one day would no longer be doing so after a complete revolution, one sidereal day, later. The earth must spin a little more before our observer is facing the sun again at the next noon. The earth moves through about 1° each day in its journey round the sun, so that it must rotate about its axis this extra degree before noon. It spins 360° in a day; therefore the extra bit is 1/360 of a day—or about four minutes.

Even by the time of Ptolemy, it was clear that the apparent solar day is not constant in length. If you were to compare the time given by a sundial with that given by a watch, you would find a discrepancy, even allowing for British summer time and your longitude. In fact the lengths of the apparent solar days vary over a range of about 50 seconds. These discrepancies accumulate and then diminish in the course of a year so that the sundial gains and loses with respect to your watch in a quite complicated way. This difference, between the times given by your watch and by the sundial, is known as the equation of time; it can amount to as much as 16 or 17 minutes.

> *April the Fourth, and June the Sixth remember;*
> *August the Twentieth, and Twenty-fourth December;*
> *On these Four Days and none else in the year,*
> *The Sun and Watch both the same Time declare.*
>
> FROM WING'S *Sheet almanac*, 1752

To understand this difference, first note the seasonal variations in the difference between the sidereal and solar day, as discussed (see page 24). Two factors are at work. First, the speed of the earth in its orbit varies—it is fastest when it is at its perihelion and slowest at its aphelion (the two 'pointed' ends of its elliptical orbit when it is respectively nearest and furthest from the sun). Secondly, the earth's axis is tilted with respect to the plane of its orbit; this means that the motion of the earth in its orbit has two components, one parallel and the other perpendicular to the equator. These two components, the parallel component in particular, vary with the seasons.

To allow for this variation, the mean sun was invented. This theoretical and fictitious body moves along the celestial equator at a constant speed, taking exactly one year to move right round it. As the celestial sphere rotates about the earth, this mean sun sweeps through your meridian once each day just like the real sun but, unlike the real sun, the intervals between its successive transits are constant. This period is the basis of the unit of time called the mean solar day which is the unit used in calendars.

The sidereal day, defined by the transit of the first point of Aries, is a more regular time interval, but even this is not quite constant because of a multitude of small effects.

It is useful to distinguish several sorts of small change in the length of the day or other rotational periods. First there are secular changes; these continue over long periods of time. Secondly there are periodic changes characterized by regular increases and decreases which repeat indefinitely. Lastly there are irregular changes which are essentially unpredictable. When we calculate an average period over many rotations, the periodic and irregular changes cancel each other out leaving only the long-term secular changes.

The effects of tidal friction and similar phenomena cause a gradual decrease in the rate of rotation of the earth so that the day is subject to a secular increase of about half a millisecond (0.001 seconds) every century. Other effects cause small periodic changes, and finally there are small irregular variations of a few milliseconds. These variations could not be detected before clocks more accurate than the rotation of the earth had been invented. In time the earth will have slowed down so that it will present the same face to the moon (as the moon does to the earth today)—but not for many billions of years.

Modern estimates of the lengths of the mean solar day and the sidereal day are given in Appendix I.

🖾 The sun and the zodiac

During the day the stars are invisible, and during the night the sun is hidden from view. Nevertheless just before dawn or just after sunset it is possible to observe the rising or setting of stars which are close to the position of the sun. Such risings and settings are called heliacal risings and settings. In practice it may be easier to observe stars in the opposite direction to that of the sun as they set at dawn or rise at dusk, for there the sky is darker.

The Egyptians at one time observed the heliacal rising of the star Sirius (which they called 'Sopdet' and the Greeks called 'Sothis'), on a date which was close to that of the annual flooding of the Nile in the middle of July. Each day, after its heliacal rising, Sirius was found to be even higher in the sky at sunrise; in time, as the year progressed it was found to have set by sunrise. Would it ever rise again? In due course it reappeared after about 70 days to make its heliacal rising again. It is as if the sun were slipping behind the star in the celestial sphere. In the course of a full year, it slips behind by a full circle, almost back to where it started.

By observing heliacal risings and settings day after day you could list the sequence of stars that rise or set just before the sun and mark them on a star map. You would find that the sun seems to move in a well-determined path called the zodiac. The Babylonians, maybe as long ago as 3000 BC, noted the constellations lying in the path of the sun (or the moon, because it too moved in the zodiac). They later divided the zodiac into twelve equal parts known as the signs of the zodiac; the sun is found in each sign in turn as the year progresses, so that each sign corresponds to a particular time of year.

The Babylonians gave the names of the constellations to the corresponding signs of the zodiac. These names, whose origin is complicated and obscure, were inherited by the Greeks and our word 'zodiac' derives from the Greek 'zodion', a 'little animal'. The words passed to the Romans and, in turn, to us. The Indians, like ourselves, also use translations of the Greek names and they too may have inherited the idea from the Babylonians.

The names of the signs of the zodiac together with the approximate time of year when the sun can be found in them are shown in TABLE 2.1.

The path of the sun is called the ecliptic and this passes through the signs of the zodiac. The ecliptic defines a great circle in the celestial sphere and the sun appears to move along it as the earth moves in its orbit round the sun. The two great circles in the celestial sphere, the ecliptic and the equator, are inclined at an angle of about $23\frac{1}{2}°$, which is known as the obliquity of the ecliptic. Complicated interactions with the other planets cause this angle to vary a little with time.

The equator and the ecliptic intersect at two points called the equinoxes. These are situated at the start of the two signs of the zodiac, Aries and Libra,

TABLE 2.1 *The signs of the zodiac*

	Babylonian	Latin	Sanskrit	English	Season
♈	Hired hand	Aries	Mesha	The Ram	Spring
♉	Star	Taurus	Vrishabha	The Bull	
♊	Twins	Gemini	Mithuna	The Twins	
♋	?	Cancer	Karka	The Crab	Summer
♌	Lion	Leo	Simha	The Lion	
♍	Furrow	Virgo	Kanya	The Virgin	
♎	Balance	Libra	Tula	The Scales	Autumn
♏	Scorpion	Scorpio	Vrischika	The Scorpion	
♐	Name of God	Sagittarius	Dhanus	The Archer	
♑	Goat-fish	Capricorn	Makarus	The Goat	Winter
♒	?	Aquarius	Kumbha	The Water bearer	
♓	Tails	Pisces	Mina	The Fish	

The Latin names are translations from the Greek.

The meanings of the Sanskrit names are close to, but not always identical, to those of the Latin names. In particular Makarus is a sea serpent or crocodile rather than a goat.

Other Nations had other names, some similar in meaning to the names in the table.

The symbols in the first column are frequently used in astronomical and astrological works.

TABLE 2.2 *The houses of the sun at the equinoxes and solstices at different periods*

Position of sun	100 BC–AD 2050	2250 BC–100 BC	4400 BC–2250 BC	6650 BC–4400 BC
Vernal equinox	Pisces	Aries	Taurus	Gemini
Summer solstice	Gemini	Cancer	Leo	Virgo
Autumn equinox	Virgo	Libra	Scorpio	Sagittarius
Winter solstice	Sagittarius	Capricorn	Aquarius	Pisces

It is possible that from 6650 BC to 100 BC a new set of constellations was named and assigned to the sun at approximate intervals of 2150 years. After the third such change, all 12 of the constellations (marking 30° segments) around the ecliptic had been so assigned.

Some have attempted to discern symbolic or allegorical similarities between the names for the four houses of the sun which announce its ascent, height, descent, and depth in its yearly journey round the ecliptic.

and they are thus also called the first point of Aries and the first point of Libra. In the time of the Babylonians, these two points on the celestial sphere lay in the constellations of Aries and Libra. Today, the signs of the zodiac no longer correspond to the constellations with the same names (see TABLE 2.2); this is because the points where the ecliptic intersect the equator have moved, because of a complication known as the precession of the equinoxes. Nowadays the sign of Aries is, confusingly, in the constellation of Pisces. Before long it will be in Aquarius.

In the course of a year the sun crosses the equator at the two equinoxes. After it has passed through the first point of Aries, or vernal equinox, at the start of spring, it climbs further and further from the equator till it reaches its highest point at the summer solstice; it is then furthest from the equator. Thereafter it begins its descent to eventually cross the equator once more at the first point of Libra at the equinox marking the start of the autumn. It continues to move south till it reaches its lowest point at the winter solstice, before returning once more to the first point of Aries. At the two solstices it appears to come to a halt before reversing its motion; the term 'solstice' means in Latin 'the sun stands still'. Note that the term 'equinox' is used to refer not only to positions on the ecliptic but also to the times of year when the sun has reached them. The position of the sun when at a solstice is termed a solstitial point.

The tropics are two small circles parallel to the earth's equator. The tropic of Cancer passes through the summer solstice, and the tropic of Capricorn through the winter solstice. At the summer solstice, the sun is directly overhead at noon at the tropic of Cancer; at the winter solstice it is overhead at the tropic of Capricorn. The word 'tropic' means 'turning'; when the sun is at the tropic of Cancer its direction of motion turns from being northward to being southward, and vice versa when it is at the tropic of Capricorn.

Two other small circles define the Arctic and Antarctic Circles; these are again parallel to the equator and have latitudes of about $90 - 23\frac{1}{2}° = 66\frac{1}{2}°$. When the north polar distance of the sun is less than $23\frac{1}{2}°$, there are days when it does not set. As you move north from the Arctic Circle, the summer sun remains above the horizon for longer and longer periods. At the north pole itself it remains above the horizon for one half of the year and below it for the other half.

As we have seen, the earth does not move at a constant speed in its elliptical orbit. A consequence of this is that the seasons are not of equal length: the times taken for the sun to move from the vernal equinox to the summer solstice, to the autumnal equinox, to the winter solstice, and back to the vernal equinox are roughly 92.8, 93.6, 89.8, and 89.0 days respectively. The consolation in the northern hemisphere is that spring and summer last longer than autumn and winter.

⚅ The precession of the equinoxes

When the Babylonians were observing the heavens and recording what they saw, Polaris was not close to the centre of the circumpolar stars; instead, another star, still visible, called Thuban (and before that still, another called Vega) were near the pole. In good time these, and Polaris itself, will return to their central position, again and again about every 25 800 years. This effect, called 'the precession of the equinoxes', is said to have been discovered by the great Greek astronomer, Hipparchus of Nicaea, in 130 BC.

The effect implies that the direction of the earth's axis with respect to the fixed stars is changing. It is found that the pole or axis sweeps out, in a circular motion, describing a cone with an angular radius of about $23\frac{1}{2}°$; this rotation is due to the fact that the earth is spheroidal rather than spherical— the equatorial diameter is some 26 miles greater than the polar diameter— and that its axis is tilted with respect to the ecliptic. The sun and the moon (and to a much smaller extent the planets) each exert a gravitational attraction which is greater for the bulge nearer to them than for the further bulge. This produces a torque which causes the earth's axis to precess like a toy gyroscope.

One consequence of this precession is that the points where the equator intersect the ecliptic (the equinoxes) gradually move along the ecliptic—by about 50″ ($=360/25\,800°$) of arc each year, or by 30° in about 2150 ($=25\,800/12$)

years. At the time when the Babylonians named the constellations the first point of Aries was actually in the constellation of Aries. Today, as we have seen, it is in the constellation of Pisces and moving towards Aquarius. TABLE 2.2 shows the houses of the sun (the constellations marking its position) at the two equinoxes and the two solstices during several 2150-year periods.

The precession of the equinoxes is not the only motion suffered by the pole of the earth. It also undergoes a small rhythmic wobble or nutation over a period of about 18.6 years, which is caused by the gravitational attraction of the moon as the orientation of its orbit changes. One consequence of this is that the length of the sidereal day varies very slightly; another is that the actual rate at which the equinoxes precess is not constant throughout the year—being greater at the solstices; yet another is that the points at which the ecliptic crosses the equator vary slightly. This has the further consequence that the interval between two equinoxes—the year—increases and decreases through a range of about 4.7 minutes over the 18.6-year cycle.

⚅ The year

The passage of the seasons is marked by many meteorological and biological signs which have been noted from time immemorial. Equally obvious, at least in temperate latitudes, are changes in the behaviour of the sun. These are observed to be cyclic and to be synchronized with the meteorological and biological changes; the period of the cycle is the year.

At the two equinoxes the sun rises and sets due east and west, and the day and the night are almost equal in length. As the sun moves north towards the summer solstice, the days lengthen, the sun rises and sets at more northerly points on the horizon, and the length of the shadow of a gnomon at noon shortens. At the summer solstice, the sun rises and sets at its most northerly position, the day is longest, and shadows at noon are shortest. Thereafter, as the sun moves south, the days shorten and the shadows lengthen till the winter solstice is reached. Then, the trend reverses once more as the sun returns to the equinox. Less obvious is the movement of the sun through the signs of the zodiac and all that this implies.

One effect of the gradual change in the obliquity of the ecliptic is to alter the precise direction of the setting sun at the solstices. This was different in 3000 BC (when Stonehenge was being built) from what it is now.

For many reasons it became important to know the number of days in a year. In principle the average of the number of days between each recurrence of an annual event will yield an estimate of the length of the year. *The Times* publishes a report of the date on which the cuckoo (a migrant bird) is first heard in England. Reports from Amersham from 1925 to 1938 yielded an average interval of 365 days. The more variable the date of the event, the more years you need to average to obtain a value to a desired precision.

For more precise determinations, it is important to have an easily identifiable point in time which marks the end of one year and the start of the next. Several are available. One method is to observe the lengths of the shadow of a gnomon, day after day, and to note the days when it is shortest at noon, when it falls due north. These days mark the summer solstice, so that a year may be measured by the number of days between two such solstices. This was the method of choice of the Chinese astronomers, who built ever higher and higher gnomons to improve the precision. It was possible to interpolate between readings taken at noon on successive days to produce an estimate accurate to better than a day.

Within the tropics, there is a day each year when the sun passes directly overhead, through the zenith at noon. On such a day there are no noontime shadows. This fact was used by people living near the equator, such as the Incas of South America and the Javanese, to determine the length of the year. They noted the days when the sun could be seen through a vertical tube—a zenith tube.

Another method, more suitable in higher latitudes, is to observe the positions on the horizon where the sun rises and sets. At the two solstices, the sun rises and sets at its most northerly and southerly points on the horizon. At the two equinoxes it rises and sets due east and west, midway between its southernmost and northernmost points. To make this method work, it is necessary to stand at the same point at many dawns and dusks, from day to day, and year to year, noting the exact points on the horizon where the sun rises and sets. The natural thing to do would be to build an observatory, set with stones along which sightings could be made. Possibly Stonehenge and other neolithic stone circles and megaliths were used for this purpose. This method runs into difficulties north of the Arctic Circle since the sun does not set there on the summer solstice or rise on the winter solstice. Nobody who needed to measure the length of the year has ever lived south of the Antarctic Circle.

Yet another method, much favoured by the Egyptians is to count the days between two successive heliacal risings of the star, Sirius. This star is not in the zodiac and, because of the precession, the mean interval between such risings depends slightly on one's latitude. At Memphis it is about 365.2507 days.

It should be noted that two of these methods give an estimate of the tropical year, whereas the third does not take into account the precession of the equinoxes and so yields a slightly shorter year, close to what is called the sidereal year. The tropical year is the time required for the sun to return to the first point of Aries, whereas the sidereal year is the time required for it to return to the same point in the celestial sphere. The two differ because of the precession of the equinoxes which entails that the equinoxes move relative to the celestial sphere.

All these methods are hampered by the possibility of bad weather on the crucial days, particularly in northern climes. The second method works best near the equator where the period of twilight is shorter. All three methods have one important defect—they cannot give the length of a year to an accuracy of better than a day. In time it was realized that 365 days was too short and 366 was too long. It was necessary to count the number of days over a period of several years and to take an average.

A more sophisticated method is to determine the moment at which the sun is in the plane of the equator; this is the instant of the true equinox. Such a method was described by Ptolemy in his *Almagest*: a brass ring (an armillary or bracelet) is carefully aligned parallel to the equator—one way of doing this is to sight a long rod in the direction of the pole and mount the ring perpendicular to it; at the moment of the equinox, providing this occurs during daylight and the weather is fine, the sun is in the plane of the equator and the shadow of one side of the ring falls exactly on the other side. The year is the period between two such equinoxes.

Although accurate clocks were not available in Ptolemy's day, it was possible to tell whether the event happened in the morning, near noon, or in the evening. This, together with the date, fixed the instant of the equinox to within an interval of six hours. It was necessary to time equinoxes separated by several years and take an average to get a more accurate length of the year. In Egypt, the use of the wandering calendar to record the number of days between the two determinations was particularly convenient since there were no leap years or other complications to allow for.

You may care to measure the length of the year yourself using this method. I suggest a metal ring with a diameter of about a metre. Slits and pinholes in it will facilitate the determination of the moment the sun crosses the equator (avoid looking at the sun itself). With luck you should be able to get the time to within an hour or so, but bad weather, the possibility of an equinox happening at night, and the need to average over several years suggest you should embark on this enterprise early in your life.

It is important to note that the interval between successive true equinoxes varies by about 18 minutes on account of nutation and interactions with the other planets. This variation was not apparent to the ancients. It should also be noted that this effect must be taken into account when defining the sidereal day.

Astronomers now define a mean tropical year (as they define a mean solar day) by using a theoretical first point of Aries, or mean equinox, which precesses at a uniform rate along the ecliptic. The interval between two passages of the sun through this determines the mean tropical year. During the first millennium BC several astronomers recorded the times of equinoxes in this manner, but it was left to Hipparchus to take the credit for the first accurate estimate of the length of the tropical year. He took the determination of the equinox carried out by Aristarchus of Samos in the fiftieth year of the first Calippic cycle, and one of his own in the forty-third year of the third cycle. Since the cycle was of 76 years, the two determinations were 145 years apart. He found that the interval between the two determinations was half a day shorter than if the length of the year was exactly $365\frac{1}{4}$ days. The discrepancy was thus 0.5/145 of a day or about 5 minutes. The error should not have exceeded 6/145 hours or about $2\frac{1}{2}$ minutes (or 0.0017 days).

The determination was repeated many times as the years passed: by Ptolemy (who, it is claimed, 'cooked' his result to agree with that of Hipparchus); by the Arabs; by the Chinese; by the Indians; by the Danish astronomer, Tycho Brahe; and by others. TABLE 2.3 shows some

TABLE 2.3 *Determinations of the mean length of the tropical year as expressed in mean solar days*

Author	Place	Date	Estimate	True	Error
?	Babylon	*c.* 700 BC	365.24579	365.24236	0.00344
Hipparchus	Egypt	150 BC	365.2466	365.24233	0.0043
San Tong	China	8 BC	365.25016	365.24231	0.0078
Da Ming	China	AD 462	365.2496	365.24229	0.0073
Aryabhata	India	AD 476	366.2589	365.24229	0.0167
?	Mexico	AD 700	365.2420	365.24227	−0.0002
Da Yen	China	AD 724	365.2441	365.24227	0.0019
Al-Battani	Arabia	AD 900	365.24056	365.24226	−0.00170
Al-Zarqali	Arabia	AD 1270	365.24264	365.24223	0.00041
Guo-Shoujing	China	AD 1280	365.2425	365.24223	0.0002
Ulugh Beg	Samarkand	AD 1400	365.24253	365.24223	0.00030
Copernicus	Europe	AD 1500	365.24256	365.24222	0.00034
Xing Yunlu	China	AD 1620	365.24219	365.24222	−0.00003
Tycho Brahe	Europe	AD 1600	365.24219	365.24222	−0.00003
Modern		AD 2000		365.242199	

An error of 0.00001 corresponds to about one second. The year, like the day, is not quite constant; it has decreased by about 0.00014 days (12 seconds) since the time of Hipparchus of Nicaea.

determinations and the errors in them. Modern values are obtained using accurate clocks.

Improvements in timekeeping have shown that the length of the mean solar year is not quite constant; it is subject to a small secular decrease of about half a second per century. It has thus decreased by about 15 seconds during the last 3000 years. Accurate modern values for the lengths of the tropical and sidereal years and the way they vary with time are given in Appendix I.

Ⅺ The moon and the month

The moon reflects sunlight to us and this enables us to see her as she revolves in orbit about the earth. It is easy to detect the movement of the moon in the celestial sphere within an hour or so because she moves eastward even faster than the sun. But the most important thing about her is the way she waxes and wanes.

When the moon is in conjunction and is thus directly between the earth and the sun, we cannot see the face illuminated by the sun. As she continues in her orbit, however, the crescent of the new moon appears; this waxes or grows wider as she continues to move, till she is in opposition, with the earth between her and the sun and her full face illuminated. Then she wanes, and the portion of her face that is exposed to the sun, and which we can see, diminishes. Eventually she disappears for a while at conjunction. The cycle is repeated indefinitely.

The average interval between two successive new moons is her synodic period, and a single cycle is sometimes called a lunation. The actual length of successive lunations can vary by up to seven hours because of complicated interactions between her and the earth and the sun and, to a lesser extent, with the other planets.

The moon also moves relative to the stars, and the time she takes to return to the same part of the celestial sphere defines her sidereal period; this is shorter than her synodic period.

More relevant to primitive calendars is the interval between the moment of conjunction and the instant at which the new moon first becomes visible or the old moon disappears. This is different at different latitudes; at Babylon (latitude 32.5°), it can vary between 16.5 and 42 hours in the course of the year. This is partly a consequence of variations in the speed of the moon in its orbit, and partly due to variations of the moon's distance from the ecliptic.

The mean synodic period may be measured by counting the days in a large number of lunations and taking an average. This is facilitated by the fact that the earth, moon, and sun are exactly in line (as we Copernicans would say) during an eclipse. Thus, counting the number of days between eclipses and

TABLE 2.4 *Determinations of the mean length of the synodic period of the moon expressed in mean solar days*

Calendar	Place	Date	Estimate	Error (seconds)
Naburinos	Babylon	*c.* 500 BC	29.530614	2.6
Cidenus	Babylon	383 BC	29.530594	0.52
Ching Chhu	China	AD 237	29.530598	0.86
Yuan Chia	China	AD 443	29.530585	−0.26
Da Ming	China	AD 463	29.530591	0.26
Modern estimate			29.530588	

1 second is 0.000012 days.

dividing the result by the number of new moons observed would give an accurate estimate of the average duration of a lunation. Some results for this measurement are shown in TABLE 2.4. The mean synodic period of the moon is subject to a small secular increase of about 19 milliseconds per century. Modern estimates of the synodic and sidereal periods of the moon are given in Appendix I.

The azimuth (or compass bearings) of the moon at moonrise and moonset changes throughout the lunation, just as those of the sun at sunrise and sunset change during the course of a year. The angle between the extreme positions of the moonrises (or moonsets) depend on your latitude, and, as we shall see, it varies with time. These extreme positions are known as the major and minor standstills.

The plane of the orbit of the moon is not parallel to the ecliptic but at an angle of about 5° to it. If this were not the case there would be an eclipse of the sun at every conjunction and an eclipse of the moon at every opposition. If we suppose that the zodiac defines a band of sky about 5° on either side of the ecliptic, the moon will be found to move in it.

Twice each orbit, the moon passes through the plane of the ecliptic at two nodes (see FIG. 2.6). The line joining these nodes rotates within a period of about 18.61 years—a phenomenon known as the regression of the nodes. A consequence of this is that the angle between the equator and the orbit of the moon is sometimes greater and sometimes less than the obliquity of the ecliptic, according to the position of the nodes in their 18.61-year cycle. This also means that the extreme azimuths of moonrise and moonset are sometimes

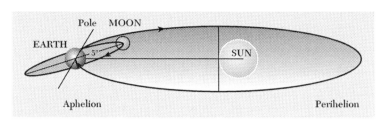

FIG. 2.6 Orbits of the earth and moon.

greater (at a major standstill) and sometimes less (a minor standstill) than those of sunrise and sunset.

If the moon is simultaneously at a node at the moment of conjunction or opposition, there will be an eclipse which will be repeated after a slightly shorter interval of about 18.03 years (or 18 years and 11 days) called the Saros, which contains 223 lunations. This fact was known to some of the ancient astronomers who used it (as can you) to predict eclipses. Eclipses, being rare events and of little economic importance, have not been used in the construction of calendars. But they can confirm the day of the start of a month.

Another aspect of the movement of the moon in the celestial sphere has been used by Indian astronomers for calendric purposes. They adapted an idea which originated in Babylon. They defined a 'tithi' as the time taken for the angular separation between the sun and the moon to change by 12°. The movement of the moon in its orbit is not uniform. However, since at each full moon, the sun and moon are in opposition with an angular separation of 0°, the mean tithi is 1/30th (360/12) of the mean synodic period or about 0.98435 days or 23 hours 37 minutes and 28 seconds. Calendars were constructed in which the basic time unit is the tithi.

Ⅱ The planets

Five planets—Mercury, Venus, Mars, Jupiter, and Saturn—are visible to the naked eye and were known to the ancients. The word 'planet' means 'wanderer' and the ancients referred to the sun and moon as well as the other five as planets. Sometimes we call them wandering stars.

The planets rise and set like the fixed stars, but they also move within the band of the zodiac like the sun and moon. Two of them, Mercury and Venus, are nearer the sun than the earth and are always seen close to the sun (when they are not hiding behind it) and thus only near sunset or sunrise. The other three can appear anywhere in the zodiac at any distance from the sun.

TABLE 2.5 *The five planets known to the ancients*

Name	Orbital period	Sign
Saturn	29 years	♄
Jupiter	12 years	♃
Mars	687 days	♂
Venus	224 days	♀
Mercury	88 days	☿

The sign is often used in astronomical works to symbolize the planet. The order in which the planets are listed is in decreasing order of their orbital periods; this happens to coincide with their supposed distances from the earth according to the Ptolemaic system. The ancients had no conception of the orbital periods themselves.

Their orbital periods are shown in TABLE 2.5. The ancients had no conception of these periods since they believed that the planets moved in a complicated manner on spheres with the earth at their centre. They could only determine the mean synodic periods of the planets—the average times taken for a planet to return to the same conjunction or opposition. These depend as much on the movement of the earth in its orbit as that of the planet.

The planets have played little role in European or Middle Eastern calendars but were of interest, Jupiter in particular, in the Chinese, Indian, Mayan, and other American calendars. The orbital period of Jupiter was nearly 12 solar years (11.86) and its progress was used to count off the years. Nevertheless the seven moving luminaries—the sun, moon, the five planets—had an astrological role and influenced the week as we know it, as explained in Chapter 21.

☲ A brief history of astronomy

Qualitative knowledge of the movement of the stars, the sun, the moon, and the planets in the celestial sphere is ancient. At various times within the first millennium BC, if not before, the great ancient civilizations of China, India, Babylon, Central America, and Greece began to make serious observations.

Among the earliest surviving records are the great numbers of clay tablets that have been unearthed from the cities of Babylonia. The earliest are dated to about 2000 BC but it was not until about 400 BC that they record sophisticated measurements of the positions of stars and other bodies. The

Babylonians could predict eclipses and they measured the year; although the inspiration for these astronomical labours may have had much to do with astrology, it may be said with some justification that Western science began at this time. The Egyptians, a practical people, built their pyramids accurately aligned North–South, but there is no evidence that their knowledge of astronomy approached that of the Babylonians.

After the campaigns of Alexander the Great in about 331 BC, Babylonian astronomical knowledge spread far and wide—to the East, to Persia, and to the Mediterranean. It inspired the Greeks and the Indians; it may have reached China. The city of Alexander with its famous Library was founded in Egypt, under the dynasty of the Ptolemys; astronomy flourished there and elsewhere in the Greek or Hellenic world for several hundred years.

Sufficient ancient documents survived or were later copied or collected by the Arabs to show the progress and discoveries made by the Hellenic astronomers. A few highlights must suffice. Aristarchus of Samos (c. 280 BC) tried to measure the distance of the sun; Eratosthenes of Alexandria (276–194 BC) measured (and got them nearly right) the diameter of the earth and the angle between the ecliptic and the equator; Hipparchus of Nicaea (c. 110 BC) measured the year and is credited with the discovery of the precession of the equinoxes (though it is in fact more likely that this honour belongs to the Babylonian astronomer, Kiddanu). Claudius Ptolemeus of Alexandria (c. AD 137), usually known as Ptolemy, compiled a catalogue of a thousand stars arranged in 48 constellations. He described his grand conception of the earth-centred universe in his famous work, *Mathemateke syntaxeos biblia* (or the *Syntaxis* for short), in thirteen volumes; this became known to the Greeks as the *Megiste syntaxis* (*The greatest syntaxis*).

After the collapse of Roman civilization, and the start of the Dark Ages, Western involvement in astronomy declined. Many Greek scholars emigrated to Persia where there was an interest in their writings. Later, the torch of astronomy was taken up by Indian and Islamic scholars. These latter preserved what they could of the ancient writings and built the great library and observatory at Baghdad. They developed the astrolabe and other instruments, they refined the star tables bequeathed to them by Ptolemy, and made new measurements of their own.

We are indebted to the Islamic scholars for translating the *Syntaxis*, which they called *Al Megast*, into Arabic in AD 827. In the late twelfth century it was translated into Latin under the title of *The almagest*. Its star catalogue, in Books VII and VIII, was the basis of the Alfonsine tables, published by Alfonso the Wise (1252–84), King of Castile, in 1277 and first printed in 1483. These tables provided the astronomical data used by Clavius and Gregory XIII for the reform of the calendar in 1582. Through them classical and Arabic astronomical knowledge began to be be re-established in Christendom.

The Alfonsine tables were followed by the Prutenic tables, compiled and published by Erasmus Reinhold (1511–53) in 1551 and based on the work of Copernicus (1473–1543). These were superseded by the Rudolfine tables, compiled by Kepler (1571–1630) from new and better observations made by Tycho Brahe (1546–1601), and published by him in 1627 in honour of their patron, the Holy Roman Emperor, Rudolf II. These were not superseded for a century. The painstaking observations of Tycho Brahe and the revolutionary ideas of

FIG. 2.7 The surviving part of the meridian arc of the quadrant of Ulugh Beg (1394–1449), at Samarkand, which was about 40 metres radius, set in the meridian. Source: Dr David Gold.

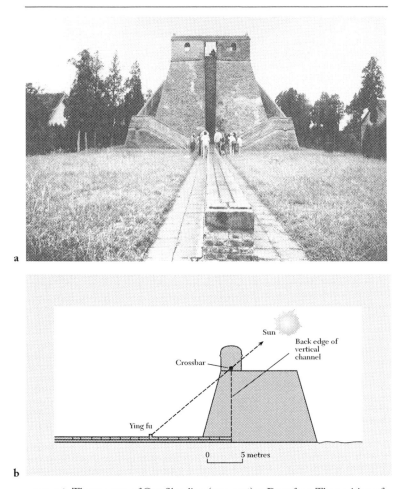

FIG. 2.8 The gnomon of Guo Shoujing (1231–1316) at Dengfeng. The position of
the shadow of the crossbar—which was about 9 metres above the horizontal scale
which was in the meridian—could be read with fair accuracy and used to determine
the day when the shadow was shortest. Photo: Robin Rector Krupp,
Griffith Observatory; Drawing: Griffith Observatory.

Copernicus led Kepler to his laws of planetary motion. These, in turn, led Newton to his celebrated theory of gravitation and planetary motion which was modified by Einstein only in the twentieth century; but these matters have no bearing on the calendar.

Ancient astronomical instruments were rudimentary and the telescope was not invented till much later, around AD 1608. Ptolemy described several ancient instruments which could be used to determine the altitude and bearing of heavenly bodies or the arc between two of them: the gnomon (essentially a vertical post) which had been used from very ancient times to measure the lengths and directions of shadows; the armillary, with its brass rings inscribed with angle marks and equipped with sights; the quadrant, and the closely similar triquetrum and astrolabe, which were used for measuring stellar altitudes. These instruments were improved by the Islamic scholars, and Ulugh Beg (1394–1449), grandson of the notorious Timur (Tamberlane), built a 180-foot quadrant in Samarkand (FIG. 2.7) before he was murdered by his son. Tycho Brahe had a quadrant with a 19-foot arm; each minute of arc was marked, and with it he could measure the altitude of a star at its culmination to 10″ of arc. Similar developments took place in China, where astronomy was profoundly influenced by the arrival of the Jesuit missionaries in the early seventeenth century. One of the first of these was Matteo Ricci (1552–1610) who was a friend of Galileo and had studied under Clavius (see Chapter 19).

The measurement of time has also played an important role in the development of astronomy, and that will be discussed in Chapter 3.

CHAPTER 3

Time and the Clock

⌛ Time and its measurement

Time which stretches from the remote past into the unforseeable future may bemuse philosophers, but we are only concerned with its measurement. You and I can remember at least some of our past experiences and, with greater or lesser difficulty, we can arrange them in temporal order—a task made easier if we know the times and dates of events. We may then note the minute, the hour, the day, the year of events as they happen. Their order is then obvious— a matter which may be of crucial importance (for example, did you arrive at the scene of the crime before or after the murder?).

To measure time, we need a clock. Clocks tick, endlessly and repetitively. This regular ticking presupposes some cyclic motion, an oscillation, or a vibration which is the essential feature of a clock. We (or better, the clock itself) can count these ticks, giving each a number in both numerical and temporal order. If you record the number of the tick which occurs just before each event of interest, you can easily arrange the events in order, and take the number of ticks between any two events as the duration of the interval between them. It may even be possible, if the ticks are not too close together, to estimate the time to a fraction of a tick, perhaps by noting the progress of some continuous change (such as the movement of a hand on a dial) that begins anew at each tick.

Three matters have to be agreed upon before we can use a clock to convey useful information to other people about the order of events or about durations. First, we must agree on a standard clock. It is not profitable to ask if the intervals of time between each tick are all the same; one tick is just like another, and it is natural to assume that they occur at equal intervals. Any clock that is not in step is out of order—like the new recruit on the parade ground. Secondly, we must agree on the phase; sometimes a tick is so indistinguishable from the following one that there is no well-defined instant within the cycle at which one tick ends and the next begins: we must agree on some recognizable instant. Thirdly, we must agree on an epoch—the moment at which the counting of the ticks is presumed to have begun.

The first clocks to be used were based on the daily motion of the sun across the heavens, the waxing and waning of the moon, and the yearly cycle of the sun around the zodiac; these provided us with standards of the day, month, and year. Others composed variations on this theme. The Indian astronomers developed a scheme whereby the passage of the moon against its background of stars was used for timekeeping. The moon cycles through its phases in the synodic period of 29.53 days; they defined a unit of time, the tithi, as $\frac{1}{30}$th of this (or a little less than a day). Both the tithi and the movement of the sun in the zodiac have been used by Indians for calendrical purposes.

⚏ The day

The most obvious natural cyclic event is the alternation of day and night, and this provides the primary unit of time of most calendars. It corresponds to the period of rotation of the earth about its axis or, from our human viewpoint, the time taken by the sun to return to the same place after completing its journey across the sky. The passage of the days is marked, except at certain times of the year in the polar regions, by the alternation of night and day, and these provide the ticks of this celestial clock.

The first problem in using the day is to decide when it begins and ends—its phase—so that events may be assigned to a day without ambiguity. Some nations decided to begin their day at dusk, and some at dawn: the Egyptians chose dawn; the Babylonians chose dusk; the Chinese and the Romans chose midnight, as we do ourselves today. Until 1925 astronomers began their day at noon but have now reverted to midnight. The different times at which people reckoned their day to start necessitates care when a day in one calendar is to be identified with a day in another.

By the time of Ptolemy in the second century, it had been noted that the days measured from noon to noon were not the same length as the interval between successive transits of some star—the sidereal day. What is worse, the comparison showed that the length of the solar day varied with the seasons. It became necessary to define a mean solar day, as explained in Chapter 2.

We usually begin the counting of days anew each month, but it is sometimes convenient to count them continuously from a fixed epoch. Astronomers do this by counting the days from Greenwich noon on 1 January 4713 BC in the Julian calendar; a number, the Julian day number, is thus assigned to each day. They go further and specify the exact time by adding the fraction of the day; 6 p.m. GMT on 1 January 2000 will have a Julian day number of 2 451 545.25. Note that the Julian day starts 12 hours after the midnight at Greenwich at which the corresponding day starts.

⧗ The divisions of the day

Anyone can estimate the time of day, even the Hopi Indians; these people were once attributed with a lack of any conception of time by a book-bound scholar, who had never met them. The Hopi can point to the sun in the sky and indicate the time of day like anyone else. Such divisions of the day are not precise and are often given names; we still do this, distinguishing dawn, morning, noon, afternoon, dusk, evening, and night.

Early attempts to subdivide the day more precisely resulted in 'hours' whose duration (perhaps $\frac{1}{12}$th part of the period from dawn to dusk) depended on the season, and which were called 'temporal hours'; the hours were longer in the summer than in the winter. We owe the division of the whole period (from one dawn to the next) into 24 temporal hours to the Egyptians. Temporal hours were most inconvenient for astronomers, and in the second century BC, Hipparchus divided the day into 24 equinoctial hours for astronomical purposes; these were based on the length of the hour on the day of an equinox, when the day is of the same length as the night. Ordinary people, however, continued to use temporal hours for more than a 1000 years. All 24 equinoctal hours are of equal length and independent of the seasons. Following the Babylonians, Ptolemy went on, in his *Almagest*, to divide each equinoctal hour into 60 minutes.

The canonical hours of Christian monasticism marked the times when devotions were sung in praise of God—at approximately three-hour intervals; they were Matins, Lauds, Prime, Tierce, Sext, Nones, Vespers, and Compline. Nones was sung at the ninth hour, at noon—hence that word's derivation. The practice of praying every three hours is of Jewish origin but had become established in the Christian church by the sixth century. It was the duty of the sexton to call the monks to prayer at the appointed times. Particular trades, to whom time is important, divide the day in other ways; for example, mariners divide their day into watches and ring bells to mark the start of each.

It is difficult to make clocks in which the lengths of the hours in the day and the night are different and adjustable. Accordingly, the equinoctial hour started to replace the temporal hour from the fourteenth century, after mechanical clocks had started to be made in Europe. Early clocks struck each hour, giving the time of the clock (or the time 'o'clock', as the time reckoned in equinoctial hours by the clock came to be called).

At first, clocks tolled 24 times a day. The difficulty of counting up to 24 'dongs' lead to the division of the day into two periods of 12 hours each. These were distinguished as being *ante meridiem* (a.m., before noon) or *post meridiem* (p.m., after noon). Much later, compilers of timetables,

TABLE 3.1 *Divisions of time*

Nation	Divisions	Modern equivalence
Babylonia	1 day = 12 beru	
	1 beru = 30 ges	2 hours
	1 ges = 60 gar	4 minutes
Egypt	1 day = 24 'hours'	
India	1 day = 30 muhala	
	1 muhala = 2 ghati	48 minutes
	1 ghati = 60 palas	24 minutes
	1 pala = 60 vipala	24 seconds
China	1 day = 12 shichen	
	1 shichen = 2 xiaoshi	2 hours
	1 ke = 0.01 day	14.4 minutes
Israel	1 day = 1080 chalakim	
	1 chalak = 76 rêgaïm	80 seconds
Republican	1 day = 10 heures décimales	
France	1 heure = 10 minutes décimales	2.4 hours
	1 minute = 10 secondes décimales	14.4 minutes
Modern	1 day = 24 hours	
	1 hour = 60 minutes	
	1 minute = 60 seconds	

The Babylonians are said to have divided the day into 24 parts at a later date. The Babylonian division shown above originated with the Sumerians.

the military, and others who feared confusion, reverted to the 24-hour system.

Other peoples subdivided the day into different numbers of units, which occasionally play a role in calendars. Some of these units are listed in TABLE 3.1. This table demonstrates a marked tendency to use subdivisions that are multiples of 12 or 30. This possibly reflects an analogy between the day, the month, and the year; the year was naturally divided into 12 since there were approximately 12 lunations in an astronomical year; likewise there were about 30 days in a lunation. These ratios precisely match the Sumerian and Babylonian division of the day into 12 beru and each beru into 30 ges. The subdivision into 60 parts reflects the Babylonian sexagesimal system; 60 is also the lowest common multiple of 12 and 30. The French revolutionaries were adamantly decimal, and the Chinese also liked decimal subdivisions.

⊠ The 24-hour day

The Egyptians were the first to divide the day into 24 parts as we do today. As we shall see, the Egyptian wandering year had 12 months, each of 30 days, and an extra 5 epagomenal days; the months were divided into three decades of ten days apiece, which thus divided the year into $36\frac{1}{2}$ periods.

At one time, Sirius (or Sopdet as the Egyptians called this star) was observed to rise just before dawn; this was its heliacal rising. As the days went by it would be found higher and higher in the sky at dawn until, in time, it had set in the West before sunrise. It then remained invisible for 70 days till once more it made its heliacal rising, a year after the cycle began. Meanwhile other stars (or groups of stars) were seen to rise just before dawn. The Egyptians noted the 36 stars that rose just before dawn on the first day of each of the decades; these came to be called decans. We have not been able to determine which stars the decans were, except for Sirius and Orion, but it is likely that they lay in a band near to the ecliptic.

In the course of a single night decans would rise in the East in sequence and could thus be used to count the 'hours' and judge the time. Since the sunrise occurred earlier and earlier as summer approached, and later and later as the days drew in, the decans were not equally spaced across the sky and so the actual length of an 'hour', the period between the rising of two consecutive decans, varied according to the seasons, being shorter in summer than in winter.

It turns out that the successive rising of about twelve decans could be observed during the hours of darkness. As the seasons changed, the 12 decans would be different, according to the decade of the calendar. To tell the time properly by the rising of the decans, you would have to know which stars to use. To aid the memory as to what decan rose at which hour during which decade, the Egyptians drew charts which we call diagonal star calendars. These are often depicted on the ceilings of Egyptian tombs from the Middle Kingdom and on the lids of coffins from about 2100 BC where they were presumably intended to be of assistance to the departed in the afterlife.

These star calendars consisted of a table with 36 columns, one for each decan, and 12 rows in each column, one for each 'hour' that could be counted during the night. The entry in a particular row and column indicated the decan that last rose at the corresponding 'hour' in the corresponding decade. The construction was simple, for in each successive column all the decans in the previous columns were moved down one row. The bottom decan was discarded and the next new decan introduced into the top row. Special provisions were made for the five epagomenal days at the end of the diagonal calendars.

One defect in this system of decans arose from the fact that the wandering Egyptian calendar was not properly in tune with the astronomical year; this meant that the start of the calendar year (and of each decade) slowly drifted with respect to the astronomical year and the decans, so that any particular diagonal clock would only be useful for a limited period. Attempts were later made to tell the time at night using other methods such as observing the transits of stars.

Later, the 36 decans were adjusted so that they divided the ecliptic into 36 equal portions of 10° each; each sign of the zodiac spanned 30° or three decans. Decans became important in astrology; as such they travelled to India and then back to Europe via Islam. They thus appear in several late mediaeval works of art and play an important role in Indian calendars.

The Egyptians divided the daytime into another twelve 'hours' by analogy with the night, using shadow clocks. In these, according to one interpretation, the shadow of a horizontal bar fell on to a graduated scale carrying four marks

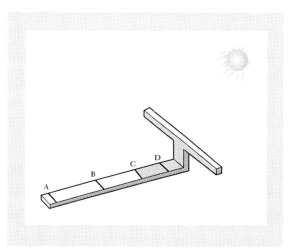

FIG. 3.1 An Egyptian shadow clock. The shadow of the bar reached mark A two 'hours' after sunrise, and marks B, C, and D at hourly intervals thereafter. At noon when the sun was overhead, the clock was reversed to point to the west.

(A,B,C,D) and orientated in an east–west direction. It was reckoned that the shadow of the bar reached the first mark (A) after two 'hours' had passed since sunrise. As the shadow reached the next three marks (B,C,D) it indicated the next three hours (up to five hours after sunrise). When the sun was at its highest, the scale was reversed through 180°, and eight hours after sunrise the shadow fell on mark A again. Another two 'hours' elapsed before the sun set. Two more hours were added to account for twilight before sunrise and after sunset, making twelve in all. The same clock could be used throughout the year, but it is obvious that these twelve daylight 'hours' were also seasonal, like the night-time hours.

The twelve daylight 'hours' and the twelve night-time 'hours' divided the day into 24 periods, although it is clear that the periods were not of equal duration and that they varied with the seasons. This scheme is the ancestor of our division of the day into 24 hours.

⚀ The measurement of time by the sun and the stars

The human heart provides a regular beat that Galileo used as a clock in his famous experiments in dynamics, but this clock could not very well be used to estimate the time of day. Initial attempts to do this depended on astronomical observations of the sun by day or the stars by night.

Many peoples developed shadow clocks and sundials. The sun rises in the east, is in the south (north of the equator) in the middle of the day, and sets in the west. This progress is marked by changes in the length and direction of the shadow of a stick. A shadow clock survives from Egypt from the second millennium BC which worked in the manner just described.

The shadow clock gave way to the more sophisticated sundial, in which the shadow of an upright rod or gnomon (a Greek word meaning 'indicator') fell on a circular scale. There are many different forms of the sundial; mention is made of one in the Bible (Isaiah 38: 8) which probably dates from about 700 BC:

> Behold, I will bring again the shadow of the degrees, which is gone down in the sun dial of Ahaz, ten degrees backward. So the sun returned ten degrees, by which degrees it had gone down.

This miraculous occurrence was a sign given by God rather than a description of some standard astronomical event.

Chinese sundials from the second millennium BC survive, as does a Babylonian one, described by a certain Berossus in about 300 BC. Sundials of essentially the same design were used for more than a 1000 years. The Greeks learnt of the sundial from the Egyptians, and the first one constructed in Rome was

FIG. 3.2 Telling the time in Borneo. Many people could estimate the time of day from the position and length of the shadow of a vertical stick used as a gnomon.

FIG. 3.3 A Roman sundial from Utica, Tunisia. Source: E T Archive.

built for Marcus Phillipus in 164 BC—but the Romans themselves made little contribution to their design.

The Arabs significantly improved the theory and design of sundials and also further developed the astrolabe, which had been invented by the Greeks and which could be used to tell the time. There is no period from antiquity till the present when sundials were not in use, and there are English examples dating from the eleventh century. By the eighteenth century they had reached a state of considerable sophistication and gnomonics provided a subject for courses at universities.

Time at night could be read by observing the rising, or more accurately the culmination, of a star—a technique used by the Egyptians. In the thirteenth

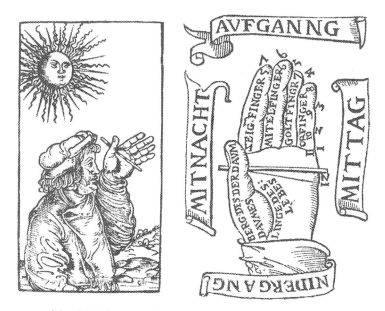

FIG. 3.4 A hand dial. A stick held in the open hand and orientated towards the pole acts as a simple sundial and casts a shadow on the fingers which can be used to tell the time.

century, possibly earlier, the nocturnal was invented. In the course of a night, stars, such as the two 'pointers' in the Great Bear (alias the Big Dipper or the Plough), move in such a way that the line joining them to the Pole Star rotates like the hour hand of a 24-hour clock. The nocturnal carried a pointer which could be aligned on such stars. A simple adjustment of a scale to take account of the date then permitted the time of night to be read off—provided the weather was clear. Nocturnals, like portable pocket sundials, were in use till the nineteenth century, when cheap watches began to supplant them.

⧗ Water clocks

The inconvenience of observations of the sun and the stars may have prompted the Egyptians to invent the water clock, or clepsydra as the Greeks called it; the word itself means 'water thief'.

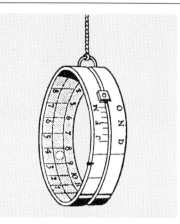

FIG. 3.5 A portable sundial described by Chaucer. The hole on the circumference casts a shadow on the scale inside the ring, which indicates the time of day. It could be adjusted both for the season and also the latitude of the user.

The earliest known Egyptian example survives from the fifteenth century BC, from the Temple of Karnak in Egypt; the Chinese were using them by the sixth century BC, if not before. In one form of the clepsydra, water drips from an upper vessel, through a small hole into a lower vessel. The depth of water in either the upper vessel or the receiver may be recorded by marks on their sides; these are graduated in 'hours'. In some Egyptian clepsydras, different sets of graduations were used at different times of year to ensure that 'water time' kept in synchrony with 'decan time'. Later, some of these clepsydras were quite elaborate, with mechanical floats and levers. One defect of early water clocks was that the rate of flow depended on the head of water in the upper vessel. This was overcome by Ctesibius of Alexandria (c. 135 BC), who was noted for great expertise in hydraulics; he arranged for the head of water to remain constant. Another difficulty is that the time for a given volume of water to flow out depends strongly on its temperature; it doubles as the temperature falls from 26 °C to 1 °C.

Similar in principle is the hourglass or sand-timer, invented in about the thirteenth century; small versions of these are still used today for timing the boiling of eggs. They require special sand, once made from ground eggshells (ordinary sand erodes the hole too rapidly). The Chinese built a clock that used mercury instead of water which might have frozen at night. Other analogue devices included graduated candles and the slow burning of incense sticks; with the latter, it was said, you could tell the time by the smell.

FIG. 3.6 Sixteenth-century nocturnal. Holding the nocturnal vertically at arms'
length, the pole star is viewed through the central hole. The radial bar is then
turned till it is aligned on the pointer stars of the Great Bear (or Dipper). The time
of night can then be read from the scale after adjusting it for the date. © Museum
of the History of Science, Oxford.

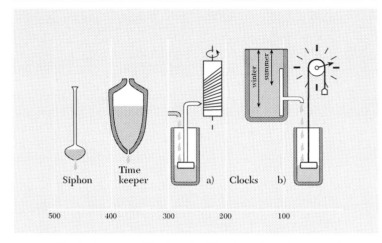

FIG. 3.7 The development of the clepsydra. The chronological scale runs from 500 BC to 100 BC.

FIG. 3.8 A seventeenth-century hourglass, attached to the pulpit of a church and used to time sermons.

FIG. 3.9 A Chinese incense seal carved from wood in about AD 1329. The single
continuous groove was filled with incense of different aromas; lit at one end, the
incense burnt for maybe 12 hours. The time could be determined from the odour of
the incense or the position of the area of burning.

Meanwhile water clocks were widely used in antiquity. The Greeks and
Romans used them to time (and limit) the length of their speeches; the Arabs
made some improvements; and mediaeval sextons used them to call the
monks to prayer. There was a street in thirteenth-century Cologne called
Urlogengasse; there could be found the guild of water clock makers.

In about 50 BC, the Greek astronomer Andronicus of Kyrrhos built the
Tower of the Winds in the Agora at Athens—the centre of political and commercial life; there what remains of it can still be seen. This octagonal tower
carried sundials on its walls and was surmounted by a wind vane; inside there
was an elaborate clepsydra, fed from a stream and adjusted to fill every 24
hours. The water rising in the tank raised a float which in turn drove an elaborate mechanism which showed the daily progress of the sun across the
heavens. Thus time could be estimated by day and by night, and boring
speeches stopped if they went on too long.

FIG. 3.10 The Tower of the Winds in the Agora at Athens, built by Andronicus of Kyrrhos in about 50 BC. Originally there were sundials on the eight walls and a wind vane on top.

⊞ Mechanical clocks

In AD 1090, Su Song, a Chinese engineer, built an elaborate water clock in which water was fed at a constant rate to scoops on a water-wheel. When a scoop was full, a latch was released which allowed the wheel to rotate till a new scoop was in position. This is a type of escapement mechanism. This device also showed the movement of the planets and various other astronomical phenomena, and was one of the first mechanical clocks to be made. The plans

FIG. 3.11 Su Song's water clock, built in AD 1090. The rotation of the water-wheel was controlled by a mechanism which released the wheel when a bucket was full. The time was indicated by the puppets, and the armillary sphere at the top moved so that a star could be kept in view as the earth rotated—as in modern astronomical telescopes.

were preserved and working models have been constructed. One is in China, but there is another in Liverpool Museum. It is said that this clock was built for astrological purposes so that the dispositions of the planets could be correctly worked out should one of the Emperor's concubines give birth during cloudy weather.

Sometime in the ninth century, the iron horseshoe was invented. The popularity of this, and of iron armour, led to more and better skilled black-smiths. Such a reservoir of skill was later harnessed to the construction of mechanical clocks.

The mechanism of a clock has three essential components. The first is a source of power: Su Sung's clock used water power; the first weight-driven European clocks date from the end of the thirteenth century; and the first spring-driven clocks were made in the fifteenth century. Much later, clocks came to be powered by electricity.

Secondly, a regulator must be provided to beat out the ticks. Su Sung's clock was regulated by the periodic filling and emptying of scoops. The first mechanical clocks in Europe were regulated by the verge escapement mechanism. This device had no natural period of its own and its accuracy was poor; sundials were required to put them right regularly. Nevertheless it was of crucial importance to the development of the mechanical clock.

Galileo (1564–1642) and Marin Mersenne (1588–1648) discovered the isochronicity of the pendulum (that is, that every swing took the same time, regardless of amplitude). They found that the pendulum swung to and fro with a natural period of its own. Christian Huygens (1629–95) contributed to the theory, and the first pendulum clock was made by Salomon Closter in 1658 according to Huygens' design. The balance spring, which regulated compact spring-driven clocks and watches, was invented either by Huygens in 1675 or Robert Hooke (1635–1702)—though by which one is controversial. In the twentieth century the frequency of the alternating current supplied throughout the civilized world provided the regulation for electric clocks. More recently this has been replaced by the quartz crystal, which vibrates at a constant rate. The final development, as we shall see, is the atomic clock—but that is not likely to become a household item.

Finally, an escapement mechanism must be provided so that the regulator can control the release of the energy which drives the mechanism. Before the invention of the verge escapement in the thirteenth century, people toyed with inadequate mechanisms involving friction. Sometime later, the anchor escapement, geared to the pendulum, was invented—possibly by William Clements in 1675. Clocks using this new mechanism kept time accurate to within about ten seconds in a day.

The first known European clock is said to have been made in 1283 for Dunstable Priory in England, but the oldest surviving mechanical clock, dating from 1386, is in Salisbury Cathedral, also in England. The accuracy of such clocks, which used the verge escapement, was no better than about 15 minutes a day, and they thus required frequent correction. They were also unreliable—but if you were rich enough to own one, you could probably afford a resident clock maker to maintain it. Their accuracy could be compared with that of the later and popular cuckoo clock, the first of which was made in Bavaria in about 1740.

The history of the clock in the eighteenth and nineteenth centuries is largely a record of the improvement of the escapement mechanism, the balance wheel, and the pendulum. New and ingenious escapement mechanisms were invented; the various sources of error in the pendulum and the balance wheel were analysed, and mechanisms to alleviate them were developed. During these two centuries, the error in clocks and watches decreased from ten or more seconds a day to a few hundredths of a second.

A landmark was the design and construction of the marine chronometers of John Harrison (1693–1776). These eventually won him a prize from the Board of Longitude at London. This body had an interest in navigation for which accurate timekeeping at sea was crucial. Despite the support of George III, the Board kept on delaying the payment of the award of £20 000—a vast sum in those days—even though John's best watch kept time to about 0.06 seconds a

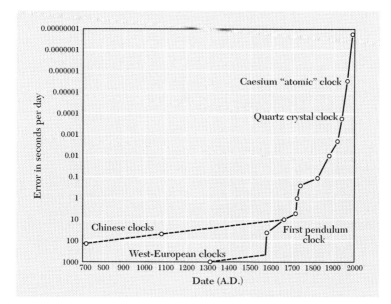

FIG. 3.12 The accuracy of clocks improved in western Europe from about 15 minutes error a day in the fourteenth century to a minute fraction of a second in the twentieth century.

day; it is with some relief that one reads that they paid up in the end, albeit reluctantly and when John was an old man.

It is ironic that no sooner had the mechanical clock and watch been perfected, after the expenditure of centuries of ingenuity and effort, than they should have been rendered obsolete within a few years by the development of the more accurate and much cheaper quartz clock. Only the best and most expensive mechanical clocks can compete with these for accuracy.

⌛ Dials, calendar clocks, and orreries

The first European mechanical clocks were used to ring a bell at regular intervals—the sexton found it eased his workload—and the English word 'clock' derives from the Latin word for a bell. In fact, a clock was originally a device that rang bells, whilst a watch was something you watched.

The first dial (with only one hand) was fitted to such a clock in 1344 by Jacopo Dondi of Padua. Jacopo's son, Giovanni, went on to build an extraordinary astronomical clock, his astrarium; this took him 16 years and he finished it in 1364. It showed both solar and sidereal time, sunrise and sunset, the motion of the planets, the date of Easter, the phases and the progression of the nodes of the moon. It was not to be equalled for 500 years; its eventual fate is unknown and the last report on it, in 1529 by Charles V at Pavia, tells us that it was badly corroded. Nevertheless Giovanni left detailed drawings of it which permitted several modern reconstructions. One is housed in the Smithsonian Institute at Washington. Meanwhile, many other, though less comprehensive, astronomical clocks were built in various cathedrals, notably at Strasbourg in France.

Many long-case or grandfather clocks show the phases of the moon and the date, and calendar watches, which not only tell us the second, the minute, and the hour but also the day of the month, the day of the week, and the month, are now common. The better ones know how many days there are in each month and make adjustments for leap years. It is unlikely that the watch (if that is the right word) made by the Swiss firm of Patek Phillipe, will be surpassed. It weighed about $2\frac{1}{2}$lb and had 24 hands. Besides the usual things expected of a calendar watch, it indicated the date of Easter, the phases of the moon, and the times of sunrise and sunset. It sold for $3 million in 1989. It was the dinosaur of the watchmaker's art; today, a simple electronic device could do all these things for a fraction of the cost.

An astronomical clock is, as it were, a working model of the solar system, operating in what is known in computer circles as real time. More explicit models of the sun, the planets, and maybe their moons as well, are provided by orreries—though these do not usually run at the same rate as the real solar system. In 1712, Charles Boyle, fourth Earl of Orrery (1676–1731), bought such a model of the solar system from a London manufacturer, John Rowley; such devices have been known as orreries ever since. Today, one can purchase a simple clockwork orrery in plastic for a modest price.

John Rowley's model was, however, not the first of its kind. Cicero (106–43 BC), the Roman author, describes a similar device made by Archimedes, and in 1900, fishermen from the island of Antikythera off the south coast of Greece, dredged up a corroded object which on inspection turned out to consist of bronze gearwheels. These were later shown to be part of a device which could be used to determine the position of the sun and moon in the zodiac, and which mechanized the metonic cycle of 19 years with 235 lunar months. It would have been cranked by hand. Since then several more fragments of geartrains have been found; these date from the fifth century and performed similar functions. There is some evidence that knowledge of such devices passed to the Arabs via Persia and then, though this is more speculative, back to Europe, to inspire early clock makers such as Giovanni Dondi; others say that their inspiration came from China.

⚅ Local time

So far we have talked of the time of day in a rather vague manner, and have merely supposed that the sun crosses the meridian at noon. However, the meridian of an observer depends on his longitude; it follows that each observer has his own time-scale according to his location. My noon in London (longitude 0°) is some five hours earlier than that in New York (longitude 74°), for the earth takes 24 hours to spin through 360°, rotating through 15° every hour.

In years past this did not matter, but as soon as long-distance travel by stagecoach, ship, or train arrived, it created severe problems for compilers of timetables and the like. In international treaties of 1883 and 1884, the countries of the world agreed to establish a series of time zones. The same time would be used throughout each zone, instead of the local time of each place within it. These zones formally encompass $7\frac{1}{2}°$ of longitude on either side, or a series of meridians spaced at intervals of 15° starting at the prime meridian at Greenwich, England. Thus zone Z extends from longitudes $7\frac{1}{2}°$ west to $7\frac{1}{2}°$ east, and Greenwich Mean Time (GMT) is honoured throughout it. The next zone to the west extends from $7\frac{1}{2}°$ west to $22\frac{1}{2}°$ west, and the time therein is one hour behind GMT. Each zone is one hour behind the next zone to the east. To avoid inconvenience, the boundaries of the zones were modified where they passed over land, so that the whole of most smaller countries falls within a single zone. The instant of noon differs by an hour from one side of a time zone to another, and designers of sundials must take account of this.

As one moves to the west from London, one passes through 12 time zones before reaching longitude 180° west; this is 12 hours behind Greenwich. Likewise, if one moves east though 12 zones, one reaches the same longitude, but now this is 12 hours in advance of Greenwich. The international date-line is situated at this longitude—step over it from west to east and the date goes back a day; 1 January 2000 would become 31 December 1999; step in the other direction and you lose a day. The time of day does not change as you step one way or the other, only the date, since the line is situated at the centre of a time zone. Should this bother you, consider two brothers who set out from the international date-line and travel westward together. Peter adjusts his calendar watch every time they move into a new time zone by winding it back an hour, but Paul leaves his watch well alone. When they get back to their point of departure on the date-line, after circumnavigating the globe, Paul is pleased to see that his watch tells the right time and date, but Peter is surprised to find his to be a day slow; we may reassure him that all is in order, for it was he who kept on fiddling with his watch as he went round the world, so winding it back a total of 24 hours.

The idea of a date-line had in fact been discussed, by those who knew the world was round, since the fourteenth century at least. Nicholas Oresme (1323–82), Bishop of Lisieux, wrote in 1355: 'One ought to assign a definite place where a change of the name of the day would be made'. Nevertheless, Magellan's crew were much perplexed when they reached Cape Verde in 1522 on a Wednesday to find the inhabitants treated it as a Thursday. Phileas Fogg in Jules Verne's *Round the World in Eighty Days* travelled from west to east, resetting his watch as he went. When he returned home he forgot to wind his watch back and nearly lost his bet.

Some countries introduce a leap hour at certain times of the year and remove it later. The idea behind these 'daylight saving hours' is to advance the hour of dawn in the winter so that farmers and others whose work is geared to daylight can live their lives in phase with the rest of us. It was first mooted by Benjamin Franklin in 1784, but actually introduced during the First World War as an energy-saving measure.

⚅ Time in the twentieth century

In the course of this century it was found that, in comparison with atomic clocks, the day varied slightly in length. It thus became desirable to abandon the day as the standard of time, and the second, originally one 86 400th part of a mean solar day, was redefined, arcanely, in 1967 by the General Conference of Weights and Measures to give the Système Internationale (SI) second as: 'the duration of 9,192,631,770 periods of the radiation corresponding to the transition between the two hyperfine levels of the ground state of the Caesium-133 atom'.

The standard clock is now based on the caesium atomic clock. This contains caesium vapour which absorbs microwave radiation, generated electronically, and tuned till it is of just the right frequency to be absorbed by the caesium. The vibrations of the microwaves are counted electronically. Astronomers have used, since 1972, a time-scale based upon the average of several such clocks in a number of countries and call it International Atomic Time (TAI).

It is still desirable, for many civil purposes, to maintain a time-scale, now called universal time (UT), which keeps in step with the rotation of the earth. This is carefully monitored to determine the length of the sidereal day, which is used to calculate mean solar time; it is synchronized with the mean sun at noon at the prime meridian at Greenwich. It has certain irregularities that can be allowed for, and corrected versions of universal time called UT1 and UT2 are calculated. Even so, small discrepancies between UT2 and TAI arise unpredictably. It was thus decided to institute yet another time-scale— Co-ordinated Universal Time (UTC). The unit of time of this scale, the second, is the SI second as used in TAI, so that a clock registering UTC runs at the same rate as a TAI clock. Universal time (UT2) nevertheless gets out of step with it from time to time. When this happens the UTC clock is reset by inserting or removing a leap second; in this way, 12 noon on the UT2 scale is guaranteed to be within 0.9 seconds of 12 noon on the UTC scale. There are other time-scales which are used for astronomical calculations, such as ephemeris time—but these need not detain us.

Co-ordinated Universal Time (UTC) is the time broadcast, since 1972, by radio stations across the world and is popularly known as Greenwich Mean Time (GMT). The day of UTC starts at midnight.

CHAPTER 4

Writing and Libraries

⌘ In the beginning

Some 12 000 years ago, the glaciers were starting to recede for the last time. A few thousand years later, at the start of the Neolithic Age, man had learnt to grow crops and domesticate animals, rather than lead a hand-to-mouth existence, hunting and gathering food. His tribe and community still remained small but where the land, irrigated by great rivers, was particularly fertile, the supply of food increased. The size of communities grew, and some 6000 years ago the first cities and civilizations were established: in the land between the Tigris and the Euphrates called Mesopotamia, in Egypt by the Nile, in northern India in the Indus valley, and in China by the Yellow River. There was, in these places, sufficient surplus food to support an army (for protection or depredation) and workers (to build dams and canals and monuments to gods or kings). Travellers and merchants moved from one centre to another, spreading news of any discoveries. The keeping of records and in turn writing, became a necessity; with writing, the calendar became possible.

⌘ The calendar and writing

Our knowledge of ancient calendars is fragmentary and haphazard. The information that we have comes from a variety of sources and rarely from writings dedicated to the subject. We must be content with contemporary snippets written on papyrus, engraved in stone, impressed on clay tablets, painted on the walls of tombs, and baked on to fragments of pottery.

A spoken sentence is no longer said than gone, but writing as we know it today enables anything we can say to be preserved; a written sentence substitutes a permanent linear sequence of symbols in space, which can last for a thousand years, for an ephemeral sequence of sounds in time. Archaeologists have unearthed memorials from tombs, lists of kings, manuals of arithmetic, sets of accounts, records of astronomical observations, monuments to important events, astrological notes, oracles, even calendars—to name but a few. From such sources it has been possible to piece together the details of ancient calendars and relate them to our own. Gaps and uncertainties in our knowledge still remain however; some of the ancient scripts have not yet been deciphered.

⊠ The nature of writing

Before writing there were pictures; 40 000 years ago early man was depicting important things and events on the walls of caves, where they can still be seen today in south-western Europe at Lascaux and elsewhere. The habit never died out; rock paintings were being made in historical times by American Indians and others.

As well as lifelike pictures of animals there were more abstract designs. Maybe these were the precursors of property marks or memory aids and the like. Before writing people needed to learn their history, their myths, and their stories by heart. Sometimes they would make drawings to remind themselves of the sequence of events to relate; in these, a drawing may stand for a whole paragraph of the narrative. Likewise, a mark branded on a beast proclaims its owner. Perhaps in this way simplified drawings came to stand for things (pictographs). Later the idea was extended so that marks stood for ideas (ideographs) and for words in the spoken language (logographs). The difficulty with ideographic writing is the vast number of symbols that have to be learnt; Chinese is said to have 50 000 ideographs, but the advantage is that they can be understood by people speaking a variety of languages.

The next step in the evolution of writing involved the use of the rebus principle. This word, coined in the seventeenth century, derives from the Latin phrase '*non verbis sed rebus*' or 'not by words but by things'. A halfway stage is illustrated in the Yoruba language: a boy might send a girl six shells; the meaning of this gift derives from the Yoruba word 'efa' which means both the number 'six' and 'I am attracted'. The principle is still used in children's amusements and in heraldry. Modern examples might include a picture of a car followed by one of a cat to signify the word 'carpet', or the American road sign 'Xing'.

Finally, true phonetic writing was invented. First came syllabaries, as used by the Japanese, who imported Chinese characters and used them to stand for syllables in their own language which had the same sound as the corresponding Chinese word. Syllabaries require far fewer symbols—about 80 in many languages. Even more economical are consonantal scripts and alphabets which only require 20 to 30 characters. Modern examples of consonantal scripts are Arabic or Hebrew; in these languages vowels are indicated by diacritic signs. Alphabetical scripts, used in European languages, in which vowels are also allocated letters, include the Greek and Cyrillic scripts as well as our own Latin alphabet.

It is as well to appreciate that no phonetic script is able to capture all aspects of the spoken language—the inflexions, the contextual modifications of sounds, and so forth. It is also a curious fact that most practised readers appear to read words spelt out in an alphabet as if they were logographs, recognizing the word as a whole rather than reading the individual letters and constructing the word from their sounds.

The first people to develop true writing were the Sumerians who were dominant in Mesopotamia in the fourth millennium BC. They wrote by impressing clay tablets with special styluses. This writing, originally picto-graphic, became the stylized cuneiform script taken up by the Babylonians and others, and it became the language of diplomacy in the ancient world. The Egyptians likewise developed a pictographic script at the end of the fourth millennium. Later, further scripts emerged from other ancient civilizations: in the Indus valley, in China, in the Mediterranean countries, and, later still (*c*. 800 BC), in Central America.

Our European alphabets are believed to descend from a consonantal script invented by the Phoenicians in the eighth century BC. This was adapted by the Greeks who used some of the signs for vowels; they took it to Italy in the eighth century BC, where it was modified and used first by the Etruscans and then by the Romans. The first Roman alphabet contained 21 letters; others were added over the centuries.

All of these scripts contained both logographs and phonetic symbols. In fact it seems that no writing system has been purely phonetic or purely logographic. Even modern English contains the logographic numerals, to say nothing of mathematical symbols, marks on clothes to indicate how they should be cleaned, road signs, the zodiacal signs, and the signs denoting the planets and gender. This list can be extended indefinitely. In recent years logographic icons have invaded the computer world in GEM and Windows.

There are several examples of scripts invented during the last century for use with minority languages; a recent inventory lists 242 ancient and modern scripts from around the world. This does not include the multitude of type-faces invented since printing came to the West from China in the fifteenth century.

Almost every imaginable surface has been used for writing on; engravings in stone and impressions in clay tablets are among the more durable. Papyrus made by the Egyptians from the sheets of the inner stem of a certain reed, beaten out flat and glued together, is known from about 3000 BC. Parchment made from treated animal skins was probably invented by Eumenes II (197–158

BC) at Pergamum in Asia Minor—the site of an ancient library. It is said to have been imported into Greece from the ancient city of Byblos in the Lebanon; there it was known as 'biblion' (giving us the word 'bible'). Most likely, paper was invented by Cai Lun in China in AD 105. The manufacture of paper gradually spread West, reaching Baghdad in about 793, England in 1494, and Philadelphia in the New World in 1690.

Writing has a direction; we write horizontally from left to right with lines running from top to bottom, but other people have written in every possible way: left to right and bottom to top; (in Boustrophedon) left to right and right to left in alternate lines; circles; and spirals to mention but a few.

Of the 242 scripts already mentioned, 16 remain undeciphered. These include the pictorial Hittite, the script used by the early Indus Valley civilization, several scripts from the Far East, and others from the Mediterranean region (such as Linear A). Since these remain opaque, we have little knowledge of the calendars used by the people who wrote in these scripts.

Nevertheless, many of the ancient writings known to us have been successfully deciphered, generally by using examples of the message in both the unknown script and in a more familiar script, or, alternatively, some knowledge of the language in which the script was written and which may still be spoken today. Thus the Rosetta Stone, inscribed with the same message in Greek and two Egyptian scripts, allowed Jean-Francois Champollion to decipher ancient Egyptian writing. The decipherment of the Mayan glyphs depended on early records made by Diego de Landa, Spanish Archbishop of Yucatan, of the sounds associated with them and on knowledge of local languages still spoken.

Ⅱ Ancient libraries

Both the Greeks and the Romans wrote books on history and on science. Some of these works were transcribed by Classical, Arabic, and Christian scholars, and some have survived in libraries in the Vatican, Constantinople (Istanbul), and elsewhere in Europe and Asia. Sometimes we find a complete manuscript, sometimes only a fragment, and occasionally only an allusion to the title of a lost manuscript. There are in China 26 books of history covering in great detail the events over thousands of years.

As we approach the present, more and more original documents, reports, letters, and edicts survive. At the same time commentaries on texts ancient and modern appear. The supply of material on the calendar becomes a flood.

The fate of ancient records provides an interesting though sorry story. Tomb robbing is an ancient custom and the tombs in the Egyptian pyramids were being plundered even while others were still being built. The search for

FIG. 4.1 The Rosetta stone was discovered by Pierre Bouchard, an officer in Napoleon's expedition to Egypt in 1798. It contains a decree by Ptolemy V of 196 bc which is written in Egyptian heiroglyphs, a demotic script, and Greek. It enabled Champollion to decipher ancient Egyptian writing. © The British Museum.

FIG. 4.2 Jean-François Champollion (1790–1832). He first deciphered Egyptian heiroglyphs. Photo: AKG London.

gold still goes on, and old buildings still provide a ready mine for new building materials. Stone and pottery survives, but artefacts made of organic material decay and crumble to dust. Egyptian papyri discovered by humble Egyptian peasants would be as likely to have been burnt—until a market appeared for them in the nineteenth century.

As well as the gradual wastage brought about in these ways, our knowledge of the ancient world has been impoverished by several catastrophic events—

some accidental, some deliberate. The most tragic is perhaps the destruction of the Library of Alexandria, but there were many more—more than we can detail here in full.

Before his Asian campaign, Alexander the Great (356–323 BC) had founded the city of Alexandria in Egypt when he was 25. His successors there, the Ptolemys, built their capital at Alexandria, where Ptolemy Soter, the first of that Dynasty, founded the great Museum. This resembled what we would call today a university. There were laboratories, lecture theatres, dining halls, a park, a zoo, and, above all, two libraries which were to contain half a million books. There, the scholars collected manuscripts from all known parts of the world; they commissioned a 30-volume history of Egypt based upon the temple records.

The first catastrophe occurred in 47 BC. The Alexandrians were at the time trying to maintain their independence from Rome; Julius Caesar, who had just defeated his rival, Pompey, and had pursued him to Alexandria, took over the Royal Palace, where Cleopatra introduced herself to him. Shortly after this the Alexandrians rebelled against Caesar and, during this rebellion, Caesar gave orders for the Alexandrian navy in the harbour to be burnt. The flames spread to the docks and to the Museum where one of the libraries was burnt. The final blow came in AD 391. Egypt was at that time ruled by Christian monks giving tribute to the Roman Emperor. These monks cared little for books and culture and when Christianity became the official religion of the Empire they were encouraged to destroy the old temples. The patriarch Theophilus destroyed the Temple of Serapis at Canopus which housed the other library. Thus 4000 years of history and learning of the ancient world were finally destroyed. Later there was an attempt by Albulfargus, a Christian writer, to blame the destruction of the library on the Arabs, who were said to have burnt the books to heat the water in the baths. The Arabs certainly took Alexandria in AD 642—but the ancient library was already gone by then.

Earlier, in 221 BC, in China, the Emperor Shih Huangdi gave orders for the imperial libraries to be burnt, and for the scholars who might have been able to remember their contents to be buried alive. Our knowledge of ancient Chinese history suffered a major setback. Despite this, the monumental work started by Joseph Needham, and continued by his colleagues, has given the West a great body of knowledge of Chinese science. Undoubtedly much more remains to be done in codifying the old records which survive.

In AD 529, the Emperor Justinian closed the Academy at Athens and many scholars migrated to Persia. From there, their learning and books were carried to the great Scientific Institute at Baghdad (a city founded in 762), after the Muslim conquest. This became the successor of Alexandria as a centre of learning and reached its greatest eminence in the reign of al-Ma'mun (813–33), son of Harun el-Rashid (of Arabian Nights fame). Many Greek works were translated into Arabic, as well as works from India and Persia. But in 1258 the city was ransacked by Huluka, grandson of Genghiz Khan, the Mongol predator, and the library was burnt.

Later, in the sixteenth century, when the Spanish had subjugated the Mayan people of Central America, they set about destroying every vestige of Mayan religion. This was partly understandable, since many Spaniards had been murdered in a particularly horrific way, as sacrificial victims (as was the customary Mayan way of dealing with prisoners of war). Mayan books, including their histories, were burnt and only four survived to find their way to the libraries of Europe.

Ancient records now face different dangers. Quantities of clay tablets dug up by archaeologists from abandoned cities in Mesopotamia were transported to Europe and deposited, often uncatalogued, in museums. Only a proportion have been deciphered and published; meanwhile in climates more humid than those of Iraq, the tablets crack and crumble. It is ironic that some ancient catastrophe which resulted in the old cities being abandoned and buried beneath the sands did more to preserve their libraries than has housing them in a modern museum.

A different fate has overtaken other ancient, classical, and mediaeval documents which found their way into the libraries of Europe and Asia—they became lost within them, uncatalogued and unrecorded. Maybe in future centuries scholars will succeed in searching through the tons of parchment, papyrus, and paper to find descriptions of new calendars.

Today many millions of books in our own libraries are in grave danger; having been printed on acid-laden paper, they will crumble to dust before long.

CHAPTER 5

Numbers and Arithmetic

⌛ Counting

The possibility of a calendar just did not exist before the invention of counting and a system to express numbers. Without such an intellectual tool, there is no way of assessing the days in a month or a year, and a calendar is impossible.

For calendrical purposes we have the day, the month, and the year, and the passage of time can be marked by counting these events. When man's ability to count had become sufficiently sophisticated, one supposes that people attempted to assess the number of days in a month, of months in a year, and of days in a year.

We share a capacity with many animals to assess quantity at a glance, of 'thinking unnamed numbers': two pints of ale look qualitatively different from one; even three is different from two. But when it comes to assessing the difference between 14 and 15, we have to resort to counting.

The first trick to be mastered was that of matching things off against our fingers and maybe our toes. Finger counting has a long and venerable history and is still used even today. Methods were available to count up to several thousand in a sophisticated language of body signs. These origins are manifest in our number systems that are predominantly based on 5, 10, or 20 to match the digits (even this word reminds us of its history) of one hand, two hands, or both hands and toes.

The important conceptual basis was to have a series of objects (the fingers) or gestures that are arranged in a natural or conventional order. Other objects could then be matched off in turn, so that the finger matching the last object represented the number in the pile being counted. The next stage was to evolve words to refer to the ordered objects. Finally, still in a predetermined order, the words became abstractions that could be applied to objects in any collection, and not just fingers.

In principle a different word could be employed for every number, but the difficulty in remembering them all, let alone their order, makes this method unacceptable for numbers greater than a few. Another method is to repeat a word an appropriate number of times to represent the number (as in roman

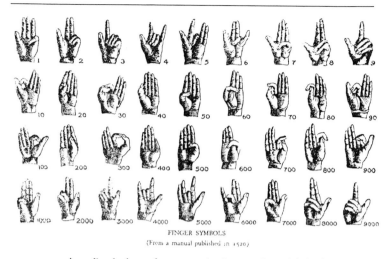

FINGER SYMBOLS
(From a manual published in 1520)

FIG. 5.1 A medieval scheme for communicating numbers with the fingers.
Source: ET Archive.

numerals); again, this is impractical for large numbers. Yet another is binary counting; the Bushmen of Botswana counted 'a', 'oa', 'oa-a', 'oa-oa', 'oa-oa-a', 'oa-oa-oa' to represent the numbers one to six. After six they gave up and said the equivalent of 'many', though in principle the system could have been extended (again with great tedium) to express any number. Computers, which do not mind the tedium, use this principle.

A great advance was the introduction of the multiplicative principle. An example of this from South America, has 'cajezeca' for four and 'teente' for five; 'cajezecaente' was then taken to mean four times five, or 20.

There is some evidence that most languages first developed counting by 5 and 20, though a survey of 307 American languages showed that about half of these used counting by 10. A hint of 20 remains in the 'quatre-vingt' of French and the 'score' of English. In time many languages evolved counting by 10 with proper use of the multiplicative principle. Such a system is used in modern English, where, for instance, fifty-six is a shortened form of five times ten plus six. Separate and ancient words are used for the next two powers of ten: 'hundred' and 'thousand'. For even larger numbers the word 'million' was coined in the fourteenth century, with 'billion', 'trillion', and so on, appearing first in a manuscript by Nicholas Chuquet dated 1485. A billion, until quite recently, meant a 'million million' in English. This usage now seems to have been superseded by the American meaning of a 'thousand million'.

The words used for numbers (and other things) in most European lan-guages, and many elsewhere, have pronounced similarities. This has led some to speculate that they have their root in an ancient and hypothetical language known as Indo-European, supposedly spoken in the fourth millennium BC, if not before. Whether or not this is true, our number words are ancient and counting itself must be even older. It has been suggested that a counting system based on 12 rather than 10 originated from this Indo-European source, and traces of this certainly exist in several European languages: the English have the dozen (12) and the gross (144), and there were 12 pence in the shilling.

> *Eena, meena, mina mo*
> *Catch a chicken by his toe*
> *If he sequeals, let him go*
> *Eena, meena, mina mo*

The first and last lines of the old children's counting rhyme are believed to be reminiscent of ancient numerals. A less politically correct version emerged

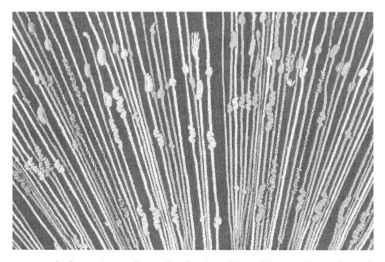

FIG. 5.2 An Inca quipu used as a tally of various objects. Clusters of knots denoted the digits of numbers. Different strings, maybe of different colours, tallied different objects. Source: E T Archive.

later in America, and returned ultimately to England. There are many other versions in various languages: An East Anglian shepherd's count has: 'Ina, mina, tethera, methera'.

Practical necessity required a method of recording numbers to aid the memory. Many people developed ways of writing them on clay tablets, papyrus, paper, slates, or with sand on a table. Others did not and never invented writing as we know it; among these were the Incas of South America. Although they never learnt to write, they could record numbers—by making knots in strings, called 'quipu'. They used the decimal system and a number such as 345 was probably recorded by three clusters of knots containing three, four and five knots respectively. Different items were recorded on strings of different colour. The number of knots in a cluster was felt with the fingers rather than seen and counted by eye. Similar systems were used in many other places and one was still in use in Hawaii in 1820.

Ⅱ Writing numbers

It is difficult to count the number of days in a year without losing count; an obvious aid is to make a notch in a stick every morning. Such would be an example of a tally stick. Several notched bones dating back many thousands of years to Palaeolithic times have been discovered, and may have been tally sticks. Wooden tallies were used by the British Exchequer until 1823 when they were cleared from the Star Chamber in the Houses of Parliament where they were stored, and burnt in the stoves which heated the building. Alas, the stoves overheated and the Houses of Parliament were burnt down. One meaning of the English word 'score' as 'cut' derives from its use in tally sticks.

When a large number of notches have accumulated on a tally stick, it becomes difficult to assess their number at a glance. One solution to this problem is to score a diagonal line through each group of five. The Chinese draw the first four strokes as the sides of a square; the fifth is put in as a final diagonal. This, or something like it, constituted the second step towards a system of written numerals. Thus, 23 might be represented by four blocks each containing five strokes completed with a diagonal, making 20, and three individual strokes. Later, the numbers five and ten were represented by separate symbols, each with a defined value: the Mayans used dots for units and bars for tens; and the Egyptians had a similar system. Such systems could be extended for numbers of any size—but again only with tedium. Later developments led to separate symbols for 5, 10, 50, 100, and so on. These schemes are additive, in that the number represented is the sum of the values of the constituent symbols; it was the basis for early roman numerals and the 'Attic' system used by some of the Greeks.

The Egyptian number system remained undeveloped and, though adequate for the practical needs of their well-known civil engineering projects, was not up to the intensive calculations required for sophisticated astronomy. In particular, their method of expressing fractions remained primitive. Every fraction had to be expressed as the sum of a set of unit fractions of the form $1/n$—though 2/3 and 3/4 were allowed as special cases. This made arithmetic with fractions difficult.

A different approach—the so-called 'Milesian scheme'—was adopted by some of the other Greek city states. This still used the additive principle and is said to have been invented by Thales of Miletus (*c.* 624–545 BC) and Anixamander of Miletus (*c.* 610–540 BC). It used the first nine letters of the Greek alphabet to represent the numbers 1 to 9; the next nine to represent 10, 20, 30, and so on, up to 90; and the next nine to represent 100, 200, through to 900—well, almost. Their alphabet contained 24 letters (alpha to omega), and they required 27; they therefore resuscitated three archaic letters (wau, koppa, sampi) for use as numerals. The system could be extended by adding dots to the letters, for instance, to denote the thousands and so forth. The number in this system could also be evaluated by adding up the individual values of each of the symbols; for example 'σμα' is equivalent to 241. Fractions were expressed as unitary fractions (indicated by primed letters) in a manner similar to that of the Egyptians; for instance, τ′ would indicate 1/300 (but β′ stood for 2/3). The system was popular throughout the Greek world up to 1453, when Constantinople fell to the Turks.

The Jews came to adopt a similar system in which the letters of the Hebrew alphabet replaced the Greek letters; numbers important in the calculations involved in their calendar were given mnemonics derived from the letters. These are examples of a so-called alphanumeric system. Such a system is even evident in fifteenth-century Europe and can be seen in *Les Très Riches Heures* (see Chapter 30). The Indian and Chinese mathematicians had developed similar systems—the Chinese by the end of the third century BC, or maybe even earlier; the Indians by the first century. These systems were decimal.

The use of alphanumeric systems permits every name or word to be assigned a value by adding up the values of its constituent letters. This practice is sometimes known as gematria. A similar pastime is the reverse: to take a name and derive from it a number. A famous example is the number 666, referred to as the mark of the beast (the Antichrist) in the Book of Revelations. Many other names have been proposed, including Nero and Adolf Hitler. If we transcribe Nero into Hebrew, and work out the number, using a alphanumeric system, we get 666. Many other games were played, including ones that established numerical relationships between people. They are not extinct, even today.

Our modern system is not strictly additive; the symbols are written in order and the value of each one depends on its position in the sequence. Thus 123 indicates 1 × 100 + 2 × 10 + 3. It is a place or positional system which was readily extended to cope with fractions, once the decimal point had been invented. It has the advantages that it is as compact as the Milesian system but requires far fewer symbols to be remembered. For it to work properly, one further invention was required: the cipher or zero.

The earliest place system we know of was used by the Babylonians in about 1800 BC; cuneiform tablets from this period give accounts of such. Later in the Seleucid period the same system was used for astronomical purposes. The Babylonians represented numbers up to 60 by two sorts of mark (wedges (T) for units and corners (>) for tens) in their cuneiform writing, in a manner resembling that of the Egyptians. Thus 43 might be written as >>>>TTT (though usually in a more compact geometrical arrangement). When a number larger than 60 was required they divided it by 60, producing a quotient and a remainder. The quotient (which represented the number of completed 60s present) and the remainder were placed side by side. If the quotient was greater than 60 they repeated the process to give three numbers, and so on, for numbers of any magnitude. We might represent the number 12345 as 3, 25, 45 to mean 3 × 60 × 60 + 25 × 60 + 45; though the Babylonians would have written 3, 25, and 45 in terms of their wedges and corners: TTT, >>TTTTT, >>>>TTTTT. This method of counting by 60s is termed sexagesimal (mathematicians would say that the base of the system is 60) and its structure is similar to our decimal (base 10) place system.

The original Babylonian system had two defects: if at any stage of division by 60, the remainder was zero, an ambiguity might arise because the absence of the number was indicated, only vaguely, by a space; only later did they use a special symbol (to mean zero) for this purpose. They could express fractions

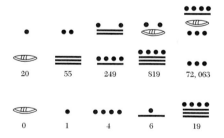

FIG. 5.3 Some numbers as written by the Maya: a dot stands for one, a bar for five, and a lozenge for zero. Numbers are written vertically in a vigesimal system (base 20).

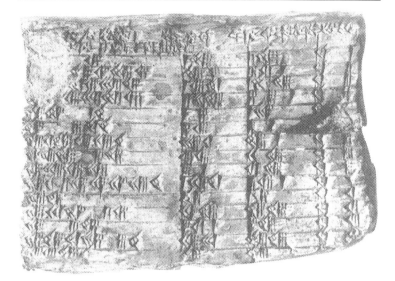

FIG. 5.4 A Babylonian mathematical text with cuneiform numbers.
Source: E T Archive.

using sexagesimals (instead of decimals), but again there was ambiguity because they had nothing analogous to the decimal point; the position of the latter had to be divined from the context of the calculation. Nevertheless, the system was convenient and was one factor in the development of Babylonian astronomy. So convenient was it that it continued to be used for many centuries and is not wholly extinct, as we shall see, even today.

The ancient civilizations of the Near East finally gave way to a new culture after AD 622, when the Arabs conquered lands stretching from India to Spain and Arabic replaced Greek as the language of civilization. But before that, the Indians developed a place system which included a sign for zero (some have credited this to Aryabhata, born AD 476). Their system was decimal, rather than sexagesimal, and it is possible that they had obtained it from the Chinese, and their zero from Babylonia. It was the start of our modern decimal place system.

The earliest Indian record of the use of this decimal place system with a sign for zero dates from AD 595; by 662 it was mentioned in the works of the Syrian bishop, Severus Sebokht. The system passed to Arabic civilization after the rise of Islam and the translation of the Indian treatises, the Siddhântas, into Arabic by Al-Fazari in about 773. Two sets of symbols were used by the Arabs for the decimal digits. One is still used by them, and the other, of gobar numerals, is derived from the Hindu symbols and was used in Spain. Arithmetic in Europe was hamstrung by the continued use of alphanumeric numbers in the East or roman numerals in the West.

From the eleventh century on, European scholars began to visit Islamic mathematicians to learn of the new numerals. Abu Jafar Muhammad ibn Musa al-Khwarizmi—Muhammad, father of Jafar, son of Musa, the Khwarizmian, 680–750 (Khwarizma is in the old Persia)—had written a treatise on Arabic numerals which survives in the form of a Latin translation dating from the twelfth century; a copy of this was found in 1857 in the library of Cambridge University. This book was the major vehicle by which the gobar arabic place system entered European civilization. The Latin form of Khwarizmi gave us the word 'algorithm'. Another book by him, *ilm al-jebr wa'l-muqabalah* (*The science of reduction and equation*) gave us algebra.

In time the Christian states began to dislodge the Moors from Spain (Toledo fell to them in 1085) and scholars flocked to Spain to learn of Arabic mathematics and science. From the end of the twelfth century on, al-Khwarizmi's treatise was translated into several European languages, and the gobar numerals and zero began to be used in Europe. Some translators managed to muddle the origin of the new numerals by attributing them to a fictitious Indian, King Algor.

Trade with the Orient began to return and merchants, and the states in which they lived, became rich; money in large quantities requires accounts, and these necessitate arithmetic. The new numerals, and the new arithmetic based on them, began to be used and taught. Leonardo of Pisa (also called Fibonacci) was one merchant who travelled widely in the Orient. In 1202 he wrote his *Liber abaci*, which described the new numerals and arithmetic and quoted al-Khwarizma. This too was of great importance in the diffusion of such knowledge in Renaissance Europe.

The new numerals were opposed at first by the Church, who perhaps saw them as products of the hated Moors. Another reason for their opposition was that they, the learned Doctors, who alone understood the old arithmetic, saw a threat to their monopoly.

Fractions remained troublesome—a difficulty overcome in the final refinement of the Indian representation of numbers. Although decimal fractions had been in use for a considerable time, and indeed the principle was known from the Babylonian sexagesimal notation, their popularization in sixteenth-century Europe is attributed to the Dutch mathematician, Simon Stevinus (1548–1620). Simon also recommended the adoption of decimal weights and measures, but it took another two centuries, till the French Revolution, for that to happen—and even at the end of the twentieth century that particular innovation is not used universally.

Ⅺ Arithmetic

One further component is required for a modern calendar—arithmetic, and the ability to add, subtract, multiply, and divide numbers; that is, the ability to compute. The oriental civilizations, including Babylon and China, developed these skills alongside their number systems, although some systems were better adapted to arithmetic than others. The Egyptians and Greeks were hampered by their ponderous methods of writing fractions, and arithmetic using roman numerals is none too easy. The Greeks themselves had a predilection for geometry rather than calculation (which was viewed as an oriental art). Nevertheless, multiplication and division with the Greek Milesian numerals can be done with reasonable ease, which perhaps accounts for its continued use in the Greek world.

From the time of the collapse of Roman civilization, there is little sign of an appreciation of mathematics in the West. One of the few spurs to the development of arithmetic was the need to compute the date of Easter. Later, as trade grew, the need for book-keeping did lead to further advances.

In the Classical world and Mediaeval Europe, paper and vellum were scarce and expensive, and arithmetic was done with the use of the abacus. The word 'abacus' is Latin but derives from the Greek word 'abax' or slab; this in turn derives from the Hebrew 'abhaq' which means dust. In the earliest times calculations were performed by drawing numerals in the sand with a stick. Later, in Roman times, a special table or counter with lines drawn on it was used; calculi (the Latin for 'stones') were placed on it and moved to keep track of the computation. The device with beads on a wire frame is later still. The abacus is still in use in many parts of the world; at the University of Uppsala its use is still taught to those interested—a skilled operator can compete with a calculator. This is not so unlikely when one realizes that the feature that limits the speed at which a calculator can be used is the speed at which you can press the buttons; beads can be moved as fast.

The new arithmetic, based on the new arabic numerals, was considerably easier to perform and was rapidly adopted by the new merchants, despite the disapproval of the Church. The earliest printed manual of arithmetic, the anonymous *Treviso arithmetic*, was published in 1478. In the same century double-entry book-keeping was invented. Capitalism had arrived, and in 1492

Columbus reached America, with its plentiful supply of gold to finance such activities.

Before the advent of these new methods, and even after, the construction of tables of the date of Easter and the like was an arduous business which only the highly educated could attempt. Today we can dispense with such tables, for the digital computer, or even the pocket calculator, can calculate the date of Easter in a trice.

Ⅺ Exclusive and inclusive, and backward and forward counting

The Romans had a practice of inclusive counting that is strange to us and which gave great trouble at the time of Julius Caesar's reform of the Roman calendar. They would have said that Monday was the second day after Sunday, rather than the first as we would say. In another example, 14 March was the twelfth day after 3 March. Their process of inclusive counting meant that they started counting on the starting day and thus counted: Sunday (day 1), Monday (day 2), so that Monday was day 2 after Sunday. In this way they succeeded in introducing leap years every three years instead of every four years as required by Sosigenes. A vestige of inclusive counting remains in the French term for a fortnight, 'quinze jours' or 'quinzième', or in the German phrase 'von heute in acht Tagen' for 'in a week's time'.

Another curiosity is the difference between specifying time as elapsed time or current time. When we specify quantities—money, weights, time—we begin our scale of measurement at zero. Thus we have no difficulty with the idea that at 2.25 p.m. more than two complete hours, but fewer than three, have elapsed since noon. On the other hand when we specify a date, the day, month, and year start at one. For example, 3 February 1995 means the third day of the second month of the 1995th year: only two complete days of February and only one complete month of 1995 have elapsed; the day which is a month before this is 3 January 1995 and a month before that it is 3 December 1994. There is no possibility of a date written to include a zero; the year AD 1 is preceded by 1 BC. This is because the date specifies the current day in the current month in the current year (rather than the time elapsed since some instant in the past); the Mayans are possibly unique in specifying dates in terms of elapsed time. The habit of digital computers starting their counting at zero sometimes gives rise to a similar confusion.

Another feature of Roman dates is that they refer to a date in terms of the number of days that must pass before the kalends, nones, or ides are reached. Thus, if the ides of March fall on 15 March, three days before that is called *ante diem III idus*—the third day before the ides, 13 March; though we would call it the second day before the ides. This practice of backward counting should be familiar to most people in the context of time: twenty to eight means twenty minutes before eight o'clock. Other examples are the countdown to an explosion or the start of a race: three, two, one, go.

⚏ Primes, seconds, and thirds

It is curious and sometimes confusing that our subdivisions of time and angle are both called minutes and seconds. The reason for this goes back to medieval Latin, and ultimately to the Babylonian sexagesimal number system. We have seen that a number was denoted in Babylonian reckoning by a sequence of 'digits', each between 0 and 59, with the place value of each successive number being 60 times less than the previous one. These 'digits' count the number of the different powers of 60 just as the digits in our decimal system count powers of ten.

This number system made arithmetical operations—addition, subtraction, multiplication, and division—relatively easy, and extensive multiplication tables were available to help. Division of A by B was often performed by multiplying A by the reciprocal of B; and archaeologists have found tables of reciprocals also.

One can only speculate as to why 60 was chosen as the base of the system. A possible reason is that in order to avoid recurring decimals for simple fractions (as in 0.33 … for $\frac{1}{3}$ in the decimal system), the base should have as many factors as possible. Of all the numbers less than one hundred, 60, 72, 84, 90, and 96 can be subdivided in as many as 10 different ways; 60 is the smallest of these. It may also be that 60 is the lowest common multiple of 12 and 30; 12 being the number of complete lunations in a year and 30 the number of days in a month (in round numbers). It is also a factor of a first approximation to the number of days in a year.

This system was used by the Babylonians for accountancy and astronomical calculations, and by succeeding generations of astronomers for several thousand years. The Christian world inherited the sexagesimal system, via the astronomers at Alexandria, from the Babylonians. Possibly to avoid the difficulty presented by the absence of a zero, medieval computists took to giving a name to each subdivision. Thus the first subdivision into 60 was described by the latin phrase 'pars minuta prima' ('the first small division'); the next was called 'partes minutae secondae' (or 'the second small division'); there were smaller divisions still—'partes minutae tertiae'; and so on. Ptolemy used the system to express fractions of a degree of angle. Its use to record subdivisions of the hour dates from the time of Hipparchus.

In time 'partes' being redundant was dropped: then the 'prima' from the first division and finally 'minute' from all except the first division were left out, leaving the familiar sequence—minute, second, third . . . The thirds have now disappeared, but were still in use in geometry books in the last century. The first, second, third divisions (and so on) were sometimes denoted by ', ", ''' (and so on); the ''' sign is still called a prime.

The system was applied to other quantities than angles and time, and to divisions other than into 60ths. Thus, Webster in 1890 mentions that 12 seconds (") make 1 inch, and 12 inches or primes (') make 1 foot, and thus 12 seconds = 1 prime. Nowadays the prime sign is attached to the number of feet and the double prime (") to the number of inches; there are no seconds any more.

It is interesting that the same principles were adapted to the decimal system and used as late as the eighteenth century, when the first digit after the decimal point was the 'prime' (which denoted the number of tenths); then came the 'second' (which denoted the number of hundredths); the 'third' (for the number of thousandths); and so on. This again may hark back to a time before the zero of our number system was introduced. The function of the zero is to act as a marker when there are no multiples or fractions of particular powers of 10 present in a number. Another way of achieving the same effect would be to name and so distinguish each such multiple as a cube, square, unit, prime, second, third, and so on.

𒐸 Numerology

There are lucky numbers and there are unlucky numbers, so we are told; whether you agree with this or not, certain numbers stand out in the various calendars of the world. Sometimes the origin of the use of these numbers is obvious and sometimes it is obscure. Here I discuss a selection of them and a few other interesting numbers, and note several curious features; the reader should not place too much weight on some of the coincidences mentioned. I have avoided gematria.

THREE (3)

Three is a number of importance in Christian theology: the Trinity. The Egyptians recognized three seasons to their year: innundation, fertility, and harvest.

FOUR (4)

Four seasons are commonly recognized: spring, summer, autumn, and winter. Leap years occur in the Julian calendar every four years, as do the Olympiads. There are four suits in a pack of playing cards.

FIVE (5)

There are five Platonic solids—polyhedra—in which all the faces are the same shape and size. There were five planets (excluding the sun and moon) known to the ancients. This coincidence appealed to the Pythagorean mind. We have five digits on each hand and foot (see also 10).

SEVEN (7)

Seven is a lucky number of great magical significance and much used in the Book of Revelations. It is the sum of the perpendicular sides of the 3, 4, 5 Pythagorean right-angled triangle (see also 12 and 50).

It is approximately the number of days in one phase of the moon. There are seven hills in Rome, seven deadly sins, seven ages of man; there were seven maids with seven mops; and Jacob served seven years—the allusions go on and on. There are seven intercalated months in a Metonic cycle, and there are seven days in the Jewish, Christian, and Islamic weeks, as well as the astrological week. But first and foremost there were seven wandering stars or planets (including the sun and moon); the astrological week is based on this.

You must agree that seven is a popular number even if you do not believe that it is lucky. The lucky quality of seven was not known in the West till the twelfth century, although it had been discussed by Ptolemy.

EIGHT (8)

The eight-year cycle, the octaeteris, was designed by Cleostratus of Tenedos to reconcile the lunations of the moon with the solar year.

TEN (10)

We have 10 fingers and 10 toes; for this reason we have the decimal system. Life would have been simpler if we had 12 twelve fingers, for 12 has more factors than 10. It is not known why evolution favoured pentagonal rather than hexagonal symmetry but even dinosaurs had five fingers and toes on their arms and legs.

The Egyptians, the Chinese, and the French revolutionaries divided their months into 10-day periods; these were the Egyptian decans. There are 10 heavenly stems in the Chinese sexagesimal counting system.

TWELVE (12)

There are 12 full lunar months in a year. The Egyptians divided their day and their night into 12 hours each; the Babylonians by analogy divided the day into 12 beru of two hours each; and the Chinese also divided the day into 12 periods. There are 12 signs of the zodiac and, in the Indian calendar, 12 sankrântis. The number of solar years in the orbital period of Jupiter is nearly 12 (11.86). There are 12 terrestial branches in the Chinese sexagesimal counting period.

Twelve was the number of inches in a foot and pence in a shilling. There are 12 objects in a dozen. These could be the vestiges of a duodecanary system of counting which has certain advantages over the decimal system now prevalent: 12 is divisible by 2, 3, 4, and 6, while 10 is only divisible by 2 and 5. Twelve is the sum of all three sides of the 3,4,5 Pythagorean right-angled triangle (see also 7 and 50).

THIRTEEN (13)

Thirteen is one of the cycles of the Mayan calendar round; these people recognized 13 gods of the heavens who battled with the nine lords of the night for our terrestial domain. There are four times 13 weeks in a year, and 13 playing cards in a suit, and 13 is a baker's dozen. There are 13 Archimedean solids which are semi-regular. There are 13 months in an embolismic year of the Jewish or Metonic calendar.

Thirteen is considered by some an unlucky number, possibly because Judas Iscariot was the thirteenth member to sit down at the Last Supper. A fear of 13 is known as trikaidekaphobia—an affliction probably dating from medieval times.) Curiously, thirteen is one of what mathematicians call 'lucky numbers'

FIFTEEN (15)

The Roman indiction, a census for tax-gathering purposes, was held every 15 years.

NINETEEN (19)

There are 19 years in the Metonic cycle, designed to reconcile the tropical year with the lunar synodic period; 235 lunar months approximate to 19 solar years. Nineteen is important in the Baha'i religion and its calendar has 19 months each of 19 days.

TWENTY (20)

We have a total of 20 digits including both fingers and toes. The Mayans used a vigesimal counting system based on 20; there were thus 20 days in their uinal. Various other people used a vigesimal system and there are traces of this in the French 'quatre-vingt' and in the English 'score'. There are also 20 hundredweight in a ton, and there were 20 shillings to the pound sterling.

TWENTY-SEVEN (27)

The number of nakshatras in the ecliptic, used to mark the passage of the moon around the ecliptic in Indian astronomy, was at one time 27 (and was also, at another time, 28). The synodic period of the moon is between 27 and 28 days.

TWENTY-EIGHT (28)

Twenty-eight represents the number of years in the solar cycle, after which, in the Julian calendar, the pattern of assignment of days of the week to the days of the year repeats. (See also 27.)

THIRTY (30)

The number of days in a lunation is, when expressed as a round number, 30. There were 30 days in each month of the Egyptian calendar and its derivatives, including the French republican calendar. By analogy the Babylonians divided their time unit, the beru of two hours, into 30 ges of four minutes each (the sun moves through about 1° in one ges). Indian astronomers divided the circle into 30 tithis, each marking the passage of the moon from one conjunction to the next.

FIFTY (50)

Fifty was the holiest of numbers to the Jews; perhaps remains of this tradition can be seen in fiftieth wedding anniversaries and jubilees, which are both golden. Fifty is the sum of the squares of all three sides of the 3,4,5 right-angled Pythagorean triangle (see also 7 and 12).

FIFTY-TWO (52)

There are 52 weeks in a year—a pure accident of astronomy; and, oddly, 52 years in a Mayan calendar round. There are 52 cards in a pack. $52 = 4 \times 13$.

SIXTY (60)

Sixty is the base of the Babylonian sexagesimal counting system; it is divisible by 2, 3, 4, 5, 6, 10, 12, 15, 20, and 30. There are other numbers with more factors but 60 is the smallest divisible by so many. The Babylonians divided their ges into 60 gar, about 4 seconds. Today we have 60 seconds in a minute and 60 minutes in an hour or degree. The Indians divided their day into 60 Ghatikas, the Ghatika into 60 Palas, and the Pala into 60 Vipalas.

Sixty is a factor of 360, an early approximation to the number of days in a year. It is also the lowest common multiple of 12 (the number of months in a lunar year) and 30 (the round number of days in a lunation). Sixty is the resonance period of the Chinese 12- and 10-year or day cycles.

SEVENTY-SIX (76)

The number of years in a Calippic cycle.

ONE HUNDRED AND FORTY-FOUR (144)

In English this is a gross; perhaps a vestige of an old Indo-European duodecanary counting system (see 12).

TWO HUNDRED AND THIRTY-FIVE (235)

The number of lunations in a Metonic cycle of 19 years.

TWO HUNDRED AND SIXTY (260)

The Mayans and other Meso-Americans counted 260 days in a Tzolkin, but this may be secondary to the fact that it results from two overlapping cycles of 13 and 20 days. 260 is the lowest common multiple of 13 and 20.

There has been much speculation about the origin of this number. It is approximately the number of days in the gestation period in humans. Some try to make out that it is approximately the period of time in days between the appearances of the planet Venus as the morning and evening star.

THREE HUNDRED AND FIFTY-FOUR (354)

The number of days in a lunar year: $354 = 6 \times 30 + 6 \times 29$

THREE HUNDRED AND SIXTY (360)

The Babylonians divided the day into 12 beru and each beru into 30 ges by analogy of the division of the year into 12 months, each of 30 nominal days. There were thus 360 ges in a day, and the sun moved round the earth by 1/360th part of the circle in a ges (about four minutes). This gave us the angular measure of the degree. Likewise the sun moved round the ecliptic through one degree in about one day: 360 was thus an approximation to the number of days in the year.

THREE HUNDRED AND SIXTY-ONE (361)

Three hundred and sixty-one is the square of 19. There are 361 regular days in the Bahá'i calendar.

THREE HUNDRED AND SIXTY-FOUR (364)

The total face value of a pack of cards is 364. Calendar reformers have noted that: $364 = 4 \times (30 + 30 + 31) = 7 \times 52 = 28 \times 13$. A year of 364 days is mentioned in the Dead Sea Scrolls and was still in use in Iceland in 1940.

THREE HUNDRED AND SIXTY-FIVE (365)

This is the nearest whole number of days in the tropical year. For that reason it was the number of days in the Egyptian vague year and the Mayan Haab. Calendar reformers are dismayed that 365 is only divisible by 5 and 73, and only slightly cheered that 366 is divisible by 2, 3, 6, 61, 122, and 183.

THREE HUNDRED AND SIXTY-SIX (366)

The number of days in a leap year. (See also 365.)

THREE HUNDRED AND EIGHTY-THREE (383)

The number of days in an abundant lunar year: $383 = 354 + 29$

FOUR HUNDRED (400)

The pattern of assignments of days of the week to days of the year in the Gregorian calendar repeats after 400 years.

FIVE HUNDRED AND THIRTY-TWO (532)

The number of years in a Dionysian or Paschal cycle.

CHAPTER 6

The Variety of Calendars

⌛ Empirical calendars

Men have ordered their affairs by the phases of the moon and the seasons for as long as records exist. Even before calendars had been invented they could have told their wives that they would be back three days after the next full moon or remarked that their son was born three winters ago. Such perhaps were the beginnings of calendars. Later, men counted the days between moons and solstices, and were better able to anticipate a day in the future or to remember the past. But they still needed to observe the moon or the solstices to determine when a new month or a year began. As astronomy developed, it became possible to foresee the new moons or the solstices with increasing precision, and calendars as we know them today began to take shape.

For as long as estimates of the length of a lunation or of a year were insufficiently precise, or the calendar inadequate to its task, the months would tend to get out of synchronization with the moon, or the seasons with the year. The remedy was to insert (or even remove) extra days in a month or a year to bring them back into register. This was intercalation (or extracalation).

We call calendars which required actual observations of the sun or the moon, empirical calendars, and note that they had several defects. First, they were dependent on the weather; storms or heavy cloud might well obscure a new moon or the rising of the sun at the summer solstice. A protracted spell of bad weather could throw the calendar into disarray. No doubt such occurrences were less likely in the deserts of Egypt or Mesopotamia than further north, but there were other problems.

As human communities increased in size and empires spread, a second defect became apparent. The time and day on which the new moons and solstices were observed would depend on where the observer stood. A man might first glimpse a new moon on one day; another man, a few miles west, would see it some minutes later, maybe after the start of the next day; after the Diaspora this discrepancy was particularly troublesome to the Jews. The whole question of exact timings became problematic. One solution was for the calendar priests, who were in charge of watching the heavens at some definitive point of observation (such as the Temple in Jerusalem), to communicate their sightings throughout the land by bonfires or messengers. But as empires grew larger, the news of the start of a new month might take days to arrive.

Yet another problem arose from the need to adjust the calendar by inter-calating extra months or days to keep it in synchrony with the heavens. The occasion at which such intercalations were made had to be left to the discre-tion of the officials or calendar priests, who were entrusted with running the calendar. These might be ignorant, incompetent, or venal people; there was money to be made or other advantages to be gained by accepting bribes to delay or advance an intercalation for political or financial reasons. By the time of Julius Caesar, the Roman calendar was seriously affected by such practices.

In time, however, precise astronomical knowledge and ingenious calendri-cal systems made it possible to design a calendar that could be calculated well in advance, and which kept in reasonable synchrony with the sun or the moon. Any intercalations required were made according to a rule and were not at the discretion of the priests.

It was at this point, to grossly oversimplify a complicated swathe of history, that the calendars of the East and of the West diverged. The calendar makers of India and China designed calendars that kept in step with the sun and the moon; they depended on calculations of the exact time of a new moon or a solstice. The lengths of the months or the years were subservient to these events. Such calendars have been called astronomical calendars. In contrast the Western nations—Egypt, Rome, Christendom, and Islam—opted for arithmetic calendars in which the lengths of the months and years were fixed according to simple but rigid rules, which were designed to keep the calendar more or less in step with the heavens, but did not require either astronomical observations or calculations.

Both the astronomical and arithmetic calendars depend on precise know-ledge of the length of a lunation or a year. Within the limitations of this knowledge the astronomical calendars are self adjusting and are always in agreement with the heavens, whereas the calculated calendars only keep in step on the average.

The basic element of almost all calendars is the day, and all longer calendri-cal periods, such as weeks, months, and years, contain a whole number of days—even if the astronomical lunation and year do not. The problem beset-ting calendar makers, whether they opt for an astronomical or an arithmetic calendar, is how to reconcile the day with the lunation and the year. Different peoples managed this in different ways, but before we turn to the different strategies that were adopted, we must first examine the day, month, and year more closely. For each of these periods of time, it is necessary to agree on the moment when each begins and to know its precise length as measured in days and fractions of a day.

⚅ The day

The day is the period defined to include a night and a day. It is mostly unproblematic to count the days as they pass, and only people who live within the Arctic Circle where the sun does not set in summer or rise in winter have difficulties. Nevertheless, when we equate a day in one calendar to a day in another, we must pay close attention to when the day starts and when it ends.

In the history of most nations, including the Greeks, the day has started either at sunrise or at sunset. For the Romans, however, it began at midnight. After the collapse of the Roman Empire, many people reverted to starting their day at sunrise; whilst today we have gone back to the Roman custom. Astronomers once preferred their day to start at noon, so that the whole of their viewing time (at night) fell in the same day, but since 1925 they have started their days at midnight. Only their Julian days, discussed in Chapter 3, start at noon.

In the absence of clocks there was very little alternative to starting the day at daybreak or sunset, despite the disadvantages. Whether the day began at noon or midnight, sunrise or sunset, the precise moment had to be determined from the sun. The instants of time when you observe these events depends on your location—they vary with longitude and also with latitude. Thus, two people living not far apart could disagree as to on which day an event took place, if it occurred close to the start of a day.

The length of the day measured by clocks is not constant and only a mean can be precisely defined. In this century it has been found that even the average length of the day is slowly increasing—but very, very slowly.

⚅ The astronomical month and year

The lunation—the period defined by the cycle of phases of the moon—is the basis of the month, and the moment when each cycle starts is a matter of convention. The start of a month was once determined by observations of the phase of the moon: some people began their months with the first sighting of a new moon; others with the disappearance of the old. Others preferred a gibbous moon, and yet others used an interpolated date of conjunction with the sun, when the moon is not actually visible.

In order to anticipate the day of the start of the month, it is necessary to know how many days there are in a lunation. The ancients found that this interval was sometimes 29 days and sometimes 30. Eventually, they accepted the fact that the exact number was not integral; we know today that it is not even constant, varying by several hours from lunation to lunation. Only the period of a mean lunation—the mean synodic period of the moon—can be stated precisely, and that too is increasing, albeit very slowly.

The astronomical year is the period in which the sun moves round the full circle of the zodiac, or, which is the same thing, the earth revolves around the sun in its orbit. Like the lunation, the moment when it starts is a matter of convention. Some have taken either the vernal or autumnal equinox as this moment; others, the summer or winter solstice. Astronomers opt for the vernal equinox. The history of the calendar is largely the history of the invention of better methods for detecting the instant of an equinox or solstice and of measuring the length of the year.

The ancients found the year to contain 365 or 366 days. Today we know that, like the lunation, it varies by several minutes from year to year; only the mean tropical year can be precisely specified, and that is slowly decreasing.

⚃ Reconciling the incommensurate

TABLE 6.1 shows modern estimates of the number of days in a lunation and a year, and the resulting number of lunations in a year. We can see that such numbers are not whole numbers. It is not convenient to have a fractional number of days in a month or a year, or a fractional number of months in a year, and the historical development of the calendar is marked by attempts to come to terms with the problems this poses.

⚃ Lunar calendars

Most early people used a lunar calendar and no doubt they began by counting the days each time a new lunation was judged to have started. If the new moon (or whatever portent indicated a new month) could not be seen because of bad weather, there was uncertainty. In time this became intolerable and formal systems in which a fixed number of days were assigned to each month were developed. Maybe extra days were intercalated or extracalated if the count got out of phase with the astronomical moon.

A glance at TABLE 6.1 suggests the obvious approximation of assigning 29 and 30 days to alternate months, and all lunar and lunisolar calendars employ variations on this theme. Taking 29 and 30 in strict alternation, the average number of days in a month is 29.5, with a discrepancy of 0.030 589 days per month; this amounts to a full day in about 32 months or a bit under three years. Such a calendar would get badly out of synchronization with the moon within a human lifetime.

TABLE 6.1 *The length in days of the lunation and astronomical year*

Period	Unit	Length	Variation
Lunation	Day	29.530 589	0.3
Year	Day	365.242 190	0.003
Year	Lunation	12.368 266	
12 lunations	Day	354.367 066	

The lunation refers to the mean synodic period of the moon, and the year to the mean tropical year; both show small secular changes. The day shows a variation in the course of a year of about 50 seconds and also a small secular change. The periods given are accurate for 1 January 2000.

It is relevant to note that one second of time is about 0.000 012 days.

One solution to this problem was to intercalate an extra day from time to time. There are two difficulties with this however. First, there is the question of when exactly to intercalate—regularly according to some rule, or after examination of the heavens to see if one were needed? Then, there is the complication that the individual lunations are not all of exactly the same length.

Here we note the approximation used in the only surviving lunar calendar of note, the Islamic calendar. This intercalates 11 extra days in each cycle of 30 years of 12 months each, to give an average month of $(29\frac{1}{2} \times 360 + 11) / 360 =$ 29.530 556 days; the synchronization is much better, and the small remaining discrepancy amounts to a day in about 2500 years. The Islamic calendar has so far been running for a little over half of this time. This rule, invented by mediaeval Arabian astronomers is simple and accurate. The period of 30 years is less than a human lifetime, so there is little difficulty in remembering how long the current cycle has been running. A more accurate rule, particularly for cycles of less than 200 years, would be to insert 29 days every 79 years. This would reduce the discrepancy some sixfold, but it is more complicated and spans longer than an average human lifetime.

It might be thought that all lunar calendars whose months began when some phase of the moon was observed, would remain in perfect synchronization with the great 'clock in the sky', in all parts of the world, at all times; but this is not so. The instant when, for example, a new moon is first observed depends on the latitude and longitude of the observer, to say nothing of the state of the weather.

☒ Lunisolar calendars

A purely lunar calendar is not acceptable to farming communities who run their lives according to the seasons, and the search started for ways of reconciling the lunation with the tropical year—that is, for a workable lunisolar calendar.

Keeping a calendar in step with both the moon and the seasons is difficult. One solution was to introduce or intercalate a thirteenth month into the year from time to time. The problem then was to decide when to do this. Some left it as an arbitrary decision of the local priest or astronomer; other people developed rules based upon simple natural or astronomical phenomena. One interesting example of the latter is the practice of the fishermen of Botel-Tobago Island (near Taiwan) who operated a lunisolar calendar. When the calendar fell too far behind the season, they failed to observe the seasonal rising of the flying fish in the month expected; then was the time to postpone their fishing season by inserting an intercalary month. There were many variations on this theme practised across the world.

The Babylonians and later the Greek astronomers, put their minds to the problem and came up with purely arithmetical rules. Over the years the accuracy of their astronomical knowledge and the adequacy of the rules improved. The aim of these rules was to provide a practical scheme for intercalating the extra month. The average length of the month should be such as to keep the months in step with the lunations, and the frequency of intercalation, such as to keep the average year in step with the seasons. It took time for a scheme which adequately satisfied both of these criteria to evolve.

The initial steps were taken in Babylon, from where development passed to the Greek world. Our knowledge of Greek calendars is all too fragmentary and often we must rely on classical authors, some writing much later. The following account must be viewed as being schematic and not necessarily historically accurate. An early Greek calendar attributed to Solon of Athens (638–558 BC) specified a year of 12 months containing alternately 30 days (full months) and 29 days (deficient months) to give a total of 354 days. The calendar was reconciled with the year by introducing an intercalated month of 30 days every other year. Thus two lunar years would contain 738 (= 2 × 354 + 30) days, whereas two tropical years contain about 730$\frac{1}{2}$ days. The average length of the month was 738/25 = 29.52 days, so that the calendar would get a day out of step with the moon after about eight years.

The discrepancy of seven and a half days was ultimately unacceptable and a little later Cleostratus of Tenedos (c. 520 BC) suggested that the intercalated month be dropped once every eight years. Thus an eight-year cycle, the octae-teris, which contained 2922 days and 99 months, was developed. Miraculously, eight solar years of $365\frac{1}{4}$ days also contained 2922 days.

The matter seemed to be settled, but, alas, the actual length of these 99 lunations was 2923.528 days. This meant that Cleostratus' calendar was a day and a half out of step with the moon at the end of eight years; after 160 years the discrepancy would be 30 days. It was suggested therefore that one of the intercalated months be dropped every 160 years. In practice the Athenians reverted to the older and unsatisfactory method of dropping the month when they felt like it, with the inevitable result of corruption and chaos in the calendar. Nevertheless, the octaeteris came to be considered a fundamental time period; for instance, the Olympic Games were held (and still are) every four years (half an octaeteris). There the matter stood till Meton and Euctemon of Athens introduced a new method in 432 BC. This method was almost certainly invented by the Babylonians and in use there by 499 BC. When contact between the East and the West was opened up by Alexander the Great, it became known in India and, later still, in China. It is, however, difficult to be certain who first discovered it or even if it was discovered independently more than once. As we shall see, the Indian and Chinese versions differ in subtle ways from the Babylonian method.

Meton introduced a cycle of 19 years—the Metonic cycle. Each of these 19 years were to contain 12 months apiece, but seven extra months were to be intercalated in the 3rd, 5th, 8th 11th, 13th, 16th, and 19th years of the cycle, to make a total of 235 months. Of these months, 125 were to be full (30 days) and the remaining 110 deficient (29 days), to give a total of 6940 days. The distribu-tion of full and deficient days was calculated as follows. First suppose that all 235 months were full (30 days each), to give a total of 7050 days; from this, 110 days must be dropped or 110 months made deficient. This was done by drop-ping every 64th day and demoting the month in which this day occurred. The average length of a month was thus 29.5319 days, whereas the astronomical average was about 29.5306. The discrepancy was only about 24 minutes in a year. Likewise, the average length of the year was 365.2632 days—about 30 minutes too long.

Meton was eventually honoured for his accomplishment by an Olympic crown and his name, and that of his co-author, Euctemon, are said to have been inscribed in letters of gold in the Temple of Minerva in Athens, together with the numbers in the calendar which gave the positions of the years in the 19-year cycle. In medieval calendars these numbers were still being written in gold, and to this day they are known as 'golden numbers'. This medieval usage dates back at least to the *Massa compoti* of d'Alexandre de Villediew, which was written in 1170. One thirteenth-century scholar wrote: 'This number excels all other lunar ratios as gold excels all other metals'. Curiously the Athenians themselves only used the Metonic calendar for a short period, but the Seleucid Empire later adapted it from Babylonian practices.

The small discrepancies in Meton's calendar were tolerated, by those who used it, for about 100 years. To correct them, Calippus of Cyzicus (c. 370–300 BC) proposed replacing Meton's cycle of 19 years by one of 76 years—the Calippic cycle—and to omit just one day (in the fourth 19-year period). The average month was now 29.5308 days and the discrepancy reduced to about 22 seconds a year. The average year became exactly $365\frac{1}{4}$ days long (about 11 minutes too long still, but the Greeks of the time believed that the true length of the year was $365\frac{1}{4}$ days). The first Calippic cycle was reckoned to have started in 330 BC.

Later still, in 143 BC, Hipparchus of Nicea (c. 180–125 BC) recommended dropping one further day every four Calippic cycles, or 304 years. This gave an average year too long by about 6 minutes and an average lunation of only half a second less than the true synodic period. This was the best that the ancients were able to do for people who wished to continue with a lunar calendar. A variation on the method is still used in the Jewish calendar and in the Christian calculation of the date of Easter. TABLE 6.2 summarizes these refinements.

At this point we may ask whether a better rate of intercalation of a thirteenth month is available. It is readily shown that the rate of intercalation (R), expressed as the average number of months intercalated in a year (this is some proper fraction), should depend only on the ratio of the lengths of a mean tropical year (Y) to that of a mean lunation (L). In fact, we can show that ideally $R = Y/L - 12$. Estimates of $Y/L - 12$ vary from 0.368270 for accurate modern values of Y and L, to 0.368530 for the old estimate of $Y = 365\frac{1}{4}$ days. The theory of continued fraction can be used to provide the following successively better approximations (whatever reasonable value of R ($Y/L - 12$) we select): 1/3, 3/8, 4/11, 7/19, 123/334. The first approximation gives us an early but crude intercalation rule; 3/8 gives us the octaeteris rule; as far as I know the 4/11 has never been entertained; 7/19 is Meton's rule; and 123/334 involves intercalating 123 months in every 334 years (but it is not really practical to use a cycle longer than the human lifetime). More detailed analysis shows that the 7/19 rule cannot be improved upon for cycles of less than 100 years.

The Indians and the Chinese also used what is in effect a 19-year cycle, but based upon more subtle astronomical considerations. (We discuss these more fully in Chapters 12 and 13.)

The Indians also used a radically different approach to the non-integral number of days in the lunation. They effectively abandoned the day and used instead the tithi. This was, on average, 1/30th part of the lunar synodic period, or 0.984353 days. The tithis were numbered consecutively; the first one of each lunar month began at the moment of conjunction. Each day was allocated the tithi which prevailed at sunrise. Not every tithi corresponded to a day; since the tithi varied in length (as does the solar day), sometimes one was skipped and sometimes two days had the same tithi.

Ⅺ Solar calendars

Those who, like the Romans, were prepared to abandon the moon and orga-nize their calendar to keep in synchrony only with the sun were presented with a simpler problem. This was first solved by the ancient Egyptians, with their 365-day wandering year. Later, the Romans used a year of 365 days with an extra day intercalated every four years. This gave a four-year cycle of 1461 days, with an average length of the year of $365\frac{1}{4}$ days—a little too long. The date on which the equinoxes and solstices occurred would advance one day after about 128 years. This was the calendar that Julius Caesar introduced on the advice of Sosigenes. It became known as the Julian calendar and was used throughout Christendom till the sixteenth century.

The discrepancy between the Julian and astronomical years eventually became unacceptable, and in the sixteenth century further refinements were introduced by Pope Gregory XIII. The intercalated day was to be dropped from the years of the Christian era which were divisible by 100, but not by 400. The resulting 400-year cycle had 146 097 days and the year contained an average of 365.2425 days. Estimates available to Gregory of the length of the astronomical year were within a few seconds of this; modern-day estimates show that it is about 26 seconds too long—but this discrepancy would not amount to a day till about the 51st century. The progress of our knowledge of the solar year is charted in TABLE 6.2.

Several other proposals have been made to reduce this remaining small dis-crepancy. The French astronomer, Delambre, suggested dropping a leap day every 3600 years—an excellent suggestion, provided someone was able to remember when exactly to do it. The astronomer Sir John Herschel (1792–1871), son of Sir William, the discoverer of the seventh planet, made a suggestion which is easier to grasp and remember, although it had been anti-cipated by Gilbert Romme in 1795 in his modification of the French revolu-tionary calendar. The proposal was to drop the leap day in years divisible by 4000. This would lead to an average length of a year of 365.24225 days, which, so he thought, was a mere five seconds too long and would lead to a discrepancy of a day only in the remote future.

We may ask whether it is possible, using all the paraphernalia of

TABLE 6.2 *The refinement of the lunisolar and solar year*

Lunisolar year

Originator	Date (BC)	Days	Years	Error	Lunations	Error
(Solon)	c. 680	738	2	3.8 d.	25	−15 m.
Cleostratus	c. 550	2922	8	11.0 m.	99	−22 m.
Meton	c. 432	6940	19	30.0 m.	235	2 m.
Calippus	c. 330	27759	76	11.0 m.	940	22 s.
Hipparchus	c. 143	111035	304	6.0 m.	3760	−0.4 s.

Solar year

Originator	Date	Days	Years	Error
(Egyptians)	—	365	1	0.25 d.
Sosigenes	46 BC	1461	4	11 m.
Clavius	1582 AD	146097	400	26 s.

Key: m. (month); d. (day); s. (second)

The errors are the difference between the mean calendar year and the mean tropical year, or between the mean calendar month and the mean synodic period.

modern mathematics and computers, to design a more accurate scheme of intercalation. The short answer is no. The problem is to discover the number of leap years required in a given period to best approximate the length of the year. Suppose that we decide to have A leap years in a period of B years. A and B must be chosen so that the ratio A/B approximates to the fractional part of a year expressed in days; this I shall take to be 0.242 189. The Gregorian calendar has A = 97 and B = 400. One may set a computer to take all values from 1 to 1000 for B, and all appropriate values of A, and calculate the ratio A/B for all such pairs of values. The best solution that can be found in this way gives A = 225 and B = 929; the difference between the assumed length of the year and the ratio A/B then amounts to a tiny 0.6 seconds a year.

However, there is a difficulty in deciding exactly when to intercalate the 225 days. Both the Julian and Gregorian calendars use very simple rules for deciding this issue. But first it is worth restating the problem. Suppose we intercalate a day regularly once every N years and then, as well, drop or insert a day K times in 100N years. For the Gregorian scheme N = 4 and K = −3, meaning that three leap days are dropped every 400 years. It then turns out that, within these constraints and examining all values of N up to 10, the method proposed by Lilio and Clavius is by far the best. The only other scheme that equals or improves on its accuracy requires a leap day every five years and the insertion of 21 more days within the space of 500 years; one way of doing this is to add another leap year every 25 years, and an extra one still every 500 years. This has several disadvantages, not the least being the possibility of having more than one leap day in some years. There is little compensating advantage, for the discrepancy is only reduced from 26 to 16 seconds a year.

In 1923, the Soviet Union adopted a somewhat more accurate, but slightly more complicated scheme. In this there are leap years every four years except in certain centurial years: the leap year is dropped in every centurial year unless it leaves a remainder of two or six when divided by nine. Thus seven out of nine of the centurial years are common. The mean calendar year is now a little under three seconds too long.

However, we now know, since the work of Simon Newcomb at the beginning of this century, that the length of the year is slowly decreasing by about half a second each century. This makes nonsense of any attempt to design a calendar that will last forever. Because of this effect the Gregorian calendar will be a day slow in about the year 4000; Herschel's adjustment would only extend this by about 1000 years. The year will presumably continue to get shorter and days will have to be dropped from the calendar year on a permanent basis from time to time—not that this need concern you or me. The effect will not be noticeable for many centuries, and at this point we will let the matter rest.

⌛ A classification of calendars

EMPIRICAL CALENDARS

(a) The start of the months or years is determined by direct observation.

(b) The number of days in the month and year is fixed save that extra days are intercalated when this is deemed appropriate.

CALCULATED CALENDARS

(a) **Lunar calendars:** no attempt is made to keep the start of the year in synchrony with the sun, but the months keep in reasonable step with the moon by intercalating days according to a rule; for example, the Islamic calendar.

(b) **Solar calendars:** the moon is ignored but the year keeps in step with the sun.

(*i*)   *Astronomical calendars:* the start of the year is determined from the calculated time of an equinox or solstice; for example, the original forms of the French Revolutionary and Bahá'i calendars.

(*ii*) *Arithmetic solar calendars:* the length of the year is adjusted by inter-calating days according to precise rules to keep it in synchrony with the sun; for example, the Julian, Gregorian, Coptic, Ethiopian calendars.

(c) **Lunisolar calendars:** the months are geared to the moon and extra months are intercalated to keep the year, on average, in synchrony with the sun.

(*i*)   *Astronomical lunisolar calendars:* these attempt to keep in synchrony with both the moon and the sun, but the start of the month or year is determined by astronomical calculation; for example, the Indian and Chinese calendars.

(*ii*) *Arithmetic lunisolar calendars:* these attempt to keep in synchrony with both the moon and the sun by rule-based intercalations; for example, the Jewish calendar and the Christian ecclesiastical calendar.

(d) **Calendars with a wandering year:** both the moon and the sun are aban-doned. Each year contains a fixed number of days—365; for example, the Egyptian civil calendar; the Mayan year.

⏳ Intercalations

We have seen how the operation of a lunisolar calendar requires the intercala-tion of an extra month from time to time. The year which contains such a month has been called 'embolismic' and those that do not, 'common'; months with more days than usual have been called 'full' or 'abundant' and those with fewer, 'deficient'. Likewise, the Roman year that contained an extra day was called 'annus bissextile', because in such years both 24 and 25 February were called VI Kal. Mart.; the first of these was called ante diem bissextum Kal. Mart or bis VI Kal. Mart.

In the Christian calendar, 29 February is the intercalated day in leap years. Leap years were so called because, as was written in the 1604 edition of the Anglican prayer book, 'On every fourth year, the Sunday Letter leapeth'. The years which are not leap years are generally called 'common' years. In the French Revolutionary calendar, as in several other calendars, the extra day came at the end of the year. The French called theirs the 'fête du Revolution', and other calendars also had special names for it.

There is a subtle difference between Anglican and Roman Catholic practices concerning leap day. In the old Roman Julian calendar, the extra day was, as we have seen, inserted between the VII Kal. Mart. and VI Kal. Mart.—that is, between 24 and 25 February. That practice was taken over by the Roman Catholic Church and continued by the English until 1662, when the extra day was moved to a place between 28 February and 1 March and called 29 February. All this makes little discernible difference, except that in the Roman Catholic practice the Sunday, or dominical letter, is changed after 24 February rather than after 29 February. This can never affect the date of Easter but it does lead to celebrations of the feast of St Matthias taking place on different dates in leap years in the two Churches. In leap years, Roman Catholics celebrate this feast on 25 February, whereas the Anglicans continue to celebrate it on 24 February. Likewise, Roman Catholics celebrate the feast of St Gabriel of the Seven Dolours one day later, on 27 February, in leap years. It is as if the leap day and the day following were counted as one.

⧗ Contrived cycles

So far we have discussed natural cycles of days based upon astronomical events and man's attempts to reconcile the lengths of the year, month, and day. Further cycles were, however, introduced by various nations from time to time.

The most ubiquitous cycle is the seven-day week which has been running in one form or another for 3000 years or more; more than 150 000 weeks have gone by. Other nations had short periods resembling the week but containing five, six, eight, or ten days. The Romans had an eight-day cycle, and the Mayan people of Central America had both a 13-day and a 20-day cycle running together.

Another cycle of some historical importance was the 15-year cycle of indiction. The pomp and splendour of Rome, as well as its army, were supported by taxes. Every 15 years, the wealth of the landowners, who paid these taxes, was assessed by an indiction. The first is said to have been held in AD 312, after Constantine had moved the capital of the Roman Empire to Constantinople, but one may extrapolate the 15-year cycles back in time; in this way the year AD 1 is generally regarded to be the fourth year of the then current cycle of indiction. The year of indiction is often taken to have begun on 1 September, but other dates have been used.

Most other ancient cultures—the Indians, the Chinese, the Mayans in particular—entertained cycles of years; some encompassed millions of years. The Chinese employed two cycles of 12 and 10 years.

⚄ Interlocking cycles

An interesting effect, which has been called a resonance, occurs when two or more cycles with different periods operate simultaneously. Any particular day (or year if they are cycles of years) has several positions as counted from the start of each of the cycles. The day is thus characterized by a set of numbers, one for each cycle. In time the same set of numbers will repeat, thus forming a longer cycle. The period of this longer period is quite simply the lowest common multiple (LCM) of the periods of its components. For instance, the Mayans used three cycles of 13, 20, and 365 days. The LCM of these is 18 980 (13 × 20 × 365/5) days, which is about 52 years. They combined the 13- and 20-day cycles into a larger 260-day cycle, usually called a tzolkin. A similar device was employed by the Chinese in numbering their years (and days). They used a cycle of 12 years in which every year of the cycle was given the name of an animal, and a cycle of 10 years named by celestial signs. Thus a year was characterized by the name of an animal and a celestial sign. The sequence of these pairs of names repeated every 60 years.

In the Christian Julian calendar, the first day of the year can fall on any of the seven days of the week, but there is a leap year every four years. This means that there is a complete cycle of 28 years before the days of the week fall on the same dates over an entire year. This cycle of 28 years is known as the solar cycle or cycle of the sun. It was invented about the time of the Council of Nicaea in AD 325, but the first cycle is regarded as having started with the year 9 BC, so that AD 1 was year 10 of the first solar cycle. Since the first day of a year may fall on any one of the seven days of the week, and be either a leap or a common year, there are 14 possible calendars. If you have not filled up your diary and are more than usually thrifty, keep it for a few years and use it again.

Following the Gregorian reform, the solar cycle is now interrupted at the end of three centuries in four, so that although there are still only 14 possible calendars, the sequence only repeats every 400 years.

The date of Easter in the Christian calendar is fixed using a variant of Meton's scheme. According to this, the full moon falls on the same day of the year every 19 Julian years (sometimes termed the lunar cycle). The lunar and solar cycles combine in the Julian calendar to give the paschal or dionysian cycle of 532 (19 × 28) years. The position of any year within the lunar cycle is known as its 'golden number'.

The Renaissance polymath Joseph Justus Scaliger (1540–1609) introduced a 'monster' cycle by combining the solar and lunar cycles, and the cycle of indiction, to give a cycle of 7980 years. This he named the Julian cycle since, to quote his own words (in translation), '. . . it is laid out for Julian years

SMILING THROUGH . . . *By LEE*

[No. 2,895] COMPLEX NEW YEAR

" Now that's a nice calendar, madam. Pre-war. But if you remember that Mondays are Sundays until Tuesday, February 29, it will do nicely for 1944."

Reprinted from the London Evening News, Dec. 30, 1943

FIG. 6.1 Most things were in short supply in Britain in 1944.

only . . .'. (There is a widespread belief that he also named it in honour of his brilliant but eccentric father, Julius Caesar Scaliger (1484–1558).) Any particular year within this Julian cycle is uniquely defined by three numbers which give it its position in the solar, lunar, and indiction cycles. Thus, the position of AD 1 in the Julian cycle is 4714; and the first year of the Julian cycle is 4713 years before this, or BC 4713.

The Julian cycle is still used by astronomers, with the small difference that the year AD 1 is preceded by the year 0, and that by the year −1, and so forth. In their scale the Julian cycle began with year −4712. In fact, they also define a Julian date—each day, which begins at noon, Greenwich Mean Time, is given a unique number; day zero is 1 January 4713 BC in the Julian calendar.

Nearly as long as the Julian cycle is the cycle implicit in the long count of the Mayas; this was 13 baktuns or about 5125 years long (1 872 000 days exactly). At the end of each great cycle the world comes to an end and is recreated. Even longer periods, up to the hablatun of about $1\frac{1}{4}$ billion years, were considered. The Chinese had similar beliefs and had numerous cycles which we consider in detail in Chapter 12. Here we note that, after a period of 31 920 years, which they called a chi, the world would come to an end only to be created anew. There are other instances of a belief in cycles of destruction and renewal.

The chi cycle did not exhaust the liking of the Chinese for large cycles. They entertained a cycle of 4617 years (which was constructed on principles similar to those of Scaliger's Julian period) and another period, 138 240 years, on whose completion all the planets were supposed to be simultaneously in conjunction. Finally, the resonance of these two gave a period of 23 639 040 years.

The Hindus had a series of long cycles based on early astronomical and cosmological speculation. One of these, the mahayuga, was of 4 320 000 years, but there is disagreement about the lengths of the shorter krta-, treta-, dvapara-, and kali-yugas. The kali-yuga is said to mark a grand conjunction of all the planets. It was argued (unconvincingly) that such an event occurred in 3102 BC, initiating an era which would last for 432 000 years. The numerological reasoning behind these numbers is unclear but could be based on 10 800—the number of muhala (see Chapter 3) in 360 days.

Some cycles of interest are listed in TABLE 6.3.

Ⅺ Regnal years and eras

The position of a year within a cycle is one way of specifying a year, but a variety of other methods have been used. You or I might use the years of salient events in our lives as landmarks, and describe other events with reference to them: 'the year I was born', 'the year I first went to America', 'the year my father died', and so forth. The method is as old as records exist; the Babylonians compiled lists of notable events, one for each year.

Events of particular importance, such as the accession of kings, acted as beacons for the naming of years. Thus an agreement might be noted as being signed in the third year of King Nebuchanezzar II. This method of specifying regnal years, sometimes called eponymous dating, has persisted into the twentieth century in England, where Acts of Parliament are dated in this way, for example, as 3 Elizabeth II (meaning the third regnal year in the reign of Elizabeth II).

Yet another method is to name rather than number the year with reference

TABLE 6.3 *Chronological cycles of years and days*

Cycles of years

Name	Length	Components	Inventor	Date	Place
Olympiad	4				Greece
Octaeteris	8		Cleostratus		Tenedos
Heavenly stems	10				China
Earthly branches	12				China
Metonic ⎤ Lunar ⎦	19		Meton		Athens
Solar	28				
Calippic	76	4 × 19	Calippus	330 BC	Cyzicus
Hyppolytic	84	3 × 28	Hyppolytus		Rome
Dionysian ⎤ = Paschal ⎥ = Victorian ⎦	532	28 solar × 19 Lunar	Dionysius Exiguus Victorius	531 AD 465 AD	Rome Rome
Julian	7980	15 indiction × 28 solar × 19 metonic	Scaliger	1583	Italy
Chi*	31920	10 × 12 × 19 × 28			China
Mahayuga*	4320000	400 × 30 × 360			India
Hablatun* *c.* $1\frac{1}{4}$ billion (460 800 000 000 days)					Maya

Cycles of days

Name	Length	Originators
Week	7	Jews (?)
Nundinae	8	Romans
Decade	10	Egyptians, French
Tzolkin	13	Mayans
Haab	20	Mayans

*The Chi, Mahayuga, and Hablatun are representative of Chinese, Indian, and Mayan cycles. These nations employed other long cycles which are multiples or fractions of the ones given, but only for mythological or cosmological purposes.

to some regularly occurring event. The Greeks of Athens dated events by mentioning the current archon—a new one was elected every year. Similarly, the Romans mentioned the current consuls—again new ones were appointed each year. So Alexandria was founded in the year of the consulship of C. Valerius, and M. Claudius. The Greeks sometimes dated events by the Olympiad; Olympic games were held every four years (as they are today following their revival in the last century). Thus Alexandria was founded in the second year after the 112th Olympiad (written as 112.2) in 331 BC. The first Olympiad is said to have been held in 776 BC.

Yet another method was to specify the year with reference to some cycle. Thus Hipparchus, when comparing the dates of the summer solstice determined by himself and by Aristarchus, several years earlier, notes that his observation was made in year 43 of the third Calippic cycle, whereas that of Aristarchus was in year 50 of the first cycle. As we have noted (see page 96), the Calippic cycle was 76 years long; the two observations were therefore 145 years apart. Since the first cycle started in 330 BC, Aristarchus made his measurement in 280 BC (when he was aged about 30), and Hipparchus made his in 135 BC (when he was about 55). The Chinese developed a system for naming the years that involved a cycle of 60 years (about the expected lifetime of a man), composed of interlocking 12- and 10-year cycles. In time however, the 60-year cycle of names returns to the start, leaving an ambiguity.

Perhaps the most important method, however, was to date all events from the year (or presumed year) of some very notable event, such as the Creation, the birth of Christ, or the flight of the Prophet from Mecca. These events mark the first year of eras. This method, in many cases, developed naturally from the practice of regnal dating. Instead of starting the count anew with each successive king, the counting continued through reign after reign, for as long as the dynasty lasted. The later Romans dated events from the foundation of Rome, though there was some disagreement as to when this took place; a popular date was 753 BC. Other eras were used by various nations at particular times, and some are listed in TABLE 6.4.

Of particular importance is the Christian era, which now pervades most of the world—despite occasional attempts to abandon it and begin a new era; one such attempt was made in the early days of the French Revolution, which gave us the Republican era (ER). The Christian era was first proposed by Dionysius Exiguus in AD 532, and the first year, AD 1, is supposed to be the year of the birth of Christ—though it is almost certainly incorrect (as I shall discuss in Chapter 16). Years after the birth of Christ are termed 'Anno Domini' (AD), 'in the year of the Lord'; years before are counted backwards and termed 'before Christ' (BC). The year before AD 1 is 1 BC, and there is no year zero—to the mild annoyance of astronomers and other mathematically

TABLE 6.4 *Eras*

Name of era	Nation/religion	Event	Year of epoch*
Era of Constantinople	Greek Church, Russia	The Creation	5508 BC
Anno lucis	Masonic lodges	The Creation	4000 BC
Anno mundi	The Jews	The Creation	3761 BC
Kali-yuga	India	Hypothetical grand conjunction	3102 BC
Olympiads	Greece	First Olympic games	776 BC
Ab urbe condita	Romans	Foundation of Rome	753 BC
Era of Nabonassar	Babylonia, Alexandria	Start of reign of Nabonassar	747 BC
Buddhist era	Ceylon, South-East Asia	Death of Buddha	544 BC
Jain era	India, Jain religion	Death of Mahavira (?)	528 BC
Era of the Seleucids	Seleucid Empire	Foundation by Seleucis Nicator	East—311 BC West—312 BC
Era of Alexander		Death of Alexander the Great	324 BC
Arascid era	Parthia		247 BC
Era of Tyre	Hellenic world		125 BC
Vikrama era	India	Victory over Saka (?)	58 BC
Era of Antioch	Antioch	Victory by Julius Caesar over Pompey	48 BC
Julian era	Rome	Reform of the calendar	45 BC
Era of Augustus	Rome	Battle of Actium	31 BC
Era of Pisa	Christian world	Birth of Christ	1 BC
Christian era	Christian world	Birth of Christ	1 AD
Saka era	India	Accession of Kaniska (?)	78 AD
Kalacuri era	India	Traikutaka dynasty	248 AD
Era of Diocletian or Martyrs	Copts, Ethiopians	Accession of Emperor Diocletian	284 AD
Gupta era	India	Gupta dynasty	329 AD

TABLE 6.4 *Continued*

Name of era	Nation/religion	Event	Year of epoch*
Era of the Armenians	Armenia	Conversion to Christianity	552 AD
Fasli era	Near East		600 AD
Harsa era	India	Harsa dynasty	606 AD
Era of the Hegira	Islamic World	Flight of the Prophet from Mecca	622 AD
Era of Yazdegerd	Persia	Accession of Yazdegerd III	632 AD
Kollam era	India	Foundation of Quilan	825 AD
Little Armenian era	Armenia		1084 AD
Era of Alfonso X	Spain		1252 AD
The Republican era	France	The French Revolution	1792 AD
Bahá'i era	Worldwide	Ali Muhommed declared Bab.	1844 AD

*The year of the epoch is given according to the Julian calendar.
Many other eras which are not included in this table have been used, particularly in India.

minded people. The terms 'BC' and 'AD' are often not used by people of different religious persuasions; for instance, the Jews refer to 'Anno Domini' as the 'Common era' (CE), and 'before Christ' as 'before the Common era' (BCE). Some writers refer to dates in the past as 'before present' (BP). Many people would write, in a consistent manner, years before and after Christ as 4 BC and 1998 AD, for instance, though it makes better grammatical sense to write AD 1998 because AD stands for 'Anno Domini' or 'In the year of the Lord'.

The start of an era may occur many years before the first date on which it is used; for instance, the Christian era began on AD 1, but was not invented till AD 532, and was not generally in use till later still. Calendar dates specifying a day before the invention of the calendar are sometimes called 'proleptic dates'.

You may see from TABLE 6.4 that some eras begin with the fanciful date of the Creation; in fact there have been over 200 different calculations of this momentous event. Many of these are based upon the ages of the patriarchs recorded in the Bible. Such calculations were abandoned in the nineteenth century when Lord Rayleigh demonstrated that the earth was every much older than any of these calculations had suggested, and its age was to be measured in millions rather than thousands of years. Many now believe that it all started with a 'big bang' some 10 or 20 billion years ago, and that the solar system coalesced out of the debris several billion years later. Curiously, the biblical estimates of the date of the Creation more or less correspond to that of the start of the earliest civilizations.

The start of other eras might signal the foundation of a new religion, whilst others could relate to a new dynasty or an important battle. A remarkable cluster of eras start in the eighth century BC, as may be seen from TABLE 6.4.

The starting date of an era is sometimes called an epoch. Epochs should be used with care when converting dates from one era to another; in some cases the years counted vary in length, and in others they start in different seasons.

CHAPTER 7

The Reform of the Calendar

⌛ Reform as sacrilege

Calendars are resistant to change. The kings of Egypt had to swear before they took office that they would not change the calendar; any change in the Jewish calendar requires a meeting of the Great Sanhedrim; the reform of the Christian calendar in 1582 required a papal bull. One possible reason for this apparent sanctity is that calendars are religious objects; they define the times of rites and fasts and festivals. Where there is a calendar there is a religion more often than not; where there is a religion, there is a calendar.

To change the calendar is therefore a sort of sacrilege and all too frequently, it would seem, its reformation has resulted in bloodshed, or, one might be tempted to infer, divine retribution: Julius Caesar was assassinated the year after he reformed the Roman calendar; Regiomontanus died the year after the Pope invited him to discuss the matter; the Gregorian reforms resulted in riots and bloodshed; Fabre d'Eglantine went to the guillotine in 1795; and Dr Bradley, the architect of British changes to the Gregorian calendar, died shortly after the Act had been passed—an event attributed, at the time, to divine judgement. Although the list seems to go on and on, some of the apparently lethal consequences of reform are obviously sheer coincidence; others less so. The backdrop of plague, assassination, war, and riot which darkens human history undermines, or so I hope, any belief that calendar reforms are in fact more perilous than any other undertaking.

⌛ Phases of reform

If calendars started as empirical lunar schemes, early reforms may be seen to pass through several distinct phases. The first was to eliminate the astronomical element, so that the calendar could be run without the constant need to regulate it by watching for the phases of the moon or the days on which the solstices or equinoxes took place. That was only possible when arithmetic was sufficiently well understood and astronomical knowledge had progressed to the point at which it was possible to predict these astronomical events with reasonable reliability.

The next phase, still not universally accepted, was to abandon any reference to the moon and gear the calendar only to the sun and the year. This step was taken by the Egyptians many millennia ago, although for religious purposes they continued to user a lunisolar calendar, as do the Jews and Christians to this day. The Moslems still use a lunar calendar.

The penultimate phase comprises attempts to improve the fit between the astronomical cycles and the calendrical cycles. A notable example of this is the Gregorian reform of the Julian calendar.

Finally, in the Age of Enlightenment, we see attempts to simplify the form of the calendar and to expunge traces of its arbitrary history.

The motivations of calendar reformers have been various and have included the need to develop a calendar useful over larger and larger empires and the desire to eliminate corrupt practices which resulted in arbitrary intercalations. The need to reconcile lunations and seasonal years grew as agriculture became the major economic activity. Until the decline of religious belief, the necessity of conserving the dates of religious rites was important, but more recently some have reformed the calendar as part of a specific attack on religion. In modern times some people have been motivated, partly by what appears to be a purely aesthetic consideration, to make the calendar more mathematically regular, and others to impose reforms for the greater good of 'big business'.

Success in reforming the calendar brings a certain fame; perhaps this goes some way in accounting for the popularity of such activities through the ages. Julius Caesar and the Emperor Augustus gave their names to two of our months; other emperors tried the same short cut to immortality, but with less success. Not many people today, let alone would-be calendar reformers, enjoy the sort of power wielded by the rulers of the Roman empire or the Christian church.

⚄ Early reforms

Some reformations have been successful, others have failed. It is recorded in the celebrated Decree of Canopus (Canopus is a coastal town some 15 miles from Alexandria) of 239 BC that the Alexandrian king, Ptolemy III, tried unsuccessfully to persuade the Egyptian priests to introduce leap days into their wandering years. The calendar reform had to wait to AD 10, when the Emperor Augustus instituted Ptolemy's reform and a leap day was appended at the end of every fourth year henceforth, as in the Julian calendar.

The Greeks kept on making alterations to their lunar calendar to better relate the length of lunations to the length of the tropical year (as discussed in

Chapter 6). Julius Caesar undertook his celebrated reform of the Roman calendar in 46 BC, but his attempts were initially thwarted by the pontifices and only completed by Augustus in AD 8; in Christendom this reform lasted for more than 1600 years.

In Arabia, before the time of Muhammad, an empirical lunar calendar was still in use and, as was usual with such arrangements bribes changed hands and the priests used their influence in the interests of their benefactors to intercalate or not to intercalate an extra month. One upshot of this corruption was that the closed season for attacking other tribes was wantonly abused (as we will discuss in Chapter 18). The Prophet, perceiving the wickedness of all this, reformed the calendar for use by his followers.

People have busied themselves in attempts to reform the Christian calendar more or less continuously since the thirteenth century. Seven hundred years ago the criticism was that the Julian calendar, and the Dionysian canon for calculating the date of Easter, were faulty: the assumed length of the year and of the lunation did not match reality adequately. Many suggestions were made; the matter was brought to the attention of the Pontifex Maximus, the Pope; several reforms were mooted but something always intervened to bring them to naught. Then, in 1582, Pope Gregory XIII carried out his celebrated reform of the Julian calendar as is described in detail in Chapters 19 and 28.

⚲ The Enlightenment and positivism

Starting in the mid-eighteenth century a new type of criticism was levelled at the calendar: its subservience to the Christian religion came under attack from the philosophers of the Enlightenment. At the start, the attack was focused on those days in the Catholic Church that were dedicated to saints and martyrs.

Several proposals to rename the days were made: Clency in 1772 replaced the saints with warriors in his 'Calendrier des heros'; Joseph Vasselier used warriors and other notables in his 'Almanach nouveau de l'an passé' of 1785; in 1788, Maréchal, the 'man without God', published his 'Almanach des honnêtes gens'—a calendar in which the days were dedicated to various benefactors of humanity. This calendar was modelled on the old Egyptian calendar with its 12 months of 30 days each and its five (or six in leap years) epagomenal days at the end. Maréchal was imprisoned for his impiety and welcomed the Revolution.

In 1793, after the Committee of Public Instruction had decided to reform the calendar, Maréchal produced another almanac, his 'Almanach des républicains pour l'instruction publique'. He kept most of the names from his earlier almanac but made several concessions to revolutionary sentiment and the vanity of the revolutionaries. At this point Maréchal took fright at the Terror and from then on kept a low political profile. His almanac evidently inspired Gilbert Romme and others who were setting up the Revolutionary calendar. Although they did not use Maréchal's system of names, they adopted its Egyptian form. The names were supplied later by Fabre d'Eglantine. The Revolutionary calendar was introduced in 1793 but only lasted a few years till its abolition in 1806 by Napoleon.

In 1849 Auguste Comte (1798–1857) published his 'Calendrier positiviste'. This is not the place to expound Comte's positivist philosophy, save to say that it encompassed lofty sentiments for the progress and betterment of the whole of mankind along rational lines. His calendar was intended as a vehicle for the inspiration of the people and to be transparently simple. Formally it broke new ground by dividing the year into 13 months, each containing 28 days or four seven-day weeks; these accounted for 364 days. The remaining day (or two days in leap years) were complementary epagomenal days placed at the end of the year; they did not belong to any week, nor were assigned a weekday name.

Comte named his 13 months after particular men who had profoundly influenced man's history; he also dedicated them to a set of abstract principles as shown in TABLE 7.1. Likewise the 52 weeks of this year were dedicated to a series of lesser worthies. The two complementary days were dedicated to all the dead and to women respectively (a curious duo which might not be thought today to be politically correct). Finally, he assigned each day to a person, in a manner analogous to the Roman Catholic dedication of the days of the year to various saints. Comte had started to run out of days in working through his list of heroes, and some days had more than one. One might expect that, as time went on, more would be added to the list and the calendar would be inundated with dedications—had it been taken seriously. This short account does not do justice to Comte's philosophy which permeates the structure of the dedications in the calendar. The reader is invited to read Comte's original account, but should note that it has been said of his literary style: 'He wrote with opium on a page of lead', but John Stuart Mill, the English political philosopher, thought highly of him.

Comte's calendar aroused considerable debate throughout Europe, though most deemed the proposal itself to be too radical. These debates and controversies continued into the next century and were partly instrumental in persuading the nations which adhered to orthodox Christianity, including the new Soviet Russia, to replace the traditional Julian calendar by the Gregorian.

The new reform movements were more concerned with the form of the calendar than with the names and dedications of the months, weeks, and days.

TABLE 7.1 *The dedications of the months in the calendar of Auguste Comte*

Month	Name equivalent	Abstract dedication	Concrete dedication
1	January	Humanity	Moses
2	February	Marriage	Homer
3	March	Paternity	Aristotle
4	April	Filiation	Archimedes
5	May	Fraternity	Caesar
6	June	Domesticity	Saint-Paul*
7	July	Fetishism	Charlemagne
8	August	Polytheism	Dante
9	September	Monotheism	Gutenberg
10	October	Women	Shakespeare
11	November	The priest	Descartes
12	December	The proletariat	Frederick
13	Final	Industry	Bichat*

*Saint-Paul was Comte's patron and mentor; Bichat was a prominent medical man of the time. One may cavil that these two are not in the same league as the others.

⌛ The numerology of reform

Before we consider modern proposals for the reform of the Gregorian calendar which still soldiers on, we must briefly consider the numerological possibilities.

The year contains 365 days or 366 in the occasional leap year. These numbers themselves present problems to the tidy minded: they do not divide into sensible factors. The possible factors of 366, 365, and, for good measure, 364, are as follows:

364	2	4	7	13	14
	182	91	52	28	26
365	5				
	73				
366	2	3	6		
	183	122	61		

FIG. 7.1 Auguste Comte (1798–1857). © Getty Images.

The most immediately obvious feature is that 365 has hardly any factors, but this has not inhibited would-be reformers.

One of the elements of the Gregorian calendar most disliked by would-be reformers is the fact that there are no fewer than 14 different calendars required to cope with the possible distribution of days in the week in common and leap years. How convenient it would be, they argue, if just one 'perpetual calendar' would suffice. This of course requires that every year begins on the same day of the week and that leap days are slipped in between two years. This in turn requires that in some way it must be contrived to have a multiple of seven days in the year.

It is therefore disappointing to them that only 364 is divisible by seven. There is thus no easy way to arrange the calendar so that each year begins with the same day of the week. The few possibilities are:

1. abandon the dream of a perpetual calendar of this sort;
2. change the number of days in the week—a resort avoided by most since the untoward experiences of the French Revolution;
3. place one or two days outside the regular seven-day cycle.

The last idea, of placing days outside the regular seven-day cycle of the week, was first proposed in 1834 by Marco Mastrofini, an Italian priest. He suggested that the 365th day be called the 'feria octava' and the 366th be placed after this in leap years and be called the 'intercalary day'. Such days have been named in various ways, though no word exists in the English language for such purposes. Some have called them 'dies non'; I shall coin a new but admittedly ugly word for these unpopular days—extrahebdomadal. It is often proposed that these extrahebdomadal days should be public holidays, excluded from the economic system and statistical consideration. A year of 364 days, designed to keep in step with the seasons and the week, requires an extra day every year and two in leap years.

Another favourite concern of reformers is the lack of regularity in the number of days in the months. Some non-Christian calendars have a tidy alternation of 29- and 30-day months, but such a pattern is absent from the Gregorian calendar; in this, some have 30 days, others 31, and poor February has only 28. Reformers argue that it would be better if the months were all of the same length. It is therefore again disappointing that the only factor close to the period of a lunation is 28, which divides into 364 to give 13 parts.

There is thus a kind of inevitability in August Comte's calendar of 13 months of 28 days or 4 weeks. It is hardly necessary however to point out that 13 is an inconvenient number: it is prime and cannot be subdivided into any set of equal periods, let alone four equal seasons. The calendar also requires one or two extrahebdomadal days.

If one cannot have equal months, maybe one can have equal seasons; in

Europe and America four seasons or quarters are recognized—spring, summer, autumn, and winter. A year of 364 days neatly accommodates four seasons of 91 days of exactly 13 weeks each. The problem then is to subdivide the seasons into months. Ninety-one is only divisible by 7 and 13 so that, however we divide it, any reasonable months will be of unequal length. One possibility is 30, 30, and 31; another is 28, 28, and 35, or some permutation of such numbers. It is interesting that a year with 364 days and four quarters each containing 30, 30, and 31 days was described in the Book of Enoch more than 2000 years ago.

Ⅹ Modern reform movements

We have seen that the first step in the modern direction was taken by Marco Mastrofini who, in 1834, put forward the idea of the extrahebdomadal day. In 1884 the Abbé Croze, chaplain to the prison at La Roquette, offered 5000 francs as a prize for the best proposal for the reform of the calendar. The Astronomical Society of France was to sponsor the competition under the supervision of the astronomer Camille Flammerion. Croze laid down two conditions for the proposals: first, every year must begin on a Sunday; secondly, there must be 12 months in the year. The first prize was awarded in 1887 to the astronomer, Gustav Armelin, and the second prize to Emil Hanin.

Armelin's year included Mastrofini's extrahebdomadal days, and the remaining 364 days were divided into four quarters each of 91 days or exactly 13 weeks. These 91 days were each to begin on a Monday and be divided into three months of 31, 30, and 30 days. One extrahebdomadal day was to be called New Year's Day and fall on 0 January, and the other, the leap day, would fall on 31 December. The jury had taken three years to award the first prize to Armelin; perhaps they were troubled by the fact that his year began with an extrahebdomadal day rather than a Sunday.

Hanin's proposal was very similar, but each quarter began with a Sunday, and the two extrahebdomadal days were placed at the end of the year. The subtle difference between these two proposals is that whereas the number of weekdays (Monday to Friday) in the three months of a quarter in Armelin's proposal are 27, 26, and 25, in Hanin's they are all the same, a more equable 26.

A further variation of this theme was made by Grosclaude (a Professor of Horology at Geneva). In his version, the number of days in the month were rearranged from 31, 30, 30 to 30, 30, 31, and each quarter began with a Monday; each month had 26 weekdays as in Hanin's version. The leap day was moved to the end of June. A further minor modification was proposed later by the English lawyer, Alexander Philip. These four calendars are summarized in TABLE 7.2.

Several other proposals were made between 1887 and the outbreak of the First World War in 1914. Enthusiasts for the reformed 12-month calendar succeeded in inducing the International Congress of Chambers of Commerce to pass a resolution in favour of reform, and the Swiss Government committed itself to diplomatic action. In 1912 the Congress endorsed Professor Grosclaude's version which was then enthusiastically taken up by Miss Elisabeth Achelis who named it the world calendar. In 1930 she established the World Calendar Association, which from 1931 to 1955 published her *Journal of Calendar Reform*. It is not easy to find copies of this journal. Most of its articles describe the triumphal progress of calendar reform or enthusiastically catalogue the resolutions passed by various bodies, the speeches made in its praise, and all the other paraphernalia of a proselytizing movement. Nevertheless, it is a gold-mine of interesting information about the calendars of various nations throughout history—although it also contains some thoroughly dotty articles. It remains the only regular publication in the English language ever to be devoted to the calendar.

Meanwhile, Comte's positivist calendar became the forerunner of another modern proposal for calendar reform—the international fixed calendar. This was designed in 1895 by Moses Bruines Cotsworth (1859–1943), an English statistician. In 1914, Cotsworth founded the International Almanak Reform League. Later, in 1922, after George Eastman (of photographic fame) had espoused the cause, Cotsworth founded the International Fixed Calendar League. Like the positivist calendar, the international fixed calendar had 13 months of 28 days each; the extra month, called Sol, was interposed between June and July. Cotsworth proposed that the months be known by number, or failing that by the name of the signs of the zodiac. The odd day out followed 28 December and was to be called the 'year day'; the intercalated day, the 'leap day', was to be inserted after 28 June. Both of these days were extrahebdomadal, but each year would begin on a Sunday. Cotsworth suggested that his fixed calendar should be introduced in the year 2000 and that, thereafter, leap years should occur every four years (except that one should be omitted once every 128 years, rather than according to the Gregorian scheme). It appears that the calendar was adopted by several commercial firms in America to simplify their internal accounting.

Those who fear the dark unseen forces may also care to note that the international fixed calendar contains no fewer than 13 Friday the 13ths. They might well think of pulling the bedclothes over their heads for the whole of the thirteenth Friday the 13th.

Between 1908 and 1947 there were 44 notes that touched upon calendar reform published in the English scientific journal, *Nature*. These include book reviews, reports of resolutions passed by various bodies, and an article or two. It suggests a widespread interest in calendar reform during this period, even if many of the notes betray an antipathy to such reforms.

There have been numerous other proposals—more than we can consider in detail here. John Robertson suggested a modification to the world calendar in which the first two months of each quarter have 28 days or four weeks, and the third has 35 days or five weeks. The 'Black plan' is similar to this but the extrahebdomadal days are allowed to accumulate till they total seven and can be intercalated as a whole week at the end of December. Broughton Richmond has proposed a metric calendar based upon 73 weeks of five days each; he also set up the International Calendar Organisation to promote his scheme. Alexander Philip proposed various minor changes, including another variant on the world calendar. Many of the protagonists have written books in which the calendars of the world are reviewed and reform is extolled.

G. N. Searle had a more radical idea. He noted that 400 Gregorian years contained 146 097 days, or exactly 20 871 weeks. He proposed that each year should normally contain exactly 52 weeks or 364 days, and to intercalate an extra 71 weeks in the 400 years to give 71 years of 53 weeks and 329 of 52. He devised a simple rule for intercalating these 71 weeks: intercalate in years ending in 5 or 0 (that is, divisible by 5) except if they end in 50 or 00 (that is, divisible by 50).

The issues of the regularity of the lengths of the months in the year and of extrahebdomadal days are, in reality, separate. One simple proposal, attributable to Alexander Philip, is to redistribute the number of days in the months as shown in the last line of TABLE 7.2. This involves minimal change to March, May, and August; each ceding a day to April and February—April gets one of them while February gets the other two. A better plan might be to give January 30 days, February 30 (or 31 in a leap year), March 30, and then, alternately, 31 and 30 for the other months.

Another proposal, the reformed Saka calendar (made by the commission set up by the Indian Government in 1952 to review Indian calendars), was to have six months of 31 days followed by six of 30, save that the first month would only have 31 days in leap years (otherwise, 30).

A related subject of contention is the date of Easter. The current practices of the Orthodox and the Catholic churches differ—the former is based on the old Julian calendar, the latter on the new Gregorian. Both are founded on agreements made after centuries of wrangling. Thus any attempt to meddle with the days of the week (and so interfere with the date of Easter Sunday) is

TABLE 7.2 *Some details of proposals for a new calendar*

Author	*	Sequence of lengths of months												Extra days	
		J	F	M	A	M	J	J	A	S	O	N	D		
Gregorian	V	31	28	31	30	31	30	31	31	30	31	30	31	—	—
Armelin	M	31	30	30	31	30	30	31	30	30	31	30	30	0 Jan,	31 Dec
Hanin	E	31	30	30	31	30	30	31	30	30	31	30	30	31 Dec,	32 Dec
Grosclaude	M	30	30	31	30	30	31	30	30	31	30	30	31	0 Jan,	0 Jul
Philip I	M	30	30	31	30	30	31	30	30	31	30	30	31	0 Jan,	31 Jun
Philip II	V	31	30	30	31	30	30	31	30	30	31	30	31	—	—

*Day of week of New Year's Day:
  M = Monday
  E = Extrahebdomadal
  V = Variable

bound to cause dissension. There are nevertheless signs that the old passions generated by this subject are abating and that reformers desiring to fix the date of Easter may have some hope of success.

⚄ The politics of reform

Even after the Reformation the Pope had the power to reform the calendar of at least Catholic Christianity; indeed, Gregory XIII's reform came, in time, to dominate the whole of Christendom, if not the world. But since the decline of religion in the West, the Pope's power to direct such changes has become more limited.

The question is: who or what has the power? Modern reformers have directed their appeals to more secular authorities. Some reformers have petitioned their own parliaments, while others have laid resolutions before the councils of the League of Nations or the United nations.

The fact of the matter is that there is no pressing need for calendar reform. Although various bodies are happy to pass abstract resolutions in favour of this reform or that, they have more urgent tasks than to court possible unpopularity by meddling with the calendar.

There are two sources of opposition to the fixed calendar, the world calendar, or indeed any similar suggestion. In the first place various religious bodies object to extrahebdomadal days. Thus Catholic and Jewish leaders point out that man is required by God to rest every seventh day. The Seventh Day Adventists have their own but similar objections.

In many parts of Europe, Roman amphitheatres are still used for theatrical displays and the like, even though the buildings may show their age and may not be wholly suited to such purposes. Nevertheless most would be aghast at the thought of pulling the old ruins down and replacing them with some new scientifically designed and purpose-built edifice. The amphitheatres have lasted for 2000 years and remind us of our past; it is a good modern building that lasts for 50. These sentiments, as is argued in several articles in *Nature*, are equally applicable to the calendar; they are of course not likely to be popular with those who prefer to forget the past and bury their heads in the future.

One of the less convincing arguments for reform is that the old Gregorian calendar, with its complications and irregularities, is less scientific than the tidier new proposal. This betrays a naive and fundamental misunderstanding of science and an admiration instead for the technological, mechanical, and non-human features of modern life. Science describes how the world is; it has no opinions of its own as to how it should be.

A great deal of support for the fixed calendar in America comes from 'big (or even little) business'. Undoubtedly the irregularities of a calendar complicate comparisons between one month and another of, for example, profit and loss. The intemperate arguments of some protagonists may belie any altruistic claims they may make; occasionally, it is pointed out to publishers that if there were 13 months in the year more magazines would be sold. A converse argument might be addressed to publishers of yearly calendars who would presumably be driven out of business should a perpetual calendar be adopted. There is no doubt that there is an economic argument for reform. The only question is: who would benefit?

There may be economic benefit in the simplification of the calendar, but there is also a substantial debit side to the argument as well. Many millions of computers and their associated programs have mechanisms for manipulating dates which assume the Gregorian calendar; the cost of replacing or altering these would be considerable. Then, there are all sorts of other devices with an in-built date or calendar—date stamps, calendar watches, car park ticket dispensers, to name but three. Would it be worth it?

Enthusiasts have managed, on several occasions, to place bills before the English Parliament, but without notable success. Several international religious, astronomical, and trade associations have convened meetings to discuss the reform. Perhaps the most important step was taken by the League of Nations which appointed a special Committee of Inquiry into Calendar Reform. This committee deliberated the matter for three years and elicited 185 proposals for reform from 38 countries. This in itself indicates the difficulty in persuading all nations to agree on any one reform—but that is what politics is about.

Although the British Parliament never legislated for a reformed calendar, it did decide to fix the date of Easter. The Easter Act of 1928 allows an Order in Council to fix the date of Easter in Great Britain as the first Sunday after the second Saturday in April. So far nothing practical has come of this.

A great deal of support for reform has emanated from the United States of America, which, in response to the report of the League of Nations, set up its own inquiry. The report eventually submitted to Congress supported reform and mentions the results of a poll of 1433 organizations—80.5 per cent supported calendar reform and, in a choice between the fixed calendar and the world calendar, 98 per cent preferred the former. But we are not told the precise wording of the questionnaire and many of the organizations consulted might be expected to have a vested interest in the matter. As far as I know there has never been a poll of the population as a whole on the subject.

As the twentieth century wore on, *Nature* continued to publish notes on calendar reform, but perhaps most idealists had other things to think about as barbarism once more began to infect Europe. During the Second World War few had time for such luxuries and, even after peace was restored, *Nature* published only a few more notes. Cotsworth had died in 1943; the *Journal of Calendar Reform* folded in 1955; in the same year, a proposal put before the United Nations was rejected by the United Kingdom and the United States as well as several other countries. The moment for calendar reform seems to have passed.

𝕏 A Martian epilogue

In almost its last issue, the *Journal of Calendar Reform* published, with its tongue in its cheek, a perpetual calendar for Mars. There were an arbitrary 12 months in a year—it being difficult to choose between Mars' two moons; eight months had 56 days apiece, and the other four had 55 days. The calendar assumed a seven-day week, giving eight weeks in each 56-day month; in the 55-day months, the last Saturday was omitted—no doubt to the chagrin of the Martian Sanhedrin.

There were 668.599051 mean Martian days in a mean Martian tropical year.

To accommodate this, there was a leap day at the end of the Martian December in three out of every five years; this was omitted every tenth century year. The calendar was thus accurate to one day in about 20 000 years. The years were counted from the same point in time as our terrestial Julian period (i.e. January 1st 4713 BC in the Julian calendar); our 1 January 1954 fell in the Martian year 3545.

PART II

THE CALENDARS OF THE WORLD

CHAPTER 8

Introduction

Modern man has been around for some 100 000 years or so. By 15 000 BC he had spread into most of the habitable world; by about 10 000 BC he had invented agriculture. We only know of the people of these remote times by their graves, their artefacts, their rock paintings, and their buildings that have survived. If they had a calendar, we can only surmise its nature from the few traces that remain.

By around 3000 BC, cities, civilization, and writing had come into being in the alluvial basins of five great rivers of the world: the Tigris and the Euphrates in Mesopotamia, the Nile in Egypt, the Indus in India, and the Yellow River in China (see FIG. 8.1). We can learn something of the calendars of these civilizations, and more of those of their descendants, from surviving inscriptions and documents.

It is conventional to divide remote human history into a number of stages; some of these are defined in TABLE 8.1.

In Part II, I cover the structure and history of a selection of calendars from all parts of the world, starting with speculative discussion of prehistoric calendars before the invention of writing, and going on to describe the calendars of the four old civilizations of Babylon, Egypt, India, and China and the later civilizations of Central America.

TABLE 8.1 *The stages of man's early history*

Stage	Dates	Level of culture
Palaeolithic	to 12 000 BC	Rough stone tools
Mesolithic	12 000–4 500 BC	Better tools
Neolithic	4 500–2 200 BC	Agriculture, polished tools
Bronze Age	2 200–800 BC	Metals, trade, stratified society
Iron Age	800 BC to present	Iron

The dates above refer to Western Europe. They vary for other parts of the world.

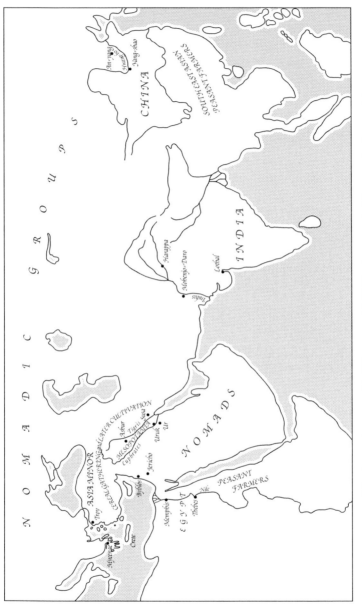

FIG. 8.1 The first civilizations of the Western world. Civilizations arose in Mesoamerica, but much later.

TABLE 8.2 *Landmarks in the history of calendars*

Date	Event	Place
2500 BC	Wandering calendar in use	Egypt
1600 BC	Oracle bones of Shang dynasty	China
432 BC	Meton describes the 19-year cycle	Athens
334 BC	Alexander begins his campaigns	Near East
239 BC	Decree of Canopus	Egypt
104 BC	Taichu calendar introduced	China
46 BC	Julius Caesar reforms Roman calendar	Rome
AD 1	Conventional date of birth of Christ	Christendom
AD 8	Augustus completes Roman reform	Rome
AD 10	Augustus reforms Egyptian calendar	Egypt
AD 325	Council of Nicaea	Turkey
AD 356	Hillel II reforms Jewish calendar	Holy Land
AD 622	Islamic era begins	Arabia
AD 1378	Start of great schism within the Roman Church	Christendom
AD 1582	Pope Gregory XIII reforms Julian calendar	Christendom
AD 1595	Jesuit missions reach China	China
AD 1752	England adopts reformed calendar	England and Colonies
AD 1792	French Republican calendar	France
AD 1917	Bolshevik Revolution	Russia
AD 1955	Report of commission of the Government of India on Indian calendars	India

I describe calendars of historical times; first, those used in various societies of the ancient European world, including those of the Greeks, the Celts, and the Romans; then, calendars set up in mediaeval times and later. These include the Jewish, Islamic, and Bahá'i calendars; the calendar of the French Revolution; and, finally, the great reform of the Roman calendar instituted by Pope Gregory XIII in the sixteenth century. Part II ends with a discussion of the origins and history of the week.

TABLE 8.2 notes a few important landmarks in the history of calendars.

CHAPTER 9

Prehistoric Calendars

⚅ Calendars before writing

Our knowledge of calendars before the invention of writing is meagre and speculative; the very existence of such calendars can only be inferred from evidence that is equivocal at best. It may be surmised that man had to take heed of the seasons and the effect of these on his food supply while he was still hunting and gathering, and the invention of agriculture can only have accentuated this. When cities and civilizations arose, the need for keeping proper records of taxes and the logistics of building works demanded writing—and the invention of calendars. Sure enough, soon after the invention of writing, calendars begin to be mentioned.

Long before the invention of writing, man was already building with stones, painting on the walls of caves, scratching marks on bones, making and decorating pots, chipping flints for weapons and tools. Some of these artefacts have survived and from some we can infer that neolithic man was interested in astronomical events. Perhaps such events had some religious significance for him; maybe they signalled an interest in a calendar. We will probably never know for certain.

Some peoples survived until the twentieth century without having invented writing, practising a Stone Age way of life that can give clues as to what others may have done much earlier.

⚅ The palaeolithic moon

In upper palaeolithic times (say from 40 000 to 12 000 BC) man was already painting and drawing pictures of animals in caves in several parts of Europe with astonishing expertise. He was also making all sorts of useful artefacts. A large number of pieces of bone and stone which bear scratch marks also made by man have survived from these times. The marks, obviously quite deliberate, take many forms and have sometimes been interpreted as decorations or perhaps palaeolithic doodles. Another possibility is that they are tally sticks of some sort. Alexander Marshack has proposed that they represent a tally of days in the lunar cycle.

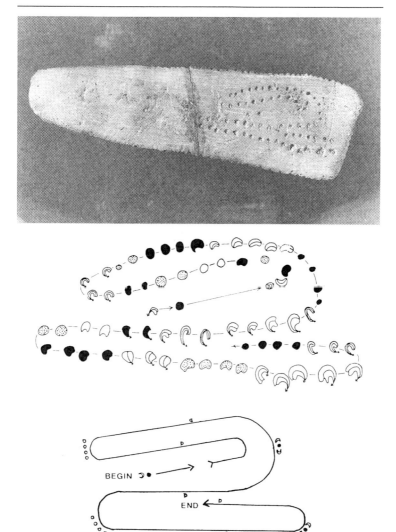

FIG. 9.1 Bone plaque from Abri Blanchard, Dordogne. This late palaeolithic piece of bone was inscribed some 30 000 years ago. Marshack supposes that the marks correspond to the days of two or more lunations and are to be read in the order indicated. Source: Marshack.

In many cases the marks are arranged linearly in rows; the marks, on examination with a microscope, frequently prove to have been made with several different stone scribes, suggesting perhaps that they were not all made at the same time. The scratches often appear to be arranged in groups of varying numbers, the groups demarcated by spacing, a change of scribe, or variation in the size, orientation, or form of the marks.

Marshack claims to have detected a periodicity of 59 in many of these rows of marks and interprets them as a tally of days grouped according to the phases of the moon. A particular set might contain a record extending over several lunations. The average synodic period of the moon is just over $29\frac{1}{2}$ days, so that new moons might be expected to occur at alternating intervals of 29 and 30 days with a periodicity of 59.

Imagine the problem of recording the progress of a lunar cycle in this way. Start on the day when the moon is invisible and make a mark. Next day perhaps, the new crescent can be seen; make another mark and yet another on each day that the moon continues to wax. Eventually it will be full—but you may not recognize this till it has started to wane, and your new group of marks will start a day late. You continue to mark till the last day of the old moon arrives—but you may not know it is the last till the day of invisibility or the appearance of the new moon. Clearly there will be deviations from strict regularity on account of these difficulties. These will be compounded by bad weather and by variations in the length of the individual lunations which may differ by several hours.

Marshack has analysed a large number of such artefacts and concluded that palaeolithic man indeed recorded the phases of the moon in this sort of way. However the pattern is often difficult to perceive and a sceptic might claim that the 'eye of faith' is often required. Nevertheless, other, more modern people, such as the Nicobar Islanders from the Indian Ocean and several North American peoples, have made similar and well-authenticated lunar tallies; there is no intrinsic reason why ancient people should not have made such counts.

If we grant that the marks do indeed represent some sort of lunar record, it is far from clear what purpose they might have served. One view is that they represent a descriptive theory of the moon that can be used to predict how it will behave in the future. Not even the ability to count is required for this, only that the character of the moon during the sequence of days should match the character of the scratch marks. Another possibility is that they were used as a clock or calendar to tell the bearer when he had, for example, to return home from the hunt; he could promise to be back 15 days after the next new moon.

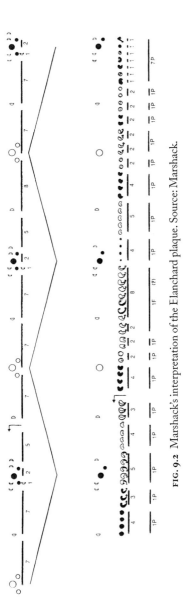

FIG. 9.2 Marshack's interpretation of the Blanchard plaque. Source: Marshack.

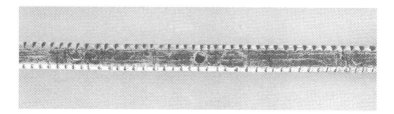

FIG. 9.3 The Winnebago calendar stick. This stick, of square cross-section, about 132 cm long, was made in 1825 by Winnebago chief Tshi-zun-hau-kau. Each side of the stick bears marks representing the days in six months; each month contains 30 or 29 days. The days of invisibility and the last crescent are marked for each month, which is divided generally into three 10-day periods. The top photograph shows the ninth month of the first year; the lower diagram clarifies the salient features. Soruce: Marshack.

Marshack's proposals are certainly interesting but cannot be ruled out as definitely correct or incorrect; the artefacts remain enigmatic and the proposal, contentious.

ⅪⅠ The megaliths of Britain and France

A recent inventory lists 368 ancient stone circles (cromlechs) in the British Isles and several more in France, notably in Brittany; there are certainly more (a figure of 900 has been given). Some have probably disappeared, and possibly a few of those listed are fakes or follies erected in recent times. There are many more large ancient stones (menhirs) which were placed in position in prehistoric times; some form tombs (cists), whilst others are aligned in rows or other patterns, or stand alone or in small groups. Carbon-14 dating of organic remains associated with many of them indicates that some were erected as long ago as 4500 BC, and some as recently as 1500 BC.

The first megaliths were built by neolithic people who inhabited the British Isles some 4000 or 5000 years ago. They had taken up farming and buried their dead in barrows, often with a wealth of goods, maybe for use in the after-life. Building was continued in the Bronze Age by the Beaker Folk, who made their presence known about 2000 BC. These buried their dead in characteristic

FIG. 9.4 Rock carving from Monterrey, Mexico, from about 2500 BC. It is believed that the marks represent days which are divided into periods which match the lunar phases.

long barrows and made distinctive pottery. All this activity took place well before the arrival of Celtic culture—perhaps in the fourth century BC. The Celtic Druids may or may not have used the circles as they found them (and there is no evidence that they did), but they certainly did not build them. Many were built as the ancient civilizations of Babylon, Egypt, and China were getting under way and the pyramids of Egypt were being constructed.

Some of the cromlechs are circular, some are elliptical, and some are of a more complicated shape. All of the arrangements surveyed could have been laid out in a simple manner using pegs and ropes and do not imply great mathematical expertise on the part of the builders. Some arrangements are based on Pythagorean triangles in which the sides are in simple ratio and one angle is a right angle. This does not mean that they knew about Pythagoras' theorem—only that they knew, for instance, that a triangle with sides of three, four, and five units contained a right angle; once or twice, they appear to have used a triangle which is not Pythagorean but whose angle is almost a right angle, for instance, with sides of eight, nine, and twelve units. Thus, although they presumably could communicate numbers up to 100 or more, and could have counted days by notching a stick, there is no evidence that they had learnt how to multiply. They left no writing, but there is an old and apocryphal story that Pythagoras (born about 582 BC) once visited the British Isles; if he did, and there is no evidence for this either, it was way after the cromlechs had been built.

The case for the astronomical significance of stone circles rests on the alignment of two or more stones with astronomical events at particular times of the year. For instance, two stones might have been aligned with the position of the setting sun on the horizon on the day of the summer solstice. The alignment of two markers acting as foresight and backsight on, say, the position of the setting sun, is more accurate and less ambiguous, the greater the distance between them. If the foresight is some mountain or other feature on a distant horizon, the alignment can be quite precise. Due account must be taken of changes in the obliquity of the ecliptic that have occurred during the last 5000 years and its effect on the position of sunsets and sunrises. The sun sets at an oblique angle (depending on latitude) to the horizon and the points on the horizon first reached by the top, middle, or bottom of the sun's disk are different; these ambiguities must also be considered.

In some instances human remains have been discovered in the vicinity of the megaliths and are roughly contemporary with them. Some were found in nearby barrows or buried, as at Stonehenge, within the confines of the circle. Calendars, we have argued, are religious objects and we must allow that beliefs about death, the alternation of the seasons, and the sun and the moon may have been welded into a coherent myth in a manner perhaps foreign to our present understanding, and that these beliefs were celebrated in some unknown way in the circles.

ⅺ Stonehenge

The most well-known and most impressive megalithic circle is Stonehenge. It is situated on Salisbury Plain in the County of Wiltshire in Southern England.

The Greek historian Diodorus Siculus, writing in about 40 BC, discusses a written description by Hecataeus, from the fourth century BC, of a magnificent circular temple to Apollo in the land of the Hyperboreans (probably the British). This has sometimes been taken to refer to Stonehenge. The site has been discussed and measured at intervals from the twelfth century on. John Aubrey (1626–97) first attributed it to the Druids—an idea enthusiastically taken up by William Stukeley (1687–1765) whose fanciful ideas, published in 1740, still attract followers of latter-day Druidism to the site to celebrate the

FIG. 9.5 Stonehenge from the air. Source: London Aerial Photo Library.

summer solstice. Stukeley, in a more sober mood, also noted that the entrance to the circle was orientated roughly in the direction of the midsummer sunrise.

Stonehenge consists of several rings of stones and other megalithic features. Its construction took place in stages from about 3100 to 1600 BC—from late neolithic times to the Middle Bronze Age. Between then and the present day extensive damage has been inflicted by those in search of building material, by the weather, and possibly by those attempting to eradicate pagan practices. Stones are known to have fallen in 1797 and 1900, and in this century several fallen stones have been re-positioned.

The outermost feature of Stonehenge is a circular ditch which was dug sometime from 3100 to 2300 BC; the earth removed was piled inside and outside the ring to form two circular banks. Inside the inner bank there are a total of 56 holes (the 'Aubrey holes') which were filled in almost as soon as they were dug (since their sides are not weathered); some contain artefacts and human remains which allow dating. At roughly the same time the 'heel stone' was set in place. There is evidence that some 300 years later a series of wooden posts were erected.

Around 2150 to 2000 BC the erection of a double crescent of vertical stones was begun—and later dismantled. These 'blue stones' weigh about four tons each and the geological evidence is that they came from the Prescelly mountains in South Wales, a direct distance of 137 miles. Some say the builders of Stonehenge transported them on rafts and sleds, whilst others claim that they were moved at a much earlier date, by a glacier.

A little later, 2100 to 2000 BC, the large stones with their lintels were erected. These are the 'sarsen stones' and are the ones that first strike the eye. The name 'sarsen' is probably a corruption of 'saracen' (a word used of infidels and foreigners at the time the stones were first so named). The stones were dragged some 17 miles from Marlborough Downs, and some weigh as much as 50 tons. The largest stones form the uprights of the five original trilithons which each comprise two uprights and a lintel. The five were arranged in a horseshoe configuration—though some are now fallen. Surrounding this is a circle of sarsen stones, also with lintels, and again many are fallen. The sarsen stones were carefully shaped and fitted together, and the effort involved in transporting, shaping, and erecting them was prodigious; it implies a well-ordered community of a great many people. Even so, some of the Egyptian obelisks erected in the sixteenth century BC, or earlier, weighed 10 times more than the largest sarsen stones.

Within the ring of sarsen stones there is another ring of blue stones which were erected later; the horseshoe arrangement of trilithons is echoed by the blue stones. These blue stones were moved to their present positions from the outer ring already mentioned.

As well as these rings, there are several free-standing stones and a number of holes (92, 94) that give evidence of once having supported standing stones (see FIG. 9.6). Many of these were given fanciful names in recent times: the heel stone, the slaughter stone, the altar stone, and the station stones (91, 93). The heel stone dates from the building of the original ditch, but the others were probably erected at the same time as the sarsen stones.

The original bank had an entrance orientated approximately in the direction of the midsummer sunrise, as has been recognized since the eighteenth century. More recently scholars have carefully investigated other alignments involving the special stones and the hypothetical stones which are believed to have once been in some of the holes. These alignments include moonrises and

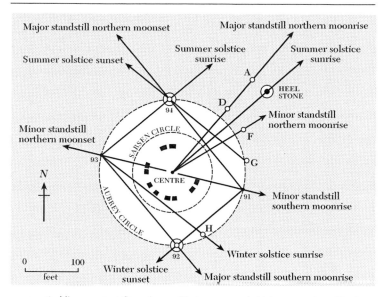

FIG. 9.6 Alignments at Stonehenge. Various stones (which are numbered in the diagram) are supposed to be aligned on various solar and lunar events. Source: Griffith Observatory.

moonsets at the major and minor standstills, and sunrise and sunset on the days of the summer and winter solstices and the equinoxes. It is hard to see that the alignments involving the moon had any calendrical significance, but it has been suggested that they were used in attempts to predict eclipses. An alternative interpretation of the alignment of the diagonals of the rectangle of the station stones (93-91 and 94-92) is that they point to the rising or setting sun on four days of the year that are roughly midway between the solstices and equinoxes: 5 February, 6 May, 8 August, and 2 November. However, if you stand by either of the surviving station stones, the other is not visible, being occluded by the sarsen stones, which appear as an impenetrable wall; they would have been no good for taking sightings of the sun or moon or anything else once the sarsens had been erected.

The alignments involving the sun may well have been used in detecting, or at least celebrating, the start or end of a neolithic year. If this was their purpose, it is clear that it cannot provide the only reason for building Stonehenge; a small number of stones would have sufficed and the prodigious effort of building the rest of the monument would have been redundant.

The most recent, and perhaps the most convincing analysis of the alignments comes from Professor North—but a full description of his conclusions requires a large volume.

ⵣ Ballochroy and New Grange

Another convincing alignment with the midsummer sunset is to be found in Scotland on the Kintyre peninsula at Ballochroy. Three megaliths are mounted on level ground close to the seashore. The centre one has been carefully dressed to give it a flat surface which is orientated towards Corra Beinn, the most northerly mountain peak of the Paps of Jura, nearly 19 miles distant. In 1800 BC the midsummer sun would have been seen to set behind Corra Beinn, passing behind the south face. Due to the shape of the mountain, the top of the setting sun would have briefly reappeared in a dip in the northern face of the mountain. This phenomenon could have been used to ascertain a precise day for the midsummer solstice. The position of the observer is not critical provided he is not too far from the stone.

This arrangement is austere, with no imposing stone rings; it would seem that the sole purpose of the stones was to fix the day of the solstice—though whether for calendrical or religious purposes or both is uncertain. A similar site at Kintraw, further north, can be used in the same way to fix the winter solstice, and there are other similar sites in Scotland and elsewhere which have been investigated in depth by Thom.

A different arrangement is seen at New Grange, some 31 miles north of Dublin in Ireland. There, there is a neolithic tomb built in about 3300 BC, several hundred years before Stonehenge was started. Above the doorway, which points to the south-east, there is a kind of window or 'roof box'. This was plugged with quartz blocks but these were easily removed. In 1969 Professor O'Kelly waited within the tomb for the sunrise on 21 December, the day of the winter solstice. As the sun rose, a shaft of sunlight penetrated the tomb from the roof box, and for 17 minutes illuminated a carving on one of the stones at the back of the tomb. Within this tomb is thus celebrated the death and rebirth of the sun (as enshrined in local tradition)—and perhaps of the occupant.

This motif of the midsummer sun shining through a window on to a special stone is found elsewhere in the world. There is an example at Macchu Piccu built by the Incas at their Andean fastness. A modern example of the principle is to be found in the war memorial in Melbourne, Australia. There, a hole in the roof directs the rays of the sun to traverse the 'rock of remembrance' at precisely 11.00 a.m. on 11 November (see FIG. 9.8).

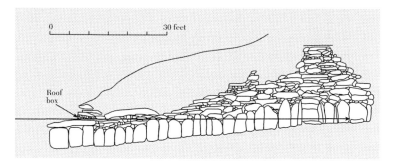

FIG. 9.7 New Grange. An elevation showing the path of the rays of the sun at sunrise on the winter solstice. They pass through the roof box on to the stone at the rear of the passage. Source: Griffith Observatory.

There is one curious feature of such solstitial alignments: the sun sets at a different position each day as it moves from solstice to equinox, back to solstice; sunrise and sunset pass every point between the solstitial extremes two times in the course of a year. As long as this idea is recognized, the setting of the sun at any fixed position on the horizon would have sufficed to determine the start of a solar year. The counting of the number of days in a year could have been done most accurately by starting when the sun rose or set due east or west at an equinox, rather than at a solstice, where it stands almost still for a few days. Nevertheless, people opted for the solstice. Presumably the solstices had some further symbolic significance of death or rebirth.

🎇 A neolithic calendar

Thom spent many years investigating alignments of megaliths and he reported that amongst about 200 in various parts of Britain, there was a tendency for them to point in several preferred directions. These included the directions of sunset and sunrise on the solstices and equinoxes, and also those of a number of intermediate dates that divide the year into 16 roughly equal periods. I have already noted that the diagonals of the rectangle of station stones are aligned approximately in four of these directions.

Thom therefore postulated a neolithic calendar in which the year is divided into 16 'months' of about 23 days each, not all equal. This division of the year is illustrated in TABLE 9.1. The calendar was constructed in such a way that alignments could serve to mark two sunrises signalling the start of two 'months', for instance, one occurring about 23 days before a solstice and one, 23 days after it.

FIG. 9.8 The war memorial at Melbourne. The rays of the sun shine through a hole in the roof on to the remembrance stone at the 11th hour of the 11th day of the 11th month. 11 November is the day dedicated to the remembrance of those who fell in the first World War. Reproduced courtesy of the Shrine of Remembrance Trustees.

TABLE 9.1 *The divisions of the hypothetical Bronze Age year*

'Month'	Length in days	Date of start	Event
1	23	20 March	Equinox
2	23	12 April	
3	24	5 May	Beltane
4	23	29 May	
5	23	21 June	Solstice
6	23	14 July	
7	23	6 August	Lughnasa
8	22	28 August	
9	22	19 September	Equinox
10	22	11 October	
11	22	2 November	Samhain
12	23	25 November	
13	23	18 December	Solstice
14	23	10 January	
15	23	2 February	Imbolg
16	23	25 February	

From TABLE 9.1 it can be seen that the start of months 5 and 13 coincide, albeit approximately, with those of the solstices, and months 1 and 9, with the equinoxes. It has been noted that the Christian festivals of Lammas, Hallowmas (or Halloween), and Candlemas, which are believed to derive from the Celtic feasts of Lughnasa, Samhain, and Imbolg, fall, again approximately, at the start of months 7, 11, and 15—an eighth of a year after the preceding solstice or equinox. Likewise the traditional (and socialist) celebration of May Day coincides with the Celtic festival of Beltane and the start of month 3. Nevertheless, although a continuous tradition originating in neolithic times and transmitted via the Celts to early Christianity is not impossible, there is little reason to connect megaliths with the Celts. Thom believed that the alignments which suggested this calendar were accurate in about 1800 BC—at least 1000 years before the Celts.

In 1808, a diamond-shaped object of beaten gold, known as the 'bush barrow lozenge', was discovered in an early Bronze Age barrow about half a mile from Stonehenge. It was found in the grave of a man, together with other

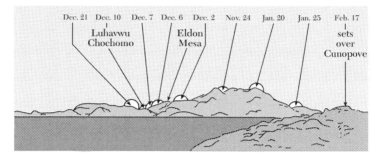

FIG. 9.9 The position of sunset at a Hopi village. The village elders could tell for which days to organize festivals by noting the position of the setting sun on the mountainous horizon. Source: Griffith Observatory, after Alexander M Stephen.

items which suggest that he was an important figure living about 1900 BC. When first found it was attached to a wooden plaque. It bears an inscribed design, which, if one axis were to be orientated due north, would indicate a series of directions. Some have speculated that these directions are those of the sunrise or sunset at the 16 points of the megalithic calendar and that the device was used, as a sort of 'mini-Stonehenge', to determine the date. Others, less convinced, have suggested other non-calendrical uses or that it was merely a decoration.

Several peoples in recent times, for instance in North America, have determined the proper day for ceremonial and economic activities by noting the position on the horizon of sunrise or sunset; rocks, mountains, and so forth served to mark the position of the sun as seen from a particular vantage point. The idea of a calendar marked by megaliths is clearly similar.

Be warned, however, if you care for truth. Thom's speculations may inflame the imagination but should not necessarily convince the mind, for his suggestion of a neolithic calendar is controversial, and the evidence sparse. Whilst the entrance to Stonehenge is certainly orientated in the rough direction of sunrise at the summer solstice, and although the alignment of other stones on important sunsets or sunrises is established, some have cast severe doubts on the statistics of Thom's inventory of alignments.

CHAPTER 10

The Calendars of Babylon and the Near East

⚏ The start of civilization

Cities and civilization probably first arose in Mesopotamia, in the region fertilized by the great rivers of the Tigris and Euphrates. Sometime in the fourth millennium BC, the Sumerians in Mesopotamia and the surrounding fertile crescent started to build cities: Eridu, Ur, Babylon, and others. This area, now part of Iraq, was ruled by a succession of dynasties from at least 4000 BC. It is convenient to refer to the inhabitants as Babylonians, though the area extended far beyond that city.

Mesopotamia is bordered to the north by Assyria, to the east by Persia, to the south by Arabia, and to the west by Asia Minor and the Levant. Tribes from these surrounding lands came to conquer and settle in this fertile region: the Sumerians, the Akkadians, the Assyrians, the Persians, to name but a few. Some salient points in the history of these invasions are as follows:

c. 5000 BC	Temples at Eridu established
c. 4000 BC	Sumerians well established in the south
c. 3500 BC	Wheeled transport appears
c. 3360–2400 BC	Wars between cities
c. 3200 BC	Sumerians invent writing
c. 2500 BC	Positional number system invented
c. 2331 BC	Sargon I established Akkadian supremacy
c. 2193 BC	Collapse of Akkadian Empire
c. 2113 BC	Ascendancy of Ur under Ur-nammu
c. 2000 BC	Ur falls to Elamites and Amorites; End of Sumerian tradition
c. 1792 BC	Accession of Hammurabi the law giver; First Babylonian Empire
1595 BC	Babylon conquered by Hittites

1570 BC	Kassite dynasty starts
1173 BC	End of Kassite dynasty
1124 BC	Nebuchadnezzor I starts second Babylonian dynasty
747–734 BC	Nabonassar
728 BC	Tiglath-Pileser III, King of Babylon
625 BC	Nabopolassar starts Chaldean dynasty
604 BC	Nebuchadnezzar, King of Babylon
597 BC	Leading Israelites deported to Babylon
539 BC	Cyrus I conquers Babylon
529 BC	Start of Persian (Achaemenian) period
331 BC	Alexander the Great enters Babylon
323 BC	Alexander dies in Babylon
321 BC	Seleucis Nicator founds Seleucid Empire
126 BC	Parthians take over Seleucid Empire
117 AD	Trajan annexes Mesopotamia for Romans
227 AD	Persian (Sassanid) Empire
637 AD	Conquest by Moslem Arabs; Abassid domination
762 AD	Baghdad founded
833 AD	Observatory at Baghdad founded
1055 AD	Turkish (Seljuk) domination
1258 AD	Baghdad sacked by Mongols
1392 AD	Timur takes Baghdad
1638 AD	Turks recover Baghdad

Writing in the form of an ideographic script was developed sometime before 3000 BC by the first city builders, the Sumerians. It later matured into the more formalized syllabic cuneiform script written by pressing cut reeds into clay tablets which were then dried. This script came to be widely used throughout the area and beyond. Clay tablets bearing accounts, astronomical tables, and much else have survived in large numbers.

The Babylonians made an important invention some time in the third millennium BC: the sexagesimal number system (explained in detail in Chapter 5). It is possible that this came about because of the need to record tax returns and the like; taxes became necessary once cities grew and large-scale civil engineering projects got under way. Later, this number system helped with the recording of detailed astronomical events; it is thus that quantitative astronomy began in Babylon. By the eighth or ninth centuries BC the Babylonians were making accurate observations of the heavens and recording their findings on innumerable clay tablets. They were inspired to do this partly because they believed in astrology (that the planets and stars were omens for our existence on earth), and partly by the need for an accurate calendar which could be used, among other things, to determine which days were propitious—or the reverse.

The various conquering peoples absorbed the culture they found in Babylonia and in turn contributed their own knowledge. In this way Babylonian learning grew and spread throughout the civilized world and beyond.

Alexander the Great took the area from the Persians before going on to India; after his death in 323 BC, Mesopotamia was ruled by one of his generals, Seleucus Nicator, who founded the Seleucid Empire in 321 BC. The Seleucids were later forced out of Mesopotamia itself by the Parthians, and their empire finally fell to the Romans, led by Pompey, in 64 BC. By then the centre of learning and science had passed to Alexandria.

The intellectual life of the area flowered again, later, under Islam, with the founding of the observatory at Baghdad in AD 833 by the Caliph al-Ma'mûn. This, together with other colleges, schools, and hospitals prospered for several centuries till the city was destroyed in 1258 by the Mongols, of whom it is said that they built a pile of 10 000 human heads in the course of their destruction.

Ⅲ The Babylonian calendar

As is appropriate in a part of the world not given to extremes in the weather, only two seasons were recognized: that of planting, and that of harvesting. The length of each season, about six lunar months, was evidently held to be significant. There is even a suggestion that instead of a year, the Babylonian calendar was structured around a period of six lunar months. The start of a season was, in later times, noted from the length of the shadow of a gnomon.

The calendar was, from the earliest times, based on the moon. The year normally contained 12 lunar months, for which the different city states had different names which were often derived from the names of festivals or tasks such as sheep shearing. In later times, from about the eighteenth century BC, during the Kassite period, Semitic names were used throughout the area; these are listed in TABLE 10.1 and were adopted by the Jewish people, probably during their captivity in Babylon. TABLE 10.1 also shows earlier Sumerian and later Seleucid names.

The day began at dusk and was divided into six watches whose length varied with the season. The months began when the new moon was first sighted so that, as long as weather conditions did not impede a sighting, the months alternated, albeit irregularly, between 29 and 30 days.

Even in Sumerian times there were attempts to keep this lunar calendar in step with seasons by intercalating, though haphazardly, a thirteenth month, later called 'iti dirig'. Records from about 2400 BC mention the practice of inserting an extra month to ensure that the barley harvest (which occurred at

TABLE 10.1 *The names of the Babylonian months*

Month	Sumerian	Semitic	Seleucid
1	Barzagga	Nisanu	Artemisios
2	Gusisa	Ayaru	Daisios
3	Sigga	Simanu	Panemos
4	Shunumun	Du'uzu	Loos
5	Nenegar	Abu	Gorpaios
6	Kinninni	Ululu	Hyperberetaios
7	Du	Tashritu	Dios
8	Apindua	Arakhsamnu	Apellaios
9	Gangan	Kislimu	Audynaios
10	Ziz	Tebetu	Peritios
11	Abbae	Shabatu	Dystros
12	Shegurku	Adaru	Xanthikos

the end of May) fell within the first month of the year, Nisanu. Later a decree issued by the Babylonian King Hammurabi, who reigned from about 1790 BC, required that an extra month, Ululu II, be intercalated between Ululu and Tashritu—but normally it would be placed after Adaru. An astronomical treatise dating from about 1000 BC, the 'mul-apin' (the plough star) tells us that if the moon was near the Pleiades on Nisanu 1, the first day of the year, there would be no intercalation in that year.

Later still, the rule for inserting the extra month became more formal. This only became possible after many years of detailed astronomical observation had established an exact value for the length of the tropical year and the average time between new moons; this measurement was facilitated, as explained in Chapter 2, by an understanding of eclipses.

This later Babylonian calendar was developed from about 500 BC and was certainly used by the Persians during their domination of Babylon after about 380 BC. It intercalated an Adaru II in years 3, 6, 8, 11, 14, and 19 of the cycle, and an Ululu II in year 17. It was described by Meton, who introduced it to the Greeks in Athens.

In the earliest times, the years were described according to notable events that took place in them. Later, this gave way to a system of regnal dating: the first year of a king would start with the next 'first of Nisanu' after his accession. A list of kings, with their dates going back to King Nabonassar in 747 BC, the so-called 'canon of kings', was constructed in the second century by Ptolemy (the author of the *Almagest*). This list has enabled Babylonian years to be correlated with years in later eras, and has been corroborated by other records discovered in this century. Other lists have enabled chronologists to extend the dating to about 2000 BC, although with increasing uncertainty before about 1500 BC.

The Babylonians held that several days in their lunar month—the 7th, 14th, 21st, and 28th—were unlucky, and that nothing important should be done on them. The 20th day, which is the 49th of the previous month ($= 7 \times 7 = 29 + 20$) was considered lucky. This practice may have given rise to the seven-day week of the Semitic people (which is discussed in Chapter 21).

The Babylonian calendar survived in various parts of the Middle East until the rise of Islam and the Arab conquests of the seventh century, when in most places it was supplanted by the less sophisticated Islamic lunar calendar—but the 19-year cycle has survived in the Jewish and Christian calendars until the present day.

🔟 The Seleucids

The Persians, under Darius III, were defeated by Alexander the Great (356–323 BC) in 331 BC, when he was 24 and before he moved on to India. After Alexander's death in 323 BC, in Babylon itself (from a fever following a drunken party), his empire came to be shared, after a few years of murderous intrigue, between his Macedonian generals. Two of these generals founded dynasties: Seleucis Nicator in Babylonia, and Ptolemy Soter at Alexandria in Egypt.

Although the Seleucids adopted the Babylonian calendar with its 19-year cycle of intercalations, they renamed the months according to their old Macedonian calendar (see TABLE 10.1) and initiated the Seleucid era, whose first year was 311 BC.

The Seleucids extended their rule towards the East, into India, as far as the Indus and Afghanistan, taking Babylonian knowledge of astronomy and the 19-year cycle with them. Use of a 19-year cycle is recorded later in China, but it is uncertain whether the Chinese invented it themselves or it reached them from India. In the West they extended to the shores of the Mediterranean. In that area a variety of calendars—some based on the Seleucid calendar, some on the Greek, some on the Egyptian, and, later, some on the Julian calendar—came to be used in the city states and vied with each other for local dominance in the succeeding centuries.

CHAPTER 11

The Egyptian Calendar

⧗ Egyptian civilization

The second civilization on earth started along the fertile banks of the Nile. Each year, water from the mountains of East Africa poured down, causing the Nile to flood its banks and fertilizing the valley with the silt washed down with it. Egyptian agriculture was dominated by this inundation.

The kings of Egypt, who became gods, ruled with a bureaucracy which, like its Sumerian counterpart, required the gathering of taxes and the keeping of records; accordingly, writing was invented—perhaps in about 3200 BC. Likewise, they developed a numbering system, which for them was also essential for surveying their fields and building works—but they never studied the heavens as intensively as did the Babylonians. Later, after the arrival of the Greeks and the foundation of the Museum at Alexandria, astronomy in Egypt did blossom.

At first there were two kingdoms, in Upper (southern) Egypt and Lower (northern) Egypt. In about 3200 BC however the southern king conquered the northern one and from that time on, for nearly 3000 years, Egypt was a united country with a continuous record of civilization ruled by a succession of Pharaohs—even though the country was, from time to time, dominated by foreign overlords.

In the fourth century BC, Ptolemy Soter took control of the country after the death of Alexander the Great (who had founded the capital, Alexandria). Later, the country was annexed to the Roman Empire.

Egyptian history is often divided into periods, and these into dynasties, as shown in TABLE 11.1—though the periods are perhaps no more precise than terms such as the Middle Ages. In later years, till the twentieth century in fact, Egypt was dominated by the following foreign powers:

330–30 BC	Ptolemeic Greeks
32 BC–AD 395	Romans
395–640	Byzantines
642–1251	Arabs

1260–1517	Mamelukes
1517–1798	Ottoman Turks
1789–1801	French
1804–82	Independent period
1882–1922	British
1953	United Arab Republic

This whole period of Egyptian history encompasses several notable events, including:

2600–1700 BC	Pyramids built
2551–2472 BC	Great pyramids of Giza
1730–1580 BC	Egypt dominated by invading Hyksos
331 BC	Foundation of Alexandria
239 BC	Decree of Canopus
47 BC	Caesar meets Cleopatra
c. AD 500	Coptic Church founded
1798–1801	Napoleon leads an expedition of discovery

The arrival of Joseph and the Israelites in Egypt, and their Exodus under the leadership of Moses, as described in the Bible, is problematic. Scholars

TABLE 11.1 *Periods and dynasties in the history of Egypt*

Date	Period	Dynasties
c. 3200–2665 BC	Protodynastic	I–II
2664–2155 BC	Old Kingdom	III–VIII
2154–2052 BC	First Intermediate	IX–XI
2052–1786 BC	Middle Kingdom	XII
1785–1554 BC	Second Intermediate	XIII–XVII
1554–1075 BC	New Kingdom	XVIII–XX
1075–525 BC	Third Intermediate	XXI–XXVI
525–332 BC	Persian	XXVII–XXXI

The intermediate periods were marked by civil war and other disruptions; the Kingdoms were relatively tranquil times.

who assume that the story of Exodus represents an historical event rather than a myth have placed it in the time of Ramasses II, in the mid-thirteenth century BC—but others favour a date some 300 years earlier.

⬚ The cycle of inundation

Egyptian agriculture and life depend on the Nile; water makes its way down from the mountains of East Africa to the Mediterranean, and each year in July the river rises and floods the fields along its banks; this irrigates the desert and deposits fertilizing silt. The date when this happens varies by several weeks from year to year, but it happens every year.

Early in their history the Egyptians noted that this flood was preceded by the heliacal rising of Sirius or Serpet, as they called this star (the Greeks called it Sothis). The heliacal rising of Serpet, or any other star, is its first appearance in the Eastern sky just before sunrise. As the days progress, the star rises a little earlier before dawn than on the night before, till eventually, after several months, it is setting at dawn. Serpet then remains invisible for 70 days and nights after which it makes another heliacal rising. This reappearance of Serpet after its 70 days journey through the 'underworld' signalled the beginning of the season that the Egyptians called 'akhet' or 'inundation' (when the Nile flooded)—the first of three seasons they recognized; the others were 'peret' or 'going forth' (the time to plant), and 'shomu' or 'deficiency' (the time to harvest as the land dried up again).

The interval between successive heliacal risings is determined by the sidereal year and the times of sunrise. It has been calculated that at the latitude of Thebes this interval was about 365.2507 days—about 12 minutes longer than the tropical year which governed the flooding of the Nile. This meant that the heliacal rising of Serpet would have been observed later and later in the tropical year, but it took about 120 years to gain a single day. The effect would have been hardly noticeable, and in the whole of the recorded history of ancient Egypt it only gained about 25 days—a time comparable with the variation in the date of the inundation.

⬚ The first lunar calendar

The early Egyptians developed a lunisolar calendar. The common year had 12 months, each month taking its name from a festival held in that month. The last month was called 'Wep-renet' or 'opener of the year', and the first was called 'Tekhy'. A thirteenth month was intercalated if the heliacal rising of Serpet occurred in the last 11 days of Wep-renet; this month was called 'Thoth' after a moon god. In this way, the year was forced into synchrony with the Sothic year.

The months began on the first day on which the old crescent was no longer visible in the East at dawn, and the day ran from sunrise to sunrise. These conventions are unusual in so far as others began their months with the first sighting of the new crescent, and their days in the evening.

It is not known when this calendar was developed; presumably it was after the Egyptians had become dependent on the inundation, but before the fourth dynasty in the middle of the third millennium BC—the date of the earliest documentary evidence.

Ⅺ The civil calendar

The Egyptians are renowned for their civil engineering. It required considerable logistic skill to organize and feed thousands of workers and to arrange for stones to be quarried and transported to the building site. This in turn required an army of bureaucrats and a system of taxation (to pay for it all), which in turn necessitated accurate records. Before these enterprises began, the variable start of a lunisolar year of variable length presented no problems, but later the bureaucracy found it inconvenient and, maybe for this reason, they invented a more regular calendar—the civil calendar.

This was founded on their observation that the period between successive heliacal risings of Serpet was usually 365 days. Each year of this new calendar contained exactly 365 days and was divided into 12 months, each of 30 days. The excess five days, called epagomenal days by the Greeks, were attached at the end of the year and were considered to be very unlucky. There was no intercalation, either of months or of days, or any attempt to keep the calendar in step with the moon; it paid only lip-service to Serpet or the sun.

Surviving records do not mention when the civil calendar was first set up but it has been suggested that the first day of the civil year originally coincided with the first day of Tekhy. Detailed calculations then point to a time of origination between 2937 and 2821 BC—but there is no corroborative evidence for this speculation.

Since the civil year of 365 days was about a quarter of a day shorter than the mean tropical year or Sothic year, it started about one day earlier with respect to the both the seasons and Wep-renet, every four years. In about 1500 years it would have fallen behind by a whole year. This gradual drift of the start of the year through the seasons may seem odd to us. The Latin term for this civil year is 'annus vagus' or 'wandering year', and it has sometimes been called, perhaps misleadingly, a 'vague year'. It had one great advantage, at least in the eyes of astronomers—it was easy to calculate the exact number of days between two dates; there was no uncertainty about whether or when an intercalation had taken place, and every year contained exactly the same number of days. For this reason astronomers continued to use it throughout the Middle Ages, and it was used even by Copernicus working in 1500.

The calendar may be correlated with the Julian calendar thanks to a remark by the Roman author Censorinus to the effect that the Julian date, 20 July AD 139, corresponded to the first day of Thoth—the first day of the civil year.

The lunar calendar, governed by Serpet, also drifted. Because of this, the heliacal rising of Serpet remained of importance and the date it was observed was announced to the people to warn them to prepare their dams and dykes for the impending flood.

At first the months were referred to as months 1,2,3, or 4 of the three seasons (Akhet, Peret, and Shomu) but later, from about 2500 BC, they received popular names (see TABLE 11.2) deriving from festivals held in those months. Each 30-day month was divided into three 10-day periods.

There was no official Egyptian era; instead, years were numbered in regnal fashion according to the reign of the current Pharaoh. A list of the Pharaohs and the length of their reigns was compiled by Manetho in about 282 BC for

TABLE 11.2 *The names of the months in calendars based on the Egyptian calendar*

Month	Egyptian	Arabic	Ethiopian	Armenian	Persian
1	Thoth	Tot	Maskarram	Navazardi	Afrudin-meh
2	Paophi	Babe	Tekemr	Hori	Ardisascht-meh
3	Athyr	Hatur	Hadar	Sahmi	Cardi-meh
4	Cohiac	Kyak	Tahsas	Dre Thari	Tir-meh
5	Tybi	Tobe	Tarr	Kagoths	Merded-meh
6	Mesir	Mashir	Yekatit	Aracz	Schaharir-meh
7	Phanemoth	Buramar	Magawit	Malegi	Maher-meh
8	Pharmouti	Baramude	Miaziah	Arcki	Aben-meh
9	Pachons	Bashans	Genbot	Angi	Adar-meh
10	Payni	Baune	Sanni	Mariri	Di-meh
11	Epiphi	Abib	Hamle	Marcacz	Behen-meh
12	Mesori	Meshri	Nashi	Herodiez	Affirir-meh
Epogomenae					
	Nisi	—	Pagumen	Avelgats	Musteraca
	Kebus	—	Quaggimi	—	—

The names used by the Parsees closely follow the Persian names, and those of the Coptic Church follow the Egyptian names.

Ptolemy II. This was, and is, of considerable help in establishing an Egyptian chronology, though there are periods in which it is unreliable.

Later, Ptolemy (the astronomer, not one of the kings) collected information about the Babylonian kings (the canon of the kings) back to Nabonassar who acceded in 747 BC. His list is preserved in several documents, but not in his *Almagest*. On the basis of this list, the era of Nabonassar (EN), which started in 747 BC, was invented. Ptolemy and other astronomers and historians used this era in specifying dates in the Egyptian year. Very remarkably, the accuracy of the dates in the canon of the kings was corroborated by information inscribed on clay tablets unearthed by archaeologists in Mesopotamia.

⚅ The second lunar calendar

The gradual drift of the lunar with respect to the civil calendars would have been hardly noticeable for a generation or two, but eventually the discrepancy became evident. The bureaucrats were unwilling to adjust their conveniently regular civil year, but on the other hand the lunar calendar was useful in that it was geared to the heliacal rising of Serpet and the inundation. Their solution to this essentially ecclesiastical problem was to invent yet another calendar—a lunar calendar geared to the civil year. The gearing was accomplished by the simple rule of intercalating a month whenever the first day of the lunar year fell before the first day of the civil year.

Thus from about the fourth century BC, or maybe earlier, there were three calendars in use: the original lunar calendar, geared to the seasons, was used for fixing the dates of religious events related to agricultural and seasonal festivals; the new lunar calendar was used for religious events related to the moon; while the civil calendar was used exclusively for administrative purposes. This multiplicity of calendars should be familiar to some of us. A Moslem living in the West will use his religious calendar for the ordering of his religious practices and the Gregorian calendar for his civil activities; he will also be aware that the Christians and Jews organize their feasts and fasts using yet further calendars.

A surviving papyrus tells us more about how the second lunar calendar was run—at any rate in later years. In 25 civil years there are exactly 9125 days, which is very nearly 309 lunations; in 25 lunar years there are 300 ($= 12 \times 25$) lunations; thus nine months needed to be intercalated in each cycle. The papyrus spells exactly how this should be done. The 309 lunations occur in just under 9125 days. The ratio of days to lunations is 29.53074 ($= 9125/309$), which may be compared with the synodic period of 29.53059. The discrepancy is only about 17 seconds a month, and the synchronization with the real moon would only deteriorate by a day in about 400 years. Thus this remarkable calendar keeps in complete synchrony with the civil year and excellent synchrony with the moon. Later, the Ptolemys introduced the custom of adding an extra day so as to keep the calendar in step with the moon. It did not, of course, keep in very good step with the rising of Serpet or the flooding of the Nile.

The 25-year period was divided into 16 small years of 12 months apiece and nine great years of 13 months, to give the required 309 lunar months. Three hundred (12 × 25) of the 309 months had alternately 29 and 30 days, giving 8850 days; the nine intercalated months of the great years had 30 days apiece, giving another 270 days. The total of 9120 was made up to 9125 by intercalating a further five days at the end of every fifth year. The months remained approximately in step with the moon and enabled the religious festivals to be synchronized with the civil year.

Later, classical writers sometimes imply that the Egyptians knew of a 'quarter day year' of $365\frac{1}{4}$ days. If true (and they never had an official calendar with such a year) they would have found that their civil year advanced in this Julian year by one day in four years. In the course of 1461 $[365\frac{1}{4}/(365\frac{1}{4}-365)]$ years the start of the civil year would return to the same Julian date. This period was called a Sothic cycle.

The calendars were regulated by religious officials who took a very conservative attitude to their charge. Each Egyptian king was required to swear before he took office that he would not attempt to alter the calendar.

Ⅺ The reform movement

After the installation of the Ptolemaic dynasty attempts were made to reform the calendar by the insertion of a leap day once every four years, in the manner introduced to Rome by Sosigenes. Thus it is recorded in the celebrated Decree of Canopus (Canopus is a coastal town some 15 miles from Alexandria) of 239 BC that the Alexandrian king, Ptolemy III (not the astronomer), tried to persuade the Egyptian priests to introduce leap days into their wandering year. The priests were happy to turn Ptolemy's little daughter (who had died in the same year) into a god and to bestow the title of 'benefactor' on the king, who was known henceforth as Ptolemy Euergetes ('do gooder'), but they were too conservative to change the calendar and merely ignored the decree. Calendar reform had to wait until later.

The Roman Emperor to be, Augustus, defeated Anthony and Cleopatra at the Battle of Actium (on the west coast of Greece) in 31 BC and annexed Egypt to the Roman Empire as his own personal estate. In 30 BC the Roman senate decreed that the Egyptian calendar should be reformed. Accordingly, Augustus ordered the introduction of Euergetes' calendar on 29 August 23 BC. A sixth epagomenal day would be inserted at the end of every fourth year in the year preceding the Julian bissextile year. The Decree of Canopus was finally put into effect.

A consequence of this reform is that the years of the modified calendar keep in exact register with those of the Julian calendar. Each Alexandrian year begins on 29 August except every four years when, in the year preceding the Julian bissextile years, it begins on 30 August. This calendar does not of course keep accurately in step with the seasons or with the Gregorian calendar. It is useful to distinguish this modified calendar from the original Egyptian civil calendar by calling the former the 'Egyptian calendar' and the latter the 'Alexandrian calendar'.

𒀭 Egyptian influence

The defining feature of the Egyptian and Alexandrian calendar is 12 months of 30 days each, and either five or six epagomenal days at the end. Both calendars were adopted in various forms by many other societies over the years; these changed the dates the years started, the names of the months, the era, or point from when the years were counted. Some of these societies are listed in TABLE 11.3.

The dates when Zoroaster flourished are uncertain, but may have been as early as about 1000 BC. The religion he founded is the ancestor of Mithraism and is still practised by the Parsees of India. A traditional story relates that the three Magi of the Gospel of St Matthew (the three wise men) were Zoroastrian princes. The original home of the religion was Persia, until it was ousted on the rise of Islam. An era was instituted in Persia on 3 March 389 BC. This used the Egyptian calendar and each of the 30 days of the month was given a name. A new era began on 16 June AD 632, on the accession of the last Sassanid King Yazdegerd III, who was then only a child. He was deposed by the invading Arabs in the same year and finally murdered in 651. The Yazdegerd calendar was anomalous in that the five extra epagomenal days were inserted, not after the twelfth month, Affirir, but after the ninth month, Aban. Later still, in the eighth century, the Zoroastrians were driven out of Persia by the Moslems to Western India, where the Yazdegerd calendar is still in use among the Parsees.

In the early years of Christianity there developed several variant theological doctrines and it was not until Christianity became the official religion of the Roman Empire that orthodoxy was established. From then on the various heresies were proscribed wherever the power of the Roman Empire and the Church held sway.

In Egypt, the country was effectively ruled by the patriarch and his followers. After the establishment of Christianity in Rome, they were emboldened

TABLE 11.3 *Nations which adopted the Egyptian or Alexandrian calendar*

Nation	Calendar	Julian date of start[†]	Era name	Usage[*]
Egypt	E	Before 4000 BC	—	Civil
		26 February 747 BC	Nabonassar[**]	Scholarly
Persia/India	E	3 March 389 BC	Zoroastrian	Parsee
Egypt	A	30 September 23 BC	(Augustan)	Civil
Ethiopia	A	29 August AD 7	Abyssynian	Christian
Coptic Egypt	A	29 August 284	Martyrs	Christian
Armenia	A	11 July 552	Great Armenian	Christian
Near East	A	24 May 600	Fasli	Civil
Persia	E	16 June 632	Yazdegerd	Parsee
Armenia	E	11 August 1084	Little Armenian	Christian
France	A	11 September 1792	Republican	Atheist

Key: A = Alexandrian (365/366 days per year)
     E = Egyptian (365 days per year)

[†] Conforming to normal historical practice, Julian dates are quoted.

[*] We have given the name of the religion under 'usage' when the calendar was used mainly for regulating religious festivals. The Republican calendar was used for civil purposes but was deliberately atheistic in intent.

[**] The era of Nabonassar was invented by Ptolemy for astronomical purposes. It was used by astronomers till the fifteenth century or later.

to burn the old pagan temples (and the Library of Alexandria). Later, encouraged by a popular feeling among the Egyptians against their Greek and Roman overlords (among other resented and discriminatory practices, native Egyptians were forbidden to live in Alexandria), some patriarchs rebelled against the established Church itself. The hero of this movement was Dioscurus, who proclaimed a new heresy—monophysism. Christians of the time were concerned with the nature of Christ: was he man or God or both? Monophysism proclaimed, to oversimplify a little, that he was all God. This heresy was promptly condemned at the Council of Chaldecon in 451; the Egyptians did not accept this and for a time there were two patriarchs at Alexandria; the first was appointed by the establishment, and the second—poor, bigoted, but popular—represented the 'Coptic' Church (the term 'Copt' meant 'Egyptian'). The schism was never healed and monophysism remains the official doctrine of the Coptic and Ethiopian Churches to this day.

The Alexandrian calendar was adopted by the Coptic Church in Egypt and the Ethiopian Church further south in Ethiopia. The Coptic Church numbered their years from the accession of the Emperor Diocletian, renowned for his persecution of early Christians. This era is called the Era of Martyrs or the Era of Diocletian. The first day of this era—first Thoth of the year 1, Era of Martyrs—corresponds to the Julian date of 29 August AD 284.

Later, the Ethiopians and the Copts were further isolated from mainstream Christianity, first by the decay of Roman civilization, and then by the rise of Islam. They held to the Alexandrian calendar, and do so to this day, whilst the rest of Christendom adopted the Julian calendar. The Copts gave their months Egyptian names, but the Ethiopians have their own names (as shown in TABLE 11.2).

An interesting variant of the Alexandrian calendar was established in Asia Minor on 24 May AD 600, just before the start of Islam. It is known as the Fasli or Soor San calendar. It has been used from time to time in various places right up to the present day. The names of the months are arabic, which in turn derive from the old Egyptian names. The years are given names according to a pattern: each year of a decade is given a name; each decade, century, and millennium is also named; the full name of a year consists of its name prefixed by those of its decade, century, and millennium. In essence the name of a year is thus indicated by a sequence of words which spells out the four digits of its decimal number.

The Armenians originally used a lunisolar calendar but adopted the Egyptian calendar in about 460 BC from the Zoroastrians, who brought it to Armenia from Persia. At one stage, it would appear, the Zoroastrians intercalated an extra month (of 30 days) every 120 years, which adjusted the average length of the year to $365\frac{1}{4}$ days. The Armenians first counted the years from the accession of their kings. In AD 522, after they had been converted to Christianity, they instituted the Great Armenian Era on 11 July 552. Attempts were made by the Anastas in the Seventh century to reform the calendar and introduce a leap day, but they came to nothing. Only later did John the Deacon establish the Little Armenian Era from 11 August 1084; this used the Alexandrian calendar, but never became popular. The Great Era continued to be used for civil purposes. Later still, in 1317, the Armenians attempted and failed to establish the Julian calendar. Attempts by the Pope in 1584 to establish the Gregorian calendar likewise failed.

Much later, in 1792, a form of the Alexandrian calendar was given a new but short lease of life by the French Revolution (as described in Chapter 20). Later still, in 1917, the Soviet Union discussed the possibility of reintroducing the French Republican calendar. They decided against it at the time, but in 1929 set up the 'eternal calendar'. Like the Egyptian calendar, this had 12 months, each of 30 days divided into six 'weeks' of five days each (as discussed in Chapter 21). The five (or six) epagomenal days were, however, distributed throughout the year. This lasted till 1934, when the Gregorian calendar (but not the week) was restored.

Two thousand years ago an observer might have concluded that the Egyptian calendar was set fair to become the dominant world calendar. This was not to be, for the influences of Rome and later of Christianity, which initially rode on the back of the Roman Empire, conspired to make the Julian calendar all but universal in Europe. The rise of Islam destroyed most of the remaining cultures where the Egyptian calendar was still used, and today it mainly persists only in Ethiopia, the Coptic Church, and among the Parsees.

CHAPTER 12

The Calendars of China and East Asia

⌛ Introduction

People have been living in settlements in China since at least 3500 BC. Although there is a traditional story that the invention of writing is as old, the earliest known writing dates from the Shang period (*c.* 1600 BC). China is ancient, vast, and varied, but a short timetable of some salient events in its history is as follows:

3500 BC	Long Shan culture
1600 BC	Bronze Age Shang dynasty
1045 BC	Zhou dynasty
771 BC	Collapse of western Zhou feudal order
453–221 BC	Warring states period
221 BC	Qin dynasty
202 BC	Han dynasty
9 AD	Wang Mang deposes Han dynasty
25	Restoration of Han dynasty
220	End of Han dynasty; Empire breaks up
589	Sui dynasty
618	Tang dynasty
907	Last Tang emperor deposed
947	Liao dynasty
960	Song dynasty
1271	Yuan dynasty (Kublai Khan)
1275–91	Visit of Marco Polo
1368	Ming dynasty
1595	Jesuit missions in China
1644	Qing dynasty (Manchu)

1911	Revolution (Sun Yat Sen)
1926	Chiang Kai Shek
1949	Communist victory

Historical records, such as the *Book of history*, survive; these give detailed records of events over several thousand years. Despite wars and invasions, a historical continuity was preserved, as in nowhere else on earth.

Religion and state ritual was directed by the emperor and his nobility, and the state calendar played a crucial part in this. Unlike other civilizations, the prophets and reformers of the seventh and sixth centuries BC—Kong-fu-zi (Confucius, 551–479 BC), Lao-tse (*c.* 604–531 BC, the founder of Taoism), and Mencius (372–289 BC)—did not engender religious zeal and did not found new religions with new rituals requiring new calendars. Their concerns were more practical than theological; ritual and astrology were left to the ruler's courts.

During the 2000 years of the Imperial Period, the calendar was taken very seriously in the Imperial Court, as epitomized by the prestigious post of Almanac Maker. He had the responsibility of preparing the almanacs which would give, besides details of the calendar and various astronomical items, the divinatory prognosis for the year and the rituals that should be performed to avoid catastrophes. Its construction must have demanded exceptional skills apart from the ability to understand the intricate details of the calendar itself: there was an ever-present danger that the Almanac Maker would be executed if he got it wrong. The civil service promulgated the almanacs throughout the Empire and took steps to ensure that counterfeit almanacs (which did not ordain proper rituals and indeed might contain counter-productive ones) were suppressed.

The Chinese counting system is decimal and this is reflected in several sub-divisions of their calendar. This liking for 10 found expression in a metaphysical fancy which supposed there to be five heavenly principles or elements: metal, wood, water, fire, and earth. Each of these could be manifest in a yin form and in a yang form, thus making 10 in all. The opposite principles of yin and yang, and their reconciliation, play an important role in Chinese thought. Yang is male and associated with heat and daytime; yin is female and associated with cold and night.

In the period between the two World Wars archaeological excavations at Anyang near the Yellow River unearthed many thousands of bones, mainly the shoulder blades of oxen and the under shells of turtles. These are the so-called 'oracle bones' which date back to the Shang dynasty. They had been used mainly for divination—a question was inscribed on the bone which was then heated until cracks appeared which were interpreted as oracular answers. A small but significant number of these inscriptions record dates and records of tribute payments. The dates have given much information about the early Chinese calendar from 1400 BC on.

The Shang oracle bones give evidence that the Chinese of the time believed that the year was $365\frac{1}{4}$ days (the 'quarter-remainder' year) and the lunation $29\frac{1}{2}$ days. This resulted in the division of the circle, not into 360 degrees in the Babylonian manner, but into $365\frac{1}{4}$ Chinese degrees or pus. Thus the sun moved round the ecliptic about one pu every day.

By this time the Chinese astronomers were making accurate observations of the heavens; oracle bones mention stars by name, and one records an eclipse of 1281 BC. Several detailed star maps survive from the fourth century BC and later, Yu Hsi (307–38) independently discovered the precession of the equinoxes.

⚏ The sexagesimal cycle

The Shang oracle bones show that, at the time they were used (round about 1400 BC) the days (in 10- and 12-day cycles) were counted over a sexagesimal period of 60 days: 60 is the lowest common multiple of 10 and 12. The days of the 10-day cycle were named the 'ten heavenly stems' and those of the 12-day cycle, the 'twelve earthly branches'. The cycle of 60 days was further divided into six decades of 10 days, each of which began with the first earthly branch known as 'zi', from the first day of the cycle.

From the Zhou dynasty onwards, the years also were counted in this fashion using the same names; in time the practice of naming the days died out, but the naming of years survived. In this century the 12 terrestial branches are prominent as names of the years: the year 2000, for example, will be the year of the dragon.

The names of the 10 heavenly stems are given in TABLE 12.1, and those of the 12 earthly branches in TABLE 12.2. Six of the animals are domestic and yin, and six are wild and yang. Likewise there are five each of yin and yang heavenly stems. In this way the years (and days) were given one of 60 combined names of which the first was jia-zi; the sequence of combined names repeats every 60 years (or days). Popular tradition dates the start of this sexagesimal cycle to 2637 BC.

The heavenly stem, H, and the earthly branch, E, of any Chinese year which starts in the spring of the Gregorian year, Y, are easily calculated from the formulae:

$$H = 1 + \text{MOD}(Y - 4, 10)$$
$$E = 1 + \text{MOD}(Y - 4, 12)$$

TABLE 12.1 *The ten heavenly stems*

Number	Chinese name	Association
1	Jia	Growing wood
2	Yi	Cut timber
3	Bing	Natural fire
4	Ding	Artificial fire
5	Wu	Earth
6	Ji	Earthenware
7	Geng	Metal
8	Xin	Wrought metal
9	Ren	Running water
10	Gui	Standing water

TABLE 12.2 *The twelve earthly branches (zhi)*

Number	Chinese name	Association
1	Zi	Rat
2	Chou	Ox
3	Yin	Tiger
4	Mao	Hare
5	Chen	Dragon
6	Si	Snake
7	Wu	Horse
8	Wei	Sheep
9	Shen	Monkey
10	You	Cockerel
11	Xu	Dog
12*	Hai	Pig

*The number 12 was significant—perhaps because there are 12 lunations in a year; perhaps because there are nearly 12 years in a Jupiter cycle; perhaps because the ecliptic was divided into 12 mansions as in the Babylonian zodiac or by the Indian sankrāntis.

where MOD($Y - 4,B$) is the remainder obtained on dividing $Y - 4$ by B (see Chapter 23). The Chinese did not count the cycles, so the reverse calculation of Y from H and E is not possible. The sexagesimal system is often useful in resolving ambiguities in dates given in the lunar solar calendar because of doubt about the number of months in a year or days in a month. This is reminiscent of the way long-count dates remove uncertainties in calendar round dates in the Mayan calendar.

The earliest record of the seven-day week dates from AD 1207 (before Marco Polo's visit), although a period equal to a quarter of a lunation may have been used in early times in the Zhou dynasty. Two sets of names for the days of the week exist: one uses the names of the planets as in the West, and the other, numbers. The latter is now standard.

🏛 The lunisolar calendar

The Shang (fourteenth century BC) oracle bones also give evidence of a lunisolar calendar which, much modified, persists to this day. This has a 12-month common year, with a thirteenth intercalated from time to time.

The *Book of history* mentions intercalated months in an anecdote about the Emperor Yao, reputed to have lived in the twenty-first century BC. For sure, from the seventh century BC, or before, the principle of intercalating seven months in a cycle of 19 years was established. This is, of course, the Metonic cycle. It is unknown whether the Chinese developed this independently or it reached them from India or Babylon or, for that matter, whether the Indians adopted it from the Chinese.

In 104 BC a new calendar (the Taichu calendar) was introduced which worked on the new principle that the thirteenth month was intercalated in a manner remarkably similar to that used in the Indian lunisolar calendar (see Chapter 13). The ecliptic was divided into 12 equal regions which are close analogues of the Indian sankrântis (and for that matter the signs of the zodiac). The passage of the sun from one region to the next defines a solar month. The lunar month was taken to begin with the first day after the new moon or conjunction. The division of the ecliptic went further than solar months, for it was also divided into 24 parts of about 15 days each. The 24 periods were given names which date from late Zhou times and which describe meteorological and agricultural events expected in them, as shown in TABLE 12.3.

When this calendar was first set up it was believed that the tropical year was, on average, $365\frac{1}{4}$ days (the so-called 'quarter-remainder' year), so that the average length of a solar month was 30.44 days. An accurate value for the average length of a lunation of 29.53 days was already available.

TABLE 12.3 *The 24 'fortnightly' periods*

Number	Name	Meaning	Gregorian date
1	Li Chun	Spring begins	February
2	Yu Shui	Rain water	February/March
3	Jing Zhe	Excited insects	March
4	Chun Fen	Vernal equinox	March/April
5	Qing Ming	Clear and bright	April
6	Gu Yu	Grain rains	April/May
7	Li Xia	Summer begins	May
8	Xiao Man	Grain fills	May/June
9	Mang Zhong	Grain in ear	June
10	Xia Zhi	Summer solstice	June/July
11	Xiau Shu	Slight heat	July
12	Da Shu	Great heat	July/August
13	Li Qiu	Autumn begins	August
14	Chu Shu	Limit of heat	August/September
15	Bai Lu	White dew	September
16	Qui Fen	Autumn equinox	September/October
17	Han Lu	Cold dew	October
18	Shuang Jiang	Frost descends	October/November
19	Li Dong	Winter begins	November
20	Xiao Xue	Little snow	November/December
21	Da Xue	Heavy snow	December
22	Dong Xhi	Winter solstice	December/January
23	Xiao Han	Little cold	January
24	Da Han	Severe cold	January/February

Since the lunation is shorter than the solar month, it happens from time to time that two lunations commence within the same solar month; when this occurs, the second is accepted as an intercalary month. This system uses the tropical year rather than the sidereal year used by the Indians; originally, the mean motion of the sun and the moon was used, but later, the true motion, which takes into account variations in the speed of the sun and moon in their orbits, was used instead. Thus, from AD 619 true new moons were used and, from AD 1645, true solar months as well.

It can easily be shown that the result of this innovation was to lead to the intercalation of, on average, about seven months in a period of 19 years—close to Metonic frequency. A scheme of this sort will, generally speaking, keep the years in synchrony with the sun, provided the assumed duration of the lunar and solar months is accurate. Unlike the systems devised by Meton, Calippus, and Hipparchus, the precision does not depend on numerological 'fiddles' of dropping a day here and intercalating a month there; it is, as it were, automatic.

The months themselves were originally given somewhat prosaic names, mostly seasonal or agricultural, but were later referred to by number. The intercalary month was given the same number as the preceding month; this doubling up of the number is reminiscent of the Roman bissextile leap day.

According to the scheme just described, the months, which began on the day of the astronomical new moon or conjunction, would have 29 or 30 days, in approximate alternation. The months were further divided into three decades of a nominal 10 days each—the 29-day months never completed their third decade. This custom of having 3 divisions of the month is reminiscent of the Egyptian calendar. At the time of the oracle bones, the month was also divided into four quarters, a practice common to many cultures.

The quarter-remainder year is, as Pope Gregory XIII realized, not adequate in the long run as the basis of a calendar and, as the discrepancies became apparent, reforms were made to the calendar. Over the centuries no fewer than 102 reforms have been recorded. Some of these have involved refinements in the length of the year and the lunation; some, improvements in the method of calculation; and others, the elaboration of new theories. Many of the reformed calendars were given names. Despite these complications, records exist which enable dates over the last 2000 years to be converted. Several sets of tables are available for this purpose.

The length of the tropical year, from which is derived the length of the solar month, is a crucial parameter in this lunar calendar. Sometime near the beginning of the third century BC it had become obvious that the quarter-remainder year was inaccurate and that it needed to be shortened a little. The favourite Chinese method of determining the length of the tropical year was to measure the interval between two winter solstices, or better still, the average of a large number of such intervals. At these solstices, the noontime shadow of a gnomon lay due north and was at its longest. From the second century BC, gnomons, two metres high, were used. Although taller gnomons would in principle yield a more accurate result, in reality the shadow would be blurred. This problem was not solved till Guo Shoujing, in AD 1281, introduced a pinhole which threw a small spot of light on to the scale. This enabled him to increase the length of his gnomon to 13.3 metres and yet read the length of the shadow with great precision.

Later, Xing Yunho (1573–1620) built a gnomon which was 20 metres tall; with it he determined a value of 365.242190 days—close to the modern value. It is interesting that he was an almost exact contemporary of Tycho Brahe (1546–1601), who used similarly large devices. These two great astronomers calculated years almost identical in length—how pleased they might have been to have met and corroborated each other's results. Perhaps their agreement had something to do with the arrival of the Jesuit missionaries, since they are known to have worked in the bureaucratic office that oversaw the calendar.

Meanwhile, other advances were being made which facilitated the measurement of the year. The problem, originally perceived by Zu Chongzhi (AD 430–510), was that the length of the shadow of the gnomon changed very little from day to day at the time of the solstice; furthermore, the exact moment of the solstice was unlikely to be at noon when the measurements were made. In response to this, he and others developed new methods which involved measuring the length of the shadow over a period of 20 days and interpolating to the instant, not necessarily at noon, when it appeared the longest. Some measurements of the length of the year are listed in TABLE 12.4.

Unlike the Indian calendar, the Chinese calendar was geared to the tropical year. In fact, the difference between the sidereal and tropical years, which is due to precession, was not suspected till about AD 7, and not unequivocally demonstrated till AD 330.

During the Zhou dynasty (1045–221 BC), the year was assumed to begin with the new moon which occurred on or before the winter solstice. The Qin Emperor, Shi Huang-ti (221–210 BC), moved it one lunation back in time, and

TABLE 12.4 *The length of the tropical year*

Date	Astronomer	Length of year in days
Fifth century BC		365.25
AD 206	Liu Hong	365.2462
AD 462	Zu Chongzhi	365.2428
AD 728	Yi Xing	365.2444
AD 1199	Yang Zhongfu	365.2425
AD 1280	Guo Shoujing	365.2425
AD c. 1600	Xing Yunlu	365.242190

this custom prevailed until 104 BC, when the Emperor Han Wu Di moved it again. The current rule is to start the year in such a way that the winter solstice occurs in the eleventh month. This means that it will usually begin at the second new moon following the winter solstice, but once in a while, at the third new moon.

The solstices and new moons were calculated or observed at Beijing, although modern Chinese communities may use other meridians; the day was considered to begin at midnight, and a month began with the day of the new moon. The construction of an almanac in advance of the days it lists clearly requires an adequate theory of the movement of the sun and the moon. For this, if no other reason, the details of the calendar varied as the theory was refined.

The Chinese learnt of the Gregorian calendar and of European astronomy from Jesuit missionaries who arrived at the end of the sixteenth century. Possibly, they played some part in the adoption of the true motion of the moon and sun in calendrical calculations. However, the Chinese continued to use their own calendar until 1912, when the Republic of Sun Yat Sen officially introduced the Gregorian calendar. In 1949, the People's Republic of China finally abolished the old calendar in favour of the Gregorian calendar. The old calendar is still used by Chinese communities around the world and, we may guess, in traditional circles in China itself.

⌛ Eras and cycles

No system of eras was ever established. The common practice was to count separately the years of the reign of each emperor; the first year of each reign was taken to be the year that started next after his accession. Lists of emperors and the lengths of their reigns are available for many centuries, allowing dates to be located with reasonable exactitude back to the eighth century BC. It was usual to mention not only the regnal year but also the sexagesimal name of the years. This usually removes any remaining ambiguity.

Nevertheless, the Chinese, like other peoples, contemplated several cycles of years. Longer cycles are resonances of shorter cycles, starting with the sexagesimal period, the lunar month, and the solar month. The longest cycle, the chi, was of 31 920 years; by a coincidence this is equal to four Julian cycles. At the end of a chi, all things come to an end and return to their original state—a notion echoed in Mayan belief (discussed in Chapter 14).

Other sets of resonances derive from the orbital periods of the planets. Like the Indians, the Chinese contemplated 'grand conjunctions' when all the planets came together after a cycle of 138 240 years. A further resonance of this gave a 'world cycle' of 23 639 040 years, the beginning of which was known as the 'supreme ultimate grand origin'.

⊠ South-East Asia

Several countries of South-East Asia owe their calendar either to the Chinese or the Indians; the Chinese and Indian lunar calendars are closely similar as I have already remarked. Thus, the Khmer of Cambodia, the Annamese, the Laotians—to name but a few—use the sexagesimal cycle to name their years. Generally, the names of the animals of the earthly branches are adopted but often translated into the local language; the names of the heavenly stems are sometimes used.

The calendars of the Khmer of Cambodia and of Thailand (or Siam), which are very alike, show a marked Indian or Hindu influence, and the months of the lunar calendar are given names derived from the Indian names shown in TABLE 13.4 (see page 178). The names of the days of the week are derived from the Sanskrit names. They number the years in three ways: according to the Buddhist era, according to the Greater era (which is identical to the Saka era of India), and according to the Lesser era which began in AD 63.

⊠ Japan

The traditional Japanese calendar, closely modelled on the traditional Chinese calendar, was adopted in the early seventh century. It employed the sexagesimal cycle of heavenly stems and earthly branches, modified slightly and translated into Japanese; the 12 earthly branch names, the menagerie, were also used to name the 12 periods into which a day was divided. The calendar itself was lunar, with a thirteenth month intercalated seven times in 19 years, but the months are given poetic, agricultural, and meteorological names. The year was divided into four seasons and into 24 periods of about 15 days, as with the Chinese calendar.

Early Japanese chronology would seem to be uncertain until the Chinese calendar was adopted in AD 660. Several systems of counting the years have been used; these include the sexagenary cycle already mentioned, the regnal year of the emperor, the Jimmu Tenno system, and the Nengo system. In the Jimmu Tenno system, the years are numbered consecutively from the accession of the first Emperor, Jimmu, in 660 BC. The nengo system was used sporadically, counting the years from some notable event; at the next notable event, the counting restarted at one.

There have been a considerable number of nengos, some lasting less than a year.

In 1873, the Japanese adopted the Gregorian calendar, together with the seven-day European week (unknown in Japan until then) and a number of other Western innovations.

CHAPTER 13

The Calendars of India

⊞ Introduction

In India (or what is now Pakistan), the Indus Valley civilizations at Harappa, Mohenjo-Daro, and elsewhere, flourished in about 2500 BC. However, their writing has not yet been deciphered, so we know nothing of their calendar. Later, Aryan invaders from the North initiated new developments in many parts of India.

From the sixth century BC, Mesopotamian astronomy and calendrical knowledge, including the tithi, began to have influence in India. Later, from about the time of Alexander, the civilizations of India and the Middle East borrowed and repaid astronomical and mathematical knowledge. The Indians learnt of the signs of the zodiac (in about AD 149) and the seven-day week; later, their number system, with its zero (sometimes attributed to Aryabhata, born AD 476) was exported to the Arab world and thence to Europe, where it became one of the key elements in the development of modern science. Some of the salient features in the history of the Indian subcontinent are as follows:

c. 2500 BC	Indus Valley civilizations at Harappa and Mohenjo-Daro.
c. 1500 BC	Aryans arrive in North India and destroy Indus Valley civilization.
c. 1500 BC	Composition of Vedas begins.
c. 700 BC	Composition of Upanishads begins.
c. 500 BC	Sinhalese Aryans reach Ceylon. Caste system established.
c. 528 BC	Death of Mahavira, founder of Jain sect.
c. 487 BC	Death of Buddha in Ceylon.
329 BC	Alexander the Great reaches India.
321 BC	Maurian Empire founded in North India by Chandragupta.
272 BC	King Asoka enthroned.
258 BC	King Asoka converted to Buddhism.
77 AD	Kanishka is Kushan emperor.

320 AD	Chandra Gupta founds Gupta Empire.
476	Birth of Aryabhata.
c. 480	Guptas overthrown by Huns.
712	Arabs conquer Sind.
846	Start of Cholas Empire.
1018	Mahmud Ghazni takes Hindu states.
1206	Sultanate of Delhi founded.
1398	Tamerlane sacks Delhi.
1498	Vasco da Gama reaches India by sea.
1526	Babur founds Moghul dynasty.
1556	Accession of Akbar.
1674	Sivaji founds Mahratta kingdom.
1690	British found Calcutta.
1707	Death of Aurangzeb; decline of Moghul power.
1787	End of war against Mahrattas.
1857	Indian mutiny.
1877	Victoria proclaimed Empress of India.
1885	Foundation of Indian National Congress.
1947	India and Pakistan independent of the British.

A survey conducted at the turn of this century reported that there were 20 calendars in use in India at the time; Nehru, writing in 1953, claimed there were over 30. The multitude of religions and sects accounts for some of this diversity: Islam and Hinduism have both waxed and waned in India over the centuries (in Islamic times the Islamic calendar was, of course, used); the Jains maintain an era in which the years are counted from the year of the death of the founder of that sect, Mahavira, in 528 BC; and the Buddhists, when that religion was popular in India, counted the years from a nominal year of the Buddha's death, 544 BC (his actual dates are uncertain, as is whether the Mahavira was his contemporary).

The East India Company and later, the British Imperial administration, used the Gregorian calendar; 10 years after independence in 1947 the Indian Government set up the reformed Saka calendar—but the Gregorian calendar is also still in use. It remains to discuss the Hindu calendars themselves.

An interest in astronomy is apparent in the earliest sacred writings of the Hindus, the Vedas, which date back to the middle of the second millennium BC. They give information about an early Vedic calendar with 12 months of 30 days each, with a common year of 360 days; this was augmented by an extra month when required. Variations on this theme are also mentioned, and there is evidence that it was of Mesopotamian origin.

Several theoretical astronomical treatises (the siddhântas), dealing with the calendar, were written from about the first century AD, but the earliest surviving dates from AD 425. They were written in Sanskrit, the classical language of India. These show obvious indications of the influence of the astronomy of Babylon and of Greece. Thus the system of time measurement divided the day into 60 ghatikas, each ghatika into 60 pala, each pala into 60 vipalas, so that an hour is 2.5 ghatikas and so forth. The names of the signs of the zodiac are mostly literal translations of the Greek or Roman names, and it is possible that the lunar calendar was based upon the Metonic cycle invented by the Babylonians.

Three of these siddhântas are particularly important for the calendar: the *Aryabhatiya*, written in the fifth century by Aryabhata himself; the *Brahmes-phuta-siddhânta* written by Brahmagupta in the seventh century; Rajamrig-anka of Bhojareda and the *Surya-siddhânta*. The *Surya-siddhânta* dates from about AD 800—but the author is unknown.

The Indian calendars are noted, at any rate by people used to occidental calendars, for being intricate, complex, and subject to numerous local variations. The last characteristic is only to be expected in a country occupying a large area and rarely united under a common government. Both a solar and a lunisolar calendar were developed and are still in use. The solar calendar is used in parts of Bengal and Madras for civil purposes, and the lunar calendar is used, mostly for religious purposes, in all parts of India—but there are exceptions to both these generalizations. There are many variants of both calendars which differ in the names of the months, the date of the start of the year, the phase of the moon on which the year or month starts, the rules for intercalation, the era used for counting the years, and so on.

𝍢 The week

The astrological week that spread through the Roman Empire in the early years of the Christian Era eventually diffused to India in the second century. The earliest record of its use is in an inscription dating from the time of the Gupta, in AD 484. The days of the week were given the Sanskrit names for the sun, moon, and planets; these names are listed in Appendix II and TABLE 21.3 (see page 283). The weekday name of the days are always given in Indian almanacs.

𝍢 The solar calendars

The solar calendar is based upon the sidereal year. The Indians have known about the precession of the equinoxes for a long time but have largely chosen to ignore it; attempts were made in the nineteenth century to reform the solar calendar to allow for precession, but they appear to have come to nought. The year thus begins when the sun reaches a certain fixed point in the ecliptic, approximately marked by a star in Pisces, about 28° before the current vernal equinox (the vernal equinox was at this point in about 6 BC). The ecliptic is divided into 12 equal regions of 30°, which are analogous to the Babylonian signs of the zodiac. The entry of the sun into each region marks the start of a new month. The names of the months in this solar calendar are usually based on the names of the zodiacal signs, as shown in TABLE 13.1; sometimes the lunar names, discussed later, are used.

The start of each region or sign was called a 'sankrânti', and the astronomical year began when the sun reached the mesha-sankrânti. For civil purposes, the first day of the year was the first day that began after the mesha-sankrânti, but there are several variations of practice, and sometimes it was one or two days after that. Likewise, each month began at the appropriate sankrânti or shortly after. The civil day began at sunrise and the days of the month were numbered in an unbroken sequence.

TABLE 13.1 *The sanskrit names of the signs of the zodiac*

Number	Sanskrit*	Latin	Time†
1	Mesha	Aries	30.9
2	Vrishabha	Taurus	31.4
3	Mithuna	Gemini	31.6
4	Karka	Cancer	31.5
5	Simha	Leo	31.0
6	Kanya	Virgo	30.5
7	Tula	Libra	29.9
8	Vrischika	Scorpio	29.5
9	Dhanus	Sagittarius	29.4
10	Makarus	Capricorn	29.5
11	Kumbha	Aquarius	29.8
12	Mina	Pisces	30.3

*The meanings of the Sanskrit names are close to, but not always identical to the Latin names (see Table 2.1, page 27). In particular, Makarus is a sea serpent or crocodile rather than a goat.

†The time is the period, measured in days, from the sankrânti at the start to that at the end of the sign. There is some small variation in these times according to the estimated length of the sidereal year given in the various siddhântas.

These solar months do not contain the same number of days, even though the sun moves the same distance in each. This is because the ellipticity of the earth's orbit implies that the sun's velocity is not constant. It may be seen from TABLE 13.1 that Dhanus is the shortest month and Mithuna, the longest. The number of days in a month may vary from 29 to 32, and will also vary from year to year and with the position of the observer (as his local time and the time of his sunrise varies).

It should not be thought that the solar calendar is empirical or demands constant observations of the position of the sun to ascertain the moment of each sankrânti. In practice, a value was assumed for the length of the sidereal year and for the intervals between each sankrânti. Different values of the length of the year were given in the several alternative siddhântas in use, with the result that the precise details of the solar year varied from region to region where the solar calendar was used. The solar calendar continually advanced on the tropical year for two reasons: first, the sidereal year is longer than the tropical year; and secondly, the values of the sidereal year used were not accurate (the lengths used and the errors they imply are shown in TABLE 13.2).

The solar year is divided into six seasons as shown in TABLE 13.3. There is now a discrepancy of about six weeks between the sort of weather indicated and the seasons themselves. This is due to the precession of the equinoxes that has taken place since the system was devised and the cumulative effect of error. (This suggests it was invented in about 500 BC.) The year is also divided into two parts: the uttarayana, which begins with the makara-sankrânti and during which the sun is moving northwards; and the dakshinayana in which the sun moves south, beginning with the karka-sankrânti.

TABLE 13.2 *The lengths of the sidereal years assumed in the siddhântas*

Siddhânta	Length*	Discrepancy[†]
Aryabhatiya	365.258681	61
Rajamrlganka	365.258691	61
Surya-siddhânta	365.258756	60
Modern sidereal year	365.256360	71
Modern tropical year	365.242193	—

*The length of the year is given in days.

[†] The discrepancy is the approximate number of years required for the nominal sidereal year to advance on the true tropical year by one day.

Finally, it may be noted that although precession of the equinoxes is ignored as far as the formal structure of the solar calendar is concerned, modern Indian almanacs record the date of the true equinoxes and solstices as being particularly auspicious occasions.

It is clear that the solar calendar is susceptible to almost infinite variations and it is difficult to derive any general method for equating a date in, for example, the Gregorian calendar, with an Indian solar date. This can only be done for a defined locality and with a suitable selection of rules for deciding the starting day of each month. Several tables are available, however, which are valid within the sort of constraints just noted.

🕮 The lunisolar calendars

The lunisolar year of India contains 12 months, marked by the phases of the moon, with an occasional thirteenth month to reconcile it to the seasons. The rules by which the thirteenth month is intercalated bear a resemblance to the Chinese system, though their effect is similar to that of the Metonic cycle discussed in previous chapters.

The months take their names from those of the nakshatras. These are 27 (or sometimes 28) stars located round the ecliptic at intervals of about 13° 20′. The moon, travelling round the ecliptic during a lunation, passes close by each one. This journey takes about 27.3 days—the sidereal period of the moon—which accounts for the number of nakshatras. These can be used to mark the days of the lunations as the moon passes by them. The nakshatras are grouped into 12 sets of two or three stars each and these indicate the passage of the sun in its yearly journey round the ecliptic. These sets thus function in a manner analogous to the signs of the zodiac and probably predate the importation of the latter from Babylon; they are sometimes referred to as the lunar zodiac. It may

TABLE 13.3 *The seasons of the solar year*

Number	Season	Meaning	Sankrântis
1	Vasanta	Spring	Mina, Mesha
2	Grishma	Hot	Vrishhabha, Mithuna
3	Varsha	Rainy	Karka, Simha
4	Sarad	Autumn	Kanya, Tula
5	Hemanta	Cold	Vrischika, Dhanus
6	Sisira	Dewy	Makarus, Kumbha

be noted that the nakshatras and the sankrântis are, as it were, interdigitated round the ecliptic.

The most common names of the lunar months are derived from the names of 12 of the nakshatras and are given in TABLE 13.4, but it must noted that variations of these names occur in different parts of India. Sometimes the solar months are also given the names in TABLE 13.4, rather than those in TABLE 13.1. The names in TABLE 13.4 superseded another set. The nakshatras, the reader may perhaps be relieved to learn, play no other role in the Indian calendar than to provide names for the months.

There are two customs concerning the start of the year. In some parts of India this is taken to be the beginning of the month Chaitra, which is in the spring; this is the Chaitradi year. The alternative is to start the year in the autumn with Karttika; this is the Karttikadi year. Similarly there are two customs concerning the start of the lunar months. One custom, prevalent in southern India is to begin the month on the day after a new moon (or conjunction); such months are amanta months, also called sukladi months, which charmingly means 'beginning with the bright fortnight'. The alternative, prevalent in north India, is to start the months at the full moon; they are then

TABLE 13.4 *The names of the lunar months*

Number	Name*	Sankrânti†
1	Chaitra	Mina
2	Vaisakha	Mesha
3	Jyaishta	Vrishabha
4	Ashadha	Mithuna
5	Sravana	Karka
6	Bhadrapada	Simha
7	Asvina	Kanya
8	Karttika	Tula
9	Margasira	Vrischika
10	Pausha	Dhanus
11	Magha	Makarus
12	Phalguna	Kumbha

*These names probably came into use in the middle of the second millennium BC and are derived from the names of certain nakshatra stars. They displaced an earlier set of names.

†The sankrântis listed are the last to be entered by the sun before the conjunction which initiates the lunation, the name of which is the one assigned.

called purnimanta or krishnadi months; krishnadi means, lugubriously, 'beginning with the dark fortnight'.

Of crucial importance is the manner in which the names of the months, shown in TABLE 13.4, are allocated to the lunations. There is a simple and ingenious rule which, when followed to the letter, not only names each lunation but automatically indicates when months should be intercalated or extracalated. This rule tells us that any lunation whose conjunction occurs when the sun is between the Mina and Mesha sankrântis should be Chaitra; any lunation whose conjunction occurs when the sun is between Mesha and Vrishabha should be called Vaisakha; and so on. The months in TABLE 13.4 are allocated to the lunation whose conjunction occurs after the sankrânti listed against it in the table.

Two interesting complexities arise. From TABLE 13.1 it may be seen that the interval between two successive sankrântis varies from about 29.4 to 31.6 days on account of the variable speed of the earth in its elliptical orbit. Likewise, the actual period between successive conjunctions of the moon varies between about 29.3 and 29.8 days on account of ellipticities in the orbits of the earth and moon, and because of complex gravitational perturbations. Most times, one single conjunction occurs between two successive sankrântis, and vice versa; that is to say, the sankrântis and the conjunctions alternate in time. But occasionally one or other of two eventualities may occur.

First, two lunar conjunctions may occur in the period between two sankrântis; when this occurs both lunar months are given the same name in accordance with the rule as illustrated in FIG. 13.1(A). This is effectively an intercalation and the first of the two like-named months is distinguished by the prefix adhika, meaning 'added', and the second by nija, meaning 'normal'.

Secondly, but much less frequently, two sankrântis may occur in a single month (between two conjunctions); that is, there is no conjunction occurring between two sankrântis. This contingency is illustrated in FIG. 13.1(B). When this happens, a month is omitted as an extracalation, again according to the rule.

The reader who is concerned by the apparently esoteric nature of these considerations may pause to ponder an analagous effect in our own calendar which arises from not too dissimilar reasons: in some months there are two full moons and in others only one; likewise in some months there are five Sundays and in some only four. The phrase 'once in a blue moon' has referred, at least since 1824, to an event which happens exceedingly rarely. More recently it has been used to refer to a month containing a second full moon—which occurs every few years. It has even been suggested (without evidence) that the latter is the original meaning of the phrase.

When a month is extracalated (and this happens in the autumn when the

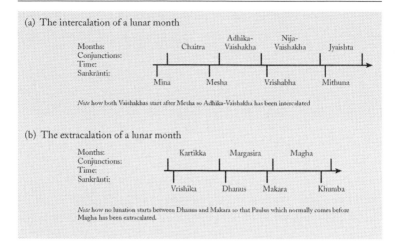

FIG. 13.1 (a) The intercalation of a lunar month in the lunar calendar. Note how both Vaishaknas start after Mesha, so Adhika-Vaishakha has been intercalated.
(b) The extracalation of a lunar month in the lunar calendar. Note how no lunation starts between Dhanus and Makara, so that Paulus, which normally comes before magha, has been extracalated.

solar months are short), it is invariably found that one or more months are intercalated earlier in the year. A single intercalation will cancel out, so to speak, an extracalation. The upshot of these intercalations and extracalations is that seven months are intercalated in a period of 19 years; the length of the lunar year oscillates by about the length of the tropical year; but the start of the lunar year never drifts far from the start of the solar year (though definitive proof of this is difficult to obtain). A slight problem arises if Chaitra itself is duplicated by an intercalation—which year does the Adhika-Chaitra belong to? Again the answer will vary, since customs differ.

This highly individual and ingenious solution to the problem of reconciling the lunar and solar years bears a superficial resemblance to the Babylonian method attributed to Meton. But although there are seven intercalations in 19 years, these figures emerge from the rule of naming, rather than the other way round, and the actual years in the cycle which receive the intercalated month vary. It thus seems at least possible that the method is indigenous—although by the time it was invented in the fifth century, or even earlier, the Indian astronomers must presumably have been aware of the Metonic cycle. It is possible that in its original form the system used mean motions of the moon; with these the length of every mean lunation was the same and there could be no extracalations. The abandonment of mean motions, if once used, probably occurred before AD 1100.

I have described a simple rule and its consequences. This is, however, not the only rule to be employed; like other aspects of the Indian calendar, there are variations. Nevertheless, all the rules employed in the different parts of India lead to intercalations or extracalations at the same points of time; all that differs are the names of the months affected.

There is one further unique feature of the Indian lunar calendar—the naming of the days in a lunar month. As already described, a lunar month began at one conjunction of the moon and ended at the next. During the intervening days the angular separation of the moon and sun waxed and waned through 360°. The Indian astronomers divided this into 30 equal segments of exactly 12° each. The time taken for the angular separation of sun and moon to change through this angle is termed a tithi or lunar day. The mean value of a tithi is the mean synodic period divided by 30, or about 0.984353 days, or 23 hrs 37 minutes—which is about 23 minutes shorter than a mean solar day of 24 hours. However, the actual time taken for the moon to pass from one tithi to the next varies from lunation to lunation, as does the actual length of a day from noon to noon. There was one further subdivision—the karana. This was equivalent to the period from conjunction to opposition of the moon, or to 15 tithis.

The days of a lunar month are assigned a tithi number which is that of the tithi in force at dawn at the start of the day. The tithi numbers in each karana run from 1 to 15, save that the last of the second Karana is number 30—but again, there are localized differences in this custom. Generally, just one tithi starts in a day and vice versa but, as we have seen, both the tithi and the solar day vary in length. It thus may happen that the numbers of the tithi valid at dawn of each day do not run consecutively: sometimes a tithi number will be repeated (adhika-tithi) and sometimes one will be skipped (kshaya-tithi). That this must happen is indicated by the fact that there are 360 tithis in 12 lunar months, but only about 354 days; six tithis must somehow be lost. The situation is analogous to the intercalation and extracalation of months necessitated by interruptions in the alternation of the lunar conjunctions and the sankrântis (as already described).

So far I have described the construction of the Indian lunisolar calendar, or at least one variant. Indian calendar makers are able to construct, before the start of a year, an almanac. These almanacs, called panchangas (five limbs), list five items for each day: the day of the week (vara), the tithi, the nakshatra, the karana, and the yoga at sunrise and sunset. The yoga is the sum of the nakshatras of the sun and moon at the time; it provides information concerning the position of the moon relative to the sun. The estimation of these items requires prior knowledge of the time at which each conjunction is going to occur and the time at which the moon passes each tithi in its journey round the sun. It is not a matter of making observations of the moon's position in the ecliptic, for it has not yet set out on its trips around the firmament when the almanac is to be constructed. Clearly, a theory of the moon's motion is required—a theory which will predict these times with adequate precision.

Although at the time of Aryabhata, in the fifth century, Indian astronomy was groping its way towards a heliocentric system, as propounded by Copernicus, there are other reasons to believe that as late as the last century, a geocentric theory was still in use for practical purposes. However, the important thing is the existence of a recipe for actually calculating the time of a conjunction. An exact calculation is complex, but relatively simple methods can produce reasonably accurate results most of the time, adequate for the calendar makers or the people who read the panchangas, but not for those who wish to reconstruct the calendars of bygone times. To explain these methods would require a consideration of astronomical and mathematical details which is beyond the scope of this book.

The philosophy underlying these Hindu calendars is very different to that behind the equivalent Christian calendar used to calculate the date of Easter and other church festivals. The date of Easter can be easily worked out by the most humble parish priest from the tables given in the Book of Common Prayer; it requires no astronomical knowledge or elaborate astronomical calculations whatsoever. On the other hand, the correct dates on which to hold a Hindu festival could, it would appear, only be worked out by a calendar maker who was well trained and equipped with sufficient astronomical lore. There is an interesting story, brought back to Europe, of an Indian calendar maker who with the aid of pebbles (calculi), was able to predict the time of an eclipse in 1824 to within four minutes. The esoteric knowledge required had, no doubt, been passed from master to pupil down the generations. Clearly, as long as the people retained an interest in their religion, the calendar makers would be assured of respect and steady employment. The Christian computists in late medieval Europe maintained a not dissimilar attitude in their opposition to the new arithmetic of the Arabs.

As with the solar calendar, the variations in the Hindu lunar calendar make futile any attempt to correlate accurately Hindu and Christian dates—though several books of tables, covering limited places and periods for which historical records are available, have been published. Likewise, mathematical methods of at least approximate validity are available. Those wishing to try their hand at making a panchanga themselves will find a reference to these in the bibliography.

⌛ Indian cycles and eras

There are innumerable systems for referring to years in the Indian calendars. One important distinction is between current and expired (elapsed) years. Indian dates frequently refer to elapsed years—a practice foreign to Westerners.

Some systems utilize long cycles and are mainly used by astronomers; among these are the yugas, in use (according to an inscription) since at least AD 634. The shortest such period is the kali-yuga, or iron age, of 432 000 solar years. I do not know how this figure of 432 000 was arrived at; 432 factors into 27 and 16. The dvapura-, triti-, and kriti-yuga are 2, 3, and 4 kali-yugas respectively. One of each of these different yugas, equivalent to 10 kali-yugas, form a mahayuga; and a thousand of these constitute a kalpa—an unimaginably long period of 4.32 billion years. The current kali-yuga began at the midnight which started on the Gregorian date of Friday 18 February 3102 BC. According to one authority it was delayed until dawn. The kali-yuga year is sometimes given in panchangas.

At the beginning of the kali-yuga, on 18 February 3102 BC, the sun, moon, and all five planets are believed to have been in the sign of Mesha—not actually in a simultaneous conjunction, but in an arrangement approaching this. This suggested an idealized set of planets which moved with uniform motion and which were all lined up on this date. The differences between these and the real bodies were defined as 'anomalies', which theoretical astronomers attempted to calculate. These lengthy ages were also incorporated into a mythological history—but that is not for this book.

Another cycle used by astronomers since the sixth century is the Jupiter cycle. The planet Jupiter, or Brihaspati, has an orbital period of a little less than 12 solar years; a twelfth part of the Jovian period—the average time it takes Jupiter to move between two sankrântis—is about 361.05 days and is called a samvatsara. These samvatsaras were given names identical to the names of the lunar months shown in TABLE 13.4. Thus, specifying the sankrânti in which Jupiter was to be found in terms of the name of the samvatsara gave a rough indication of the solar year in this cycle. This system is known from an inscription dating about AD 575, but it is now little used. Later the system was developed into a cycle of 60 years, each named. Several variations of this were and are used in different parts of India. They are mainly for astrological purposes and are similar in their degree of complexity.

TABLE 13.5 *Some eras used in India*

Era name	1st year	Event	Calendar
Kali-yuga	3102 BC	Conjunction	
Buddhist	544 BC	Nominal death of Buddha	
Jain	528 BC	Death of Mahavira (?)	
Vikrama*	58 BC	Kanishka dynasty	Lunar, solar
Christian	1 AD	Birth of Christ	Gregorian
Saka†	78 AD	Parthian dynasty of Kshasharata	Lunar
Kalacuri	248 AD	Traikutaka dynasty	
Gupta	329 AD	Gupta dynasty	
Harsa	606 AD	Harsa dynasty	
Hegira	622 AD	Flight of Muhammad	Lunar
Kollam	825 AD	Foundation of Quilan	

*Traditionally attributed to King Vikramaditi (a title rather than a name) who is said to have driven out the Sakas; but Chandra Gupta II who actually did this lived 400 years later.
† Attributed to a Saka king who lived later.

Besides these cycles, several more conventional eras have been used since the first century BC. Before that time, the years tended to be specified as regnal years whose count started afresh with each new monarch. Exceptions to this are the Buddhist and Jain eras. Some of the many eras which are or have been in use, maybe for a limited time only, are listed in TABLE 13.5. It is important to remember that the year in which a particular date occurs depends on the date of the start of the year, and that this may vary from place to place.

⌘ The Saka calendar

After gaining independence from British rule in 1947 the Indian Government commissioned, in 1952, a report on the calendars of India. Thirty calendars were in use, besides the Islamic calendar used by the Moslem community and the Gregorian calendar bequeathed by the British. In due course they instituted a new, quite formal calendar, based on the old saka era, and not tied to astronomical events, save the tropical year. This was the reformed Saka Calendar; it is similar, but not identical, to a calendar set up in the 1920s in Iran by the Shah Pahlavi.

In the reformed Saka calendar, years are counted, as in the old Saka calendar, from AD 78, which is equivalent to the first year of the new Saka era (ES). Each year contains just 12 months. Except for the first, Chaitra, the first six months have 31 days and the last six, 30 days. In Chaitra, there are 30 days in a common year and 31 in a leap year. This extra day is intercalated at the end of the Chaitra at the same frequency as in the Gregorian calendar. The months have the old lunar month names given in TABLE 13.4, and the first day of the year, 1 Chaitra, is synchronized so as to be 22 March in a common year and 21 March in a leap year. The year (Y) of the Saka era began in March of the year Y + 78 of the Christian era.

Each Saka date is equivalent to a constant Gregorian date, except in the period 29 February to 20 April (10 Phalguna to 21 Vaisakha) inclusive in leap years, when the Saka dates are one day greater. For instance, while 2 March is normally 11 Phalguna, in leap years it is 12 Phalguna.

CHAPTER 14

The Mayan and Aztec Calendars

🏛 The Maya

Man reached the Americas from Asia about 20 000 years ago. Civilization, or at least agriculture, has flourished in Mesoamerica—Belize, El Salvador, Guatemala, Honduras, Mexico—since 1500 BC, and possibly much earlier. Different peoples, speaking different languages, have successively risen and fallen from dominance. We are interested mainly in the Maya and the Aztecs, who developed an advanced civilization with writing and a calendar.

Mayan civilization was obsessed by warfare between city states and had a horrifying quality which appalled the Spanish. Prisoners caught in battle were tortured, executed, or ritually sacrificed by the excision of their still beating hearts; even the kings were required to ceremonially perforate their penises with agate knives. The Spanish, who arrived in the early sixteenth century, suppressed and decimated them in an equally brutal manner. They burnt the Mayan books, proscribed their religion, and only recorded in a fitful manner the culture they found. Nevertheless, in time the Church recanted, and many aspects of Mayan culture, including calendrical practices, survive into the twentieth century.

It was not until the nineteenth century that Mayan cities were rediscovered, and till late in this century that Mayan writing was deciphered.

The Maya recorded their history and their calendar in innumerable documents made of the flattened inner bark of a fig tree, whitened with chalk. These were written by honoured members of their aristocracy and were held to be of great value. The Spanish invaders believed them to be instruments of the devil and burnt great quantities of them. Indeed, only four Mayan books survive in the libraries of Europe, and one of those—the Dresden codex—suffered severe damage in another fire, one which was inflicted on that city in the Second World War. The wanton looting of Mayan tombs and monuments continues to this day.

Our knowledge of Mayan writing in general, and their calendar in particular, is based these surviving documents, from a large number of inscriptions remaining in the ruins of their cities, and from a partial record of the phonetic value of their syllabic writing made by a Spanish priest in 1664. These, together with knowledge of the Mayan languages as they are spoken today, has enabled scholars of recent years to decipher nearly all their writing and so to learn a great deal about their civilization. Nevertheless, the story is still an incomplete one.

Historians of Mesoamerica conventionally distinguish periods marked by cultural achievements or the domination of some culture. Some of these periods, together with other salient events in the tragic history of the area, are given in TABLE 14.1.

The Maya had a vigesimal number system of base 20 (which maybe suggests that originally they wore no socks or shoes and counted on their toes as well as their fingers) with which they could express arbitrarily large numbers. They, like the Indians and Babylonians, also invented zero as a place saver in their written numbers, and, as many other peoples did, they developed a system of communicating numbers using gestures. Many of the ideograms or glyphs that they used in writing numbers depict these gestures.

TABLE 14.1 *The chronology of Mesoamerica*

Date	Period	Achievement/culture/eve
20 000(?) BC		Arrival of man in the Americas.
3000–1800 BC	Archaic	Hunting and gathering.
1800–1000 BC	Early preclassic	Agriculture, pottery, villages.
1000–300 BC	Middle preclassic	Earliest Mayan villages.
		Olmec civilization; massive architecture; writing, calendar.
300 BC–AD 250	Late preclassic	
AD 250–600	Early classic	Dated Mayan stela.
		Domination by Teotihuacan.
600–800	Late classic	Height of Mayan civilization.
800–925	Terminal classic	Rise of Toltec civilization; collapse of Maya.
925–1200	Early post-classic	Toltec domination in Yucatan.
1200–1530	Late post-classic	Highland city states; Aztec civilization.
1492		Columbus discovers America.
1519		Cortes lands and burns his boats.
1521		Aztec civilization destroyed.
1813		Independence proclaimed by Mexico.
1952		Yuri Knorosov begins decipherment of Mayan script.

⚏ The calendar

The earliest record of a calendar survives from about 500 BC in Monte Alban near Oaxaca. This calendar employs a 260-day cycle which was commonly used by several societies—Zapotec, Olmec, Toltec, Aztec, Mayan, and more—and is still in use among the present-day inhabitants of the region.

The calendar attained its most sophisticated form among the Maya who inherited the 260-day cycle and further developed a most complicated calendar system. (A closely similar system was used later by the Aztecs.) According to it, days could be lucky or unlucky, or have other qualities; they used the calendar partly to anticipate propitious days to embark on wars and other activities. It was also used to record on stone pillars, or stelae, important events in the lives of their kings and to relate these to more mythical events of the past—sometimes the remote past. In this way the kings could associate themselves with their gods and with propitious dates. Although the Maya had considerable astronomical knowledge, not only of the behaviour of the sun and moon and their eclipses, but also of the planets Venus and Jupiter in particular, their calendar was only loosely related to such things. Nevertheless, the behaviour of the planets was deemed significant and used to fix the date of important events. One unfortunate, captured in a local skirmish, was kept alive for 12 years to be sacrificed on an inferior conjunction of Venus.

The Mayan calendar system involved two major methods of specifying a date—the calendar round and the long count. The calendar round was used to specify a date within a period of about 52 years, while the long count served to relate such dates within a longer period named a great cycle. As well as these two items of information, the full formal emblazonment of a date might record other features such as which Lord of the Night ruled the day and details concerning the phase of the moon. The complete statement of an important date was thus a portentous and long-winded affair.

The calendar round involved three interlocking cycles of 13, 20, and 365 days respectively. The lowest common multiple of these is 18 980 days, or about 52 years.

The haab

The 365-day cycle was called a haab and was similar to the Egyptian wandering year. Each haab was divided into 18 periods called uinals; each uinal had 20 days and a name. The 18 uinals were followed by five epagomenal days called uayeb to give a total of 365 days (18 × 20 + 5). Thus, a day within a haab could be specified by giving the name of the uinal and the number of the day

POP   UO   ZIP   ZOTZ   TZEC

XUL   YAXKIN   MOL   CHEN   YAX

ZAC   CEH   MAC   KANKIN   MUAN

PAX   KAYAB   CUMXU   UAYEB

FIG. 14.1 Glyphs representing the 18 uinals and uayeb of the Mayan calendar.

within it. For instance, the 27th day of a haab could be referred to as 6 Uo, where Uo is the name of the second uinal and the days of the uinal are numbered 0 to 19 (or 0 to 4 for the uayeb).

Other central American people such as the Aztecs also employed this wandering year but, although the hieroglyphs depicting the names of the 'uinals' are similar, the names themselves are very variable. Furthermore, the years of the different people were not synchronized. The Mayan and Aztec names are shown in TABLE 14.2.

TABLE 14.2 *The Mayan and Aztec names of the 20-day periods*

Number	Mayan name*	Aztec name
0	Pop	Atlcauala
1	Uo	Tlacaxipeualiztli
2	Zip	Tocoztontli
3	Zotz	Ueitocoztli
4	Tzec	Toxcatl
5	Xul	Etzalqualiztli
6	Yaxkin	Tecuilhuitontli
7	Mol	Ueitecuilhuitl
8	Chen	Miccailhuitontli
9	Yax	Ueimiccailhuitl
10	Zac	Ochpaniztli
11	Ceh	Teotl eco
12	Mac	Tepeilhuitl
13	Kankin	Quecholli
14	Muan	Panquetzalliztli
15	Pax	Atemoztli
16	Kayab	Tititl
17	Cumhu	Izcalli
Epagomenal	Uayeb	Nemontemi

*The names of the periods varied even among the Mayan peoples, and those given here were used by the Maya of Jacaltec and, for convenience, by modern scholars.

No attempt was made to intercalate in order to maintain any sort of synchronization with the seasons; instead, the cycle was allowed to run freely like the Egyptian wandering year—even though there is evidence that they had a good understanding of the actual length of the astronomical year.

The tzolkin

The Maya, the Aztecs, and others as well, also had a 20-day period; following some modern writers we will call this a 'veintena'. Each day in this cycle of 20 days received one of 20 names as shown in TABLE 14.3, and the sequence of names repeated every 20 days. Although the names themselves differed according to the people and their language, the veintena were synchronized over a large part of Central America, and there is considerable agreement con-

TABLE 14.3 *The Mayan and Aztec names of the veintena*

Number	Mayan	Aztec
1	Imix	Cipactli
2	Ik	Eecatl
3	Akbal	Calli
4	Kan	Cuetzpalin
5	Chiccan	Coatl
6	Cimi	Miquiztli
7	Manik	Macatl
8	Lamat	Tochtli
9	Muluc	Atl
10	Oc	Itzcuintli
11	Chuen	Ocomatli
12	Eb	Malinalli
13	Ben	Acatl
14	Ix	Ocelotl
15	Men	Quauhtli
16	Cib	Cozcaquauhtli
17	Caban	Olin
18	Etznab	Tecpatl
19	Cauac	Quiauitl
20	Ahau	Xochitl

cerning the meaning and origin of the names and the form of the glyphs representing them.

Several of the veintena names specify a series of animals and, curiously, this series matches another found in the lunar zodiacs of South-East Asia.

Finally, there was a cycle of 13 days which has been called a 'trecena'. Each day was given a number in the range 1 to 13 to indicate its position in the trecena.

In this way each day was given both a trecena number and a veintena name. There are 260 (20 × 13) possible combinations of these names and numbers, so that the entire sequence of names and numbers repeats every 260 days. This 260-day period is generally called a tzolkin, though this is probably not what the Maya themselves called it. A day in a tzolkin may be specified by giving the trecena number and its veintena name. Given the trecena number and the number corresponding to the veintena name of a day in a tzolkin (1 to 20), it is possible to calculate the day of the tzolkin, which is in the range 1 to 260.

The origin of the 13- or 260-day periods is not known, though there have been several conjectures and the number 13 turns up in other contexts within the calendar. The period of 260 days is close to the term of human pregnancy of nine lunations. In truth, all we do know is that their use is ancient, and was and still is widespread in the area.

⌛ The calendar round

I have described how each day belonged to a trecena, a veintena, and a haab; there are thus 94 900 (13 × 20 × 365) possible combinations. However, the combinations repeat every 18 980 days (the lowest common multiple of 13, 20, and 365) so that only a fifth of the possible combinations can ever occur. This period of 18 980 days corresponds to about 52 tropical years, and the specification of the tzolkin and haab numbers and names defines a day within this period. For instance, a day might be called 4 Ahua, 8 Cumhu which designates a day which is both day 4 of Ahau and day 8 of Cumhu.

⌛ The long count

The Maya sometimes had need to locate a date in much longer time spans. To do this, they had another system called the 'long count', which was essentially a tally of days elapsed since the start of an era—the great cycle—in the remote past. The long count was invented much, much later than the calendar round. The count was specified as a series of units, each containing various numbers of days. These units are defined in TABLE 14.4.

It is important to note that the Maya recorded elapsed rather than current time, so that the first element (day, uinal, haab, and so on) of any repeating period is counted to be zero, rather than one, as in our calendar. A date would be specified by five integers which denoted the numbers of these five units of days that had elapsed. For instance, 9.10.19.5.11 implied that 1374 951 days

TABLE 14.4 *The units used in specifying the long count*

20 kim*	=	1 uinal	=	20 days
18 uinal	=	1 tun	=	360 days
20 tun	=	1 katun	=	7200 days
20 katun	=	1 baktun	=	144 000 days

*The base unit, the kim, was one day. The Mayan counting system allowed for greater multiples of the kim than the baktun, but these were not generally used in specifying dates.

had elapsed, for $1374951 = 9 \times 144000 + 10 \times 7200 + 19 \times 360 + 5 \times 20 + 11$.

The Maya believed that time was cyclic and that at the end of a great cycle of 13 Baktuns, the world would be destroyed—only to be recreated for the next cycle.

Most scholars now agree that the current great cycle started on Wednesday, 8 September 3114 BC (Julian calendar). This date is based on a concordance of information from various sources: eclipses and other astronomical events dated in Mayan inscriptions; information from contemporary Mayan calendar priests who continue to maintain the calendar to this day; concordances between Christian and Mayan dates recorded by the early Spanish invaders and by the Mayans in writings made after the conquest. The calendar round date of this day is 4 Ahua, 8 Cumhu. The current great cycle will end on Sunday, 23 December 2012. On this day supposedly the world will end. A gloomy attitude would suggest that this is not all that unlikely.

The oldest known inscription bearing a long count date (7.16.3.2.13 6 Ben) was found outside the Mayan area. The uinal name is missing, as it usually was in early dates, but it can be interpreted as 7 December 36 BC. Later, a stela of Olmec origin bearing a similar date (7 16 6 16 18 6 Eznab) was found. It would appear that the system had been invented at least by the first century BC and by the Olmecs rather than the Maya.

⧗ Other features

The Maya never employed intercalations or paid undue attention to synchronizing their calendar with astronomical events, so the accuracy of the calendar is of no concern. The Maya knew, as we know ourselves, that the first of Pop would occur one day earlier with respect to the seasons every four haabs or so. They made adjustments from time to time of the date of such of their festivals as were considered seasonal.

I have mentioned that the full formal specification of a date on a stela involved other components besides the calendar round and the long count. One interesting element was based on the belief that each day was ruled by one of nine Lords of the Night—a practice reminiscent of the Roman astrological week with its divine planetary regents. Further elements involved the position of the day in an obscure 819-day cycle, and, more conventionally, statements concerning the phase of the moon.

Several clues point to the possibility of some sort of cross-cultural contact across the Pacific Ocean. I have already alluded to similarities between the names in the tzolkin and in lunar zodiacs from South-East Asia; likewise, some of the longer calendrical cycles used by the Chinese and occasionally by

FIG. 14.2 A Mayan stela from Guatemala. It bears the long count date 9.16.15.0.0 (or 18 February AD 766). © The British Museum.

the Maya contained the same number of years. Nevertheless, it would appear that the Mayan and other Central American calendars evolved in isolation from Old World calendars, and no artefacts from that part of the world have even been found in America; many archaeologists remain unconvinced of any contact.

CHAPTER 15

Four European Calendars

⌛ The Greek city states

Our knowledge of the calendars in use in the Greek cities is surprisingly meagre. It comes in part from classical authors such as Geminus and Censorinus, and in part from dates inscribed on tombs and the like. Certainly, a lunar or lunisolar calendar underlay Greek calendars, but the latter varied in detail from one city state to another—different names were used, the year began at different times, intercalations were managed differently, and the years were counted in various ways. It is often very unclear how one month in one state corresponded to a month in another.

The works of Homer, dating perhaps from the eighth century BC, mention the seasons of the year. A little later, Hesiod, in his poem 'Works and days', provided an almanac listing the various natural and astronomical events which occur at different times of the year. He tells us how to manage a farm and what to do in the different seasons as marked by the heliacal risings of various stars and suchlike events. He mentions the days that are lucky and those that are unlucky for various activities. He also knew that there were about 30 days between new moons and that 12 moons take us through the whole gamut of seasons.

> 'At the time when the Pleiades, the daughters of Atlas, are [heliacally] rising, begin your harvest, and plough again when they are setting.'

The names of the months differed widely from state to state but were generally based on the names of festivals held or gods honoured within them. The names of the Macedonian months were given in TABLE 10.1 (see page 148); the names of the months used in Athens in the fifth century BC were as follows:

1	Hecatombaion
2	Metageitnion
3	Boedromion
4	Pyanopsian

5	Maimakterion
6	Posideon
Intercalated	Posideon II
7	Gamelion
8	Anthesterion
9	Elaphebolion
10	Mounichion
11	Thargelion
12	Skirophorion

These names are believed to derive from festivals held in the months.

Later, some states referred to their months by number. I will concentrate on the calendar of Athens, but you should bear in mind that practices varied over the centuries, so that this account is schematic.

In Athens, the intercalated month, Posideon II (later called Hadrianion, after the Emperor), was placed after the first Posideon in the middle of the year. In other states the intercalated month was placed at the end.

The months were either 'full' with 30 days or 'hollow' with 29, and full and hollow months generally, but by no means invariably, alternated. The full months were divided into three decades of 10 days each, but the last decade of the hollow months had but nine days. These decades may have been associated with the waxing, middle period, and waning of the moon. The actual numbering of the days of a decade was idiosyncratic and variable: at one time the days of the last decade were counted backwards in Roman fashion. Certain days of each month were dedicated to a god; the seventh day, for example, was Apollo's birthday. In a curious manner, the last day of every month was at one time called the 'old and new day'.

The months were supposed to start with the new moon, but without the occasional insertion of an extra day, they would have got out of step by a day within three years. Aristophanes (*c.* 450–386 BC), in his play 'The clouds', has the gods complaining of starvation; sacrifices were not being offered on the right day and the days were all muddled. Sometimes extra days were inserted for noncalendrical reasons and, perhaps accordingly, the Athenians sometimes distinguished between the calendar just described and a lunar calendar based on sightings of the moon.

The year was presumed to begin with the summer solstice—a notion generally incompatible with the first month starting with a new moon. A simple rule proposed by Plato (428–328 BC) in his 'laws' suggested a way to reconcile these requirements: the year should begin with the first new moon after the summer solstice. There is, however, no solid evidence that this idea was taken up. Nevertheless, since there was not an integral number of lunations in the tropical year, the practice of intercalating a thirteenth month arose. Thus, the octaeteris cycle of three intercalations in eight years became established, and indeed certain festivals such as the Olympic Games and the Pythian were held at four-yearly intervals—half an octaeteris. The first Olympiads date from 776 BC.

In this way the eight-year cycle became embedded in Greek culture. A much superior rule was the 19-year cycle proposed by Meton (*c*. 432 BC) and Calippus. At most, this was used privately by people who understood its advantages and, in general, the Athenians intercalated a month when the need arose—though this resulted in an average rate of intercalation of seven months in 19 years.

The Athenian years were referred to in various ways. From early times they were associated with the archon or magistrate who held office in that year. The Athenian archons in office from 528 BC until well into imperial Roman times are mostly on record and the list can be used to decipher dates with the years specified in this manner. After the activities of Alexander the Great, the Greeks started to count the years according to the Olympic Games; for instance, the second year of the 100th Olympiad (called 100.2 OE) was 379 BC. In general, the year, P.n OE (where *n* is to be replaced by some number 1,2,3, or 4) was the year 781–4P-n BC. The practice of referring to years in this way lasted well into Christian times.

The Athenians had another way of specifying dates. The people of Athens belonged to one or another of 10 tribes (later there were 12, and then 13). Each tribe contributed 50 of its members to the ruling council of 500. The councillors from each tribe acted in turn as a standing committee; these committees were called prytanies and lasted for 35 or 36 days. Documents would be dated by the day of the prytany and the name of the tribe serving on it.

Ⅻ The celtic calendar

Some 3000 years ago, groups of people speaking variants of the ancient Indo-European language were establishing themselves in Europe and Asia. Among these were the Celts, renowned for their mastery of metal work and their artistic inspirations. In France they were known as Gauls, and in Asia Minor as Galatians. Julius Caesar and other Roman historians tell us that they were head-hunters who appalled even the hard-bitten Roman soldiery. Horrifying human sacrifices associated with the 'wicker man' are described by Caesar and Strabo. They were ruled by warrior kings and philosopher-priests, the Druids; these acted as judges in disputes and were the fount of all knowledge. They were said by the Romans to have had a considerable understanding of astronomy.

By the fifth century BC the Celts were well established in Britain and France (or Gaul). Later, during the first century BC, their various tribes were subdued or exterminated by the Romans; their religious practices were proscribed by Augustus and later emperors so that no Roman citizen was allowed to follow them. However, they did not lose their identity, particularly in the western-most parts of France and Britain—Brittany, Ireland, Scotland, Wales, and Cornwall. There they survived the threats from further waves of nomadic people—the Angles and Saxons and the Vikings—and were eventually converted to Christianity. A large number of them migrated to Brittany from Cornwall in the sixth century.

There are two branches of the Celtic language: the brythonic branch as spoken in Brittany, Cornwall, and Wales; and the Goidelic branch represented by Erse, Gaelic, and Manx, as spoken in Ireland, Scotland, and the Isle of Man. Gaelic is still spoken in Scotland and Ireland, but the last native Manx speaker died in 1977. Likewise, Celtic is still spoken in Wales and Brittany, but the last native speaker of Cornish is said (probably inaccurately) to have been Dolly Pentreath of Mousehole, who died in 1777. The Goidelic branch tend to use 'q' where the brythonic branch would use 'p'.

The Druids preferred to transmit their wisdom and knowledge orally rather than by writing. In that way they could maintain their power and prestige by keeping things secret from the ordinary folk. However, sometimes in Britain, they wrote in the ogam alphabet, which was probably invented in Ireland in the fifth century. Elsewhere, they occasionally used the Greek or Roman alphabets.

According to the Roman author, Pliny, the Celts had a calendar which involved a 30-year cycle. Further details would have been lost save for two discoveries. In 1807 two fragments of a bronze calendar were unearthed near lake Antre in the Jura, but, more importantly, 153 fragments of a shattered bronze tablet were discovered in 1897, at Coligny near Lyons in France.

The Coligny fragments presented three interrelated problems. The first was to reconstruct the tablet and arrange the pieces in their correct positions. The solution to this problem was helped by 20 edge pieces but, nevertheless, it was a jigsaw puzzle with pieces representing some 55 per cent of the area missing. The second problem was to interpolate the missing inscriptions, while the third was to interpret how the calendar worked. The result of several attempts

FIG. 15.1 The Coligny calendar. A reconstruction of the 153 surviving fragments of the bronze tablet. It is possible to discern columns of small holes into which pegs may have been inserted to mark the passage of the days. Source: Cliché Ch. Thioc, Musée de la civilisation gallo-romaine, Lyon.

to solve these problems (which agree in principle if not in detail) was a tablet measuring some five feet by three and a half feet, inscribed in the Gaulish language, but using Roman letters and numerals. It has been dated to sometime towards the end of the second century.

It turns out that the tablet represented a Celtic lunisolar calendar. It is perforated by small holes into which pegs may have been inserted to mark the passage of the days. Each day, for a total of five years, is listed and arranged in a scheme with 12 months to the year and an intercalary month inserted every two and a half years. There are thus 62 months in the five-year period. Generally, days of 29 and 30 days alternate. The names and lengths of the months in each of the five years, according to one interpretation, are shown in TABLE 15.1.

It may be seen that the entire five-year period contained 1832 days in 62 months. This gives a mean lunation of 29.548 days and a mean year of 366.4 days. Both these averages are longer than the astronomical periods they are designed to represent, so that in five years the months will be about one day behind the moon and the solstices about six days behind the sun. It is likely that some adjustments were made to correct these discrepancies every 25, or possibly 30 years.

Each month is designated as being 'matus' meaning 'good' or 'auspicious', or 'anmatus' meaning the opposite, the 30-day months are matus and the others,

TABLE 15.1 *The months of the Celtic calendar*

Number	Name*	Length in days				
		Year				
		1	2	3	4	5
	Quimonios†	30	—	—	—	—
1	Samonios	30	30	30	30	30
2	Dumannios	29	29	29	29	29
3	Rivros	30	30	30	30	30
4	Anagantios	29	29	29	29	29
5	Ogronnios	30	30	30	30	30
6	Qutios	30	30	30	30	30
	(R)antaranos†	—	—	30	—	—
7	Giamonios	29	29	29	29	29
8	Simivisonns	30	30	30	30	30
9	Equos	30	28	30	29	30
10	Elembivios	29	29	29	29	29
11	Aedrinios	30	30	30	30	30
12	Cantlos	29	29	29	29	29
		385	353	385	354	355

*The names on the calendar itself are abbreviations; the names here reconstructions of the full names.

† The intercalary months were called Quimonios and (R)antaranos (this name cannot be read in full).

anmatus. The months are divided into two 'fortnights'. The first always has 15 days and the second has 14 or 15 and is headed by 'atenovx' which means 'returning night' (presumably the fortnight in which the moon wanes).

Besides these general terms each day is given a number from I to XV and several other abbreviated designations. The latter are believed to relate to schemes for keeping track of the phases of the moon and the solstices.

The Celtic year was divided primarily into two halves, marked by Samonios and Giamonios, and also into four seasons. The first day of each season was signalled by a festival. Their names and nominal dates of celebration are shown in TABLE 15.2; they are not associated with equinoxes or solstices but fall roughly in between them. Some of these pagan festivities were taken over by the Christian Church. Several of these days still play a calendrical role in

Scotland as quarter days (see TABLE 15.3), when rents and other debts become due.

The celebration of the feast of Beltane—a fire festival—persisted into the last century in parts of Scotland. Fires were lit and practices, clearly symbolic of human sacrifice, were performed. In England, a fire festival is still widely celebrated on 5 November when bonfires (bonefires) are lit and a human effigy is burnt. The nominal excuse for this is the annual re-enactment of the punishment of Guy Fawkes. The English are exhorted to:

TABLE 15.2 *The principal festivals of the Celtic calendar*

Celtic festival*	Roman date	Christian
Beltane	1 May	(May Day)
Lughnasa	1 August	Lammas
Samhain	1 November	Martinmas
Imbolg	1 February	Candlemas

The dates of the Christian festivals do not exactly coincide with the nominal Celtic festivals. No doubt the Celtic festivals were honoured on days determined by a Celtic calendar.

*The origin of these names is uncertain except for Lughnasa which was devoted to the god Lug. The Roman name for Lyons was Lugudunum and associated with Lug.

TABLE 15.3 *English Quarter days and old Scottish Term days*

England and Wales		Scotland	
Lady Day	25 March	Candlemas	2 February
Midsummer	24 June	Whitsuntide	5 May
Michaelmas	29 September	Lammas	1 August
Christmas	25 December	Martinmas	11 November

The English quarter days are almost equally spaced throughout the year. This is not so for the Scottish term days.

Quarter days in the USA have lost contact with these Celtic festivals and were rationalized to the first days of January, April, July, and October.

Remember, remember
The fifth of November
Gunpowder treason and plot
We know no reason why gunpowder treason
Should ever be forgot

ANON., SEVENTEENTH CENTURY

Ⅺ Teutonic calendars

Another group of people, speaking Germanic languages, settled in northern Germany and Scandinavia. These are sometimes known as Teutons and include the Angles, Saxons, and Vikings. The word 'Teuton' itself derives from the name of a tribe of such people who were annihilated by the Romans in 104 BC. The Teutons are responsible for the development and use of the runic script—though the origins of this are uncertain. The order of the letters is different, so it is known as the futhark rather than the alphabet. There are few accounts of human sacrifices among the Teutons, although they were plagued by elves, norns, valkyries, and the like.

We are indebted to the venerable Bede writing in AD 725. He describes a lunar calendar and gives the names of the months used by the Angles, which were as follows:

1	Giuli	7	Lida
2	Solmonath	8	Woedmonath
3	Rhedmonath	9	Halegmonath
4	Eostromonath	10	Winterfylleth
5	Thrimilci	11	Blotmonath
6	Lida	12	Giuli

Later authorities give similar names. Of interest is the name of the first and last months, Giuli, from which comes the term 'yule-tide', during which Christmas is commemorated. The names themselves are derived from agricultural events or pagan gods.

The year, we are told, began in December at the winter solstice and, from time to time (it is not known how often) an additional month, Thrilidi, was intercalated in the summer; there were then three months bearing the name Lidi.

In other parts of Germany and Scandinavia, names like these and based on agricultural practices were used before the coming of Christianity, and even after. Some of these could have been the inspiration for the names concocted by the French Republicans for use in Germany and the Netherlands (shown in TABLE 20.1, see page 262).

⌛ The Icelandic calendar

In the most northerly parts of Scandinavia, Iceland in particular, a lunar calendar presents problems because the moon is frequently not visible for days on end. Above the Artic Circle, the sun may not rise during winter or set during summer when it becomes a circumpolar star. In these regions, calendars that operated according to different principles were developed and were still in use in Iceland in 1940.

Iceland was first colonized by the Vikings in about AD 870—though there had long been a settlement of Irish monks. Their history is recounted in the writings of the historian Ari Frodi Thorgilsson (1067–1148) who tells of the establishment of the 'althing', or parliament, in about 930. Christianity came to Iceland around 1000.

The Vikings brought to Iceland the seven-day week and an empirical lunar calendar. The year was divided into two seasons or misseri: summer and winter. In Iceland, resting on the Arctic Circle, this calendar became impossible to use in summer. Accordingly, the Icelanders abandoned the moon during the summer and took to counting the summer weeks instead. In time, the winter months were standardized at 30 days each.

The establishment of the althing necessitated fixing precise dates in advance, and the calendar was reformed accordingly. The reformed calendar contained 52 weeks or 12 months of 30 days each, plus four epagomenal days called aukenoetr, giving a year of 364 days. The summer misseri had 26 weeks and 2 days and always began on a Thursday; the winter misseri covered the rest of the year. The older practice of reckoning time by the months in winter and the weeks in summer was eventually amended so that weeks were counted throughout the year.

Although this 52-week year had an elegant simplicity, it did not keep in time with the meteorological seasons—being about one and a quarter days too short. To counteract this, the calendar was reformed again in 955 by Thorstein Surt. He introduced a supplementary week, a 'sumarauki', to be intercalated into the middle of summer in every seventh year. This gave a mean year of 365 days.

Although the Julian calendar was introduced with Christianity in 1000, the old calendar remained in use for civil purposes. In time, the inadequacy of Surt's reform became apparent, and further reforms were introduced in the next century as a result of the work of a local farmer, Oddi Helgason (popularly known as Star Oddi), who was also a competent astronomer and who managed to measure the length of the year with fair accuracy. The principle behind this reform was to adjust the rate of intercalation of sumarauki so that on average the 52-week year kept in step with the Julian calendar. The records of how this was done at the time are unfortunately unclear but the intercalation of the sumarauki was adjusted so that summer always began on a Thursday between 9 and 15 April.

CHAPTER 16

The Roman and Julian Calendars

⌛ Rome

Many people have some knowledge of Rome and Roman civilization; the following are just a few landmarks in the history of Rome:

753 BC	Traditional date of foundation of Rome.
715 BC	Traditional date of accession of Numa Pompilius, the second king.
509 BC	Rome becomes a republic led by two consuls.
458 BC	Cincinnatus becomes first dictator.
451 BC	Decemviri (10 men) set up to legislate reforms.
64 BC	Pompey takes over the Seleucid Empire.
63 BC	Julius Caesar elected Pontifex Maximus.
49 BC	Caesar crosses the Rubicon—civil war ensues.
46 BC	Caesar is made dictator for 10 years.
44 BC	Caesar is assassinated.
27 BC	Octavian become first emperor, Augustus.
4 BC	Birth of Jesus Christ.
14 AD	Augustus is deified.
285	Diocletian moves his capital to Byzantium.
325	Constantine summons Council of Nicaea.
330	Constantine makes Constantinople his Christian capital.
410	Alaric sacks Rome.
482	Split between the Churches of Rome and Constantinople.
553	Financial ruin in Rome.

⌛ The Roman republic and before

Our knowledge of the history of the Roman calendar is largely based on the works of Censorinus, who wrote in AD 238 on the division of time in his *De die natali*; of Macrobius, a grammarian and antiquarian who wrote his *Saturnalia convivia* in the fourth century; and of Plutarch (AD 46–120) and his *Moralia*, written in Greek.

The origins of the calendar used by the Romans are lost to us but it seems likely that the earliest was lunar. Macrobios and Censorinus describe a calendar supposed to have been invented by Romulus, the legendary founder and first king of Rome. Ovid and Plutarch also mention it. The year of this calendar began with March at the spring equinox and contained 304 days, arranged into 10 months; four (Mars, Maius, Quintilis, and October) had 31 days, while the six others had 30. The year ended with December which, as the name implies, was the tenth, and so the last month. The remaining period of two months or lunations is uncertain (as are other details) and mysterious.

Macrobios and Censorinus go on to describe a reform made by Numa Pompilius, the second king of Rome (who lived, so it is said, from 715 to 672 BC). Numa set up the pontifices—a college of officials headed by the Pontifex Maximus. These bore the responsibility for overseeing the bridges across the Tiber, for advising the king on religious matters, and for managing the calendar. (Bridges were, it seems, of great religious import to the Romans and the word 'pontifex' means bridge maker.) In later times, the head of the Christian Church became the Pontifex Maximus, which is why the Pope is still called the pontiff.

The legend goes on to tell us that Numa added two more months, January and February, with January following February, and reduced the number of days in the 30-day months to 29. He originally allocated 28 days to both January and February, to give a year of 354 days—plainly a lunar year. However, months with an even number of days were deemed unlucky, so he gave January one more to make a total of 355 days; such a calendar could not keep in step with the moon. He is also said to have instituted a cycle of intercalation. In this, an extra month, 'mensis intercalaris' (popularly known, so Plutarch tells us, as Mercedonius), was intercalated under the direction of the pontifices to keep the calendar in step with the seasons. The word Mercedonius derives from 'merces' or 'wages' because people were paid for their labour in this season.

There were several further reforms during the next five centuries. These included those of the King Tarquin Priscus (616–579 BC) and the Decemviri in about 450 BC. The latter was a group of 10 magistrates whose reforms, described by Ovid, included reversing the order of January and February to make January the first month of the year—Janus was the god of gateways. They may also have instituted a regular method of intercalating a Mercedonius of alternately 22 and 23 days every two years (effectively, in even years BC), after the Feast of Terminalia which was held on 23 February. The remaining five days of February followed Mercedonius. The names and lengths of the

12 standard months are shown in TABLE 16.1. They remained unchanged during the republican period, till the time of Julius Caesar.

This calendar gave a year with an average of $366\frac{1}{4}$ days and thus required occasional adjustments which were left to the discretion of the pontifices, who guarded closely the secret of exactly how they intercalated. The matter became a bone of political contention. Ovid reports that in some years, by as late as 14 February they had still not decided whether there would be an intercalation. This state of affairs was to prove to be the undoing of this calendar. Whether by stupidity, ignorance, or cupidity (they may have taken bribes to advance or postpone the start of the year), the calendar degenerated, and by 46 BC was three months in advance of the seasons—as we know from the occurrence of eclipses mentioned and dated in Roman historical works.

⚅ Numbering the years

After the fall of the Roman monarchy in 509 BC, the Romans dated events by reference to the consuls in office at the time. In 158 BC they began to take office on 1 January, and the year certainly started on that date thereafter. Later, other systems were used. One method, introduced by Varro (116–27 BC), was to date events as taking place so many years after the foundation of Rome, or 'ab urbe condita' (AUC). The practice continued well into Christian times. The difficulty with this was that there was no agreement on which year Rome was founded. In later times it was assumed, rightly or wrongly, that it was founded in 753 BC—as originally suggested by Varro. In AD 247 the Romans minted coins to commemorate the 1000th anniversary of their city.

Another system, used in Egypt, was to count the years from the accession of the Emperor Diocletian. He became emperor in AD 284 and is remembered for his persecution of the early Christians, as part of an attempt to bolster the power of the Roman state by reinstating the old Roman religion. He also made a number of important reforms.

⚅ Markets and fasts

From ancient times, the Romans held markets every eight days (every ninth day by inclusive Roman counting as described in Chapter 3). These market days were called nundinae. The period between markets was an 'inter nundium'—but it was never part of the official calendar since the pontifices were not concerned with trade. In Roman almanacs the eight days of this cycle were marked with the letters A to H. On the nundinae, country folk went to town, auctions were held, and legal disputes settled.

The days in the Roman calendar were generally also marked with letters to indicate what sort of activity was allowed on them. On days marked 'F' (dies

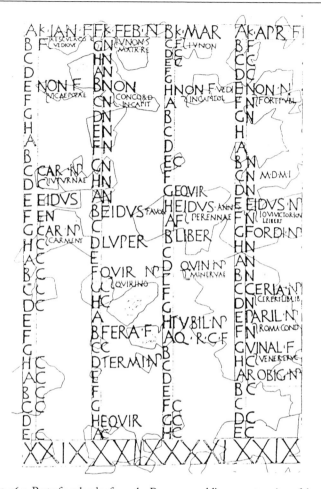

FIG. 16.1 Part of a calendar from the Roman republic; reconstruction of the shattered stone tablet.

fas) legal business, which was prohibited on 'NF' days (dies nefas), could be conducted. Assemblies could be held on 'C' days (comitialis), and public feasts, during which sacrifices could be offered, were marked 'NP' (nefas feriae publicae). So was Roman life regulated.

Sometime near the start of the Christian era, the seven-day planetary week became popular, but it did not achieve official status till the Sunday was designated a holy day in the time of the Emperor Constantine. The seven days of the planetary week were later marked in almanacs with the letters A to G. They had become the calendar letters later adopted by Dionysius Exiguus in his canon for calculating the date of Easter.

⊠ Nones, ides, and kalends

Rather than using weeks of a fixed length the Romans divided their months in an irregular way. This internal arrangement of days within the month was originally based on the phases of the moon, but it was retained long after any synchronization between the calendar months and the moon had been lost.

The three chief points of the month were the kalends, the nones, and the ides. The kalends represented the first day of a month and originally marked the appearance of the new moon; on such days a pontifex would call out and announce the start of a new month and that interest on debts should be repaid. The Latin for 'to call' is 'calere' and the first of the month became known as the kalends (abbreviated to kal.; the use of 'K' was a survival of an archaic letter), the word from which 'calendar' itself is derived. The pontifex would also announce the days of the month on which festivals were to be held and also when the nones would occur.

The ides (believed to derive from 'iduare', 'to divide') represented the full moon and occurred in the middle of the month—on the 13th or 15th day, according to the month. The ides were sacred to Jupiter who, on those particular days, provided light throughout the night by means of the moon. The nones occurred nine days before the ides—that is, by counting inclusively, in Roman style.

As well as inclusive counting, backward counting was used by the Romans when naming the days within each division (see TABLE 16.2). Thus, the day before the kalends of, for example, April (that is, the last day of March) was termed pridie (the day before) kalends—but of April, not March. The day before that (30 March) was ante diem III Aprilis; the day before that (29 March), ante diem IV Aprilis; and so on back to 15 March, which was the ides of March. The day before the ides (14 March) was pridie idus; the day before that (13 March), ante diem III Martius; and so on back to the nones on 7 March. The sixth of March was pridie Martius; 2 March, ante diem VI Martius; and 1 March, the kalends. The precise way this worked out for each of the 12 months is shown in TABLE 16.2 for the Julian months. The naming of the days of the month according to the Roman system lasted well into Christian times and is not extinct, even today.

Hints of this backward counting are also seen in some roman numerals such as 'IV' which equates to one less than five. We also use the system when we count down the seconds (days, hours, and so on) to some big event and when we specify the time as so many minutes to the hour.

In some months, the nones were on the fifth day (with the ides on the 13th), while in others they were on the seventh (with the ides on the 15th). An old rhyme was sometimes used as a mnemonic:

> *March, July, October, May*
> *Have Nones the 7th, ides the 15th day*

The distribution of these dates among the months is not as quite as arbitrary as it appears. In the old Roman calendar, in use before Julius' reform, the months with their nones on the seventh had 31 days; the others had 29 days (or 28 in the case of February). Thus the ides occurred 17 days before the kalends, except in February.

▥ The names of the months

At this point we may pause to consider the names of the months in the Roman calendar; these are shown in TABLE 16.1. You should note that the letter 'J' was a late medieval invention not used by the Romans themselves; they used 'I'. We comply with this in Latin names only, using 'J' for the modern English equivalent; thus where we write January or Julius, the Romans would write Ianuarius and Iulius; the phonetic value of this initial 'I' was probably similar to 'J' in the German 'Ja'.

The names of the last six months (Quintilis through to December) are derived from the Latin names for the numbers five to ten and are very ancient, probably antedating the foundation of Rome. Of the first six, Iunius is named after Iuno, the wife of the god Iuppiter. The origin of April is uncertain—some say it is derived from 'aprire', 'to open', others have suggested that it comes from 'Aphrilis', since the month was sacred to Venus or Aphrodite. May is named after Maia, a god of the spring (the connotation of rebirth still persists in the medieval May Day celebrations and, for that matter, in its modern political association with revolution). March is named after Mars, the god of war. The origin of February is obscure but may derive from an archaic Sabine word 'Februare' meaning 'to purify' and refer to a festival of expiation. Ianuarius is interesting; Ianus was the god of doors, and is depicted as having two faces looking in opposite directions—an appropriate stance for a god standing

TABLE 16.1 *The days in the months of three Roman calendars*

Month	Calendar (date BC)			
	Romulus (c. 753)	Numa (seventh century)	Caesar (45)	Augustus (8)
Ianuarius	—	29	31	31
Februarius	—	28	29/30	28/29
Mercedonius	—	22/23	—	—
Martius	31	31	31	31
Aprilis	30	29	30	30
Maius	31	31	31	31
Iunius	30	29	30	30
Quintilis (Iulius)	31	31	31	31
Sextilis (Augustus)	30	29	30	31
September	30	29	31	30
October	31	31	30	31
November	30	29	31	30
December	30	29	30	31
Total days	304	355/77/78	365/66	365/66

The number of days allocated to the months by Caesar is disputed. They may have been the same as ascribed to Augustus above.

on the threshold of a new year, looking forward to the new year and backward to the old. We shall see in the next section that Quintilis was renamed Iulius and Sextilis was renamed Augustus.

🜨 The Julian calendar

Julius Caesar (107–44 BC) was elected to the college of pontifices some 20 years before his murder. It was a great honour for him, and the position carried with it power and prestige. Later, after he had been elected dictator, he undertook several much-needed reforms of Roman life, and of the calendar in particular.

Caesar may have met the Alexandrian astronomer Sosigenes whilst on his Egyptian campaigns in pursuit of his rival, Pompey, in 48 BC, or maybe his mistress Cleopatra, herself an Egyptian queen, introduced him. Whether or not this is so, he knew Sosigenes by reputation and called him to Rome for consultations. Sosigenes advised Caesar to abandon any pretence to a lunar

TABLE 16.2 *The Roman Julian calendar*

Day of month	Martius Maius Iulius October*	Ianuarius Augustus December†	Aprilis Iunius September November†	Februarius§
1	Kalendae	Kalendae	Kalendae	Kalendae
2	VI	IV	IV	IV
3	V	III	III	III
4	IV	Prid. nonas	Prid. nonas	Prid. nonas
5	III	nonae	Nonae	Nonae
6	Prid. nonas	VIII	VIII	VIII
7	Nonae	VII	VII	VII
8	VIII	VI	VI	VI
9	VII	V	V	V
10	VI	IV	IV	IV
11	V	III	III	III
12	IV	Prid. Idus	Prid. idus	Prid. idus
13	III	Idus	Idus	Idus
14	Prid. idus	XIX	XVIII	XVI
15	Idus	XVIII	XVII	XV
16	XVII	XVII	XVI	XIV
17	XVI	XVI	XV	XIII
18	XV	XV	XIV	XII
19	XIV	XIV	XIII	XI
20	XIII	XIII	XII	X
21	XII	XII	XI	IX
22	XI	XI	X	VIII
23	X	X	IX	VII
24	IX	IX	VIII	VI
25	VIII	VIII	VII	V
26	VII	VII	VI	IV
27	VI	VI	V	III
28	V	V	IV	Prid. kal.
29	IV	IV	III	—
30	III	III	Prid. kal.	—
31	Prid. kal.	Prid. kal.	—	—

In leap years an extra day was inserted before VI Kal. Martias.

In the old Roman calendar, before Caesar's reform:

* these months had 31 days.

† these months had 29 days.

§ February had 28 days.

FIG. 16.2 Julius Caesar (107–44 BC). Courtesy of the Vatican Museum.

calendar and to divide the tropical year into 12 months, with no attempt to relate them to the moon.

Sosigenes suggested that instead of an intercalary month, a single intercalary day be introduced every four years, before ante diem VI kalendis Martius (that is, after 23 February, at the same point that the old mercedonius was inserted). This day was known as bis ante diem VI kal. Mar., and the year in which it occurred was referred to as a bissextile year; in these years February would have one day more, so that the average number of days in a year was $365\frac{1}{4}$.

The names of the months were to be preserved, but the number of days in most was altered. There is some contention about the lengths of the months in Caesar's reformed calendar. The traditional point of view is that their lengths alternated between 31 and 30, with January having 31 and February having 29 or 30 (as seen on page 212). However, Macrobius contradicts this supposition and, more recently, it has been proposed that the months had the same number of days as they do today.

Caesar acted on the advice given to him by Sosigenes and, in 46 BC, with the help of Marcus Aemilius Lepidus (d. 13 BC), who succeeded Caesar as Pontifex Maximus, he reformed the Roman calendar—a reform that was to last 1600 years. It involved two corrections. First, the number of days in 46 BC was augmented to bring the start of the year back into synchronization with the vernal equinox; secondly, intercalary days were to be inserted on a regular basis, as already described. Thus 46 BC contained 445 days. The extra days included a mercedonius of 23 days and two extra months totalling 67 days which were inserted between November and December. This mammoth year was known as 'the last year of confusion'. Confusing, it no doubt was, but it also signalled the end of the difficulties.

Bad feeling was engendered in Gaul where one of the governors insisted that two months' extra tax should be paid in the year of confusion—for months he called Undecimber and Duodicimber.

It is unfortunate that Julius Caesar was assassinated on the ides of March in 44 BC, and it was little consolation that the Roman senate in the same year renamed Quintilis as Iulius, our July, as a memorial to him. Caesar was unable to oversee the proper working of his new calendar and the pontifices, as usual, misunderstood their task and interpreted their orders to insert an intercalary day every fourth year as one every three years using Roman inclusive counting. Once more the calendar began to slip. This went unnoticed at first but, in 9 BC, so it is said, the Emperor Augustus decided to correct the discrepancy by omitting the next three intercalary days to make up for the superfluous ones wrongly introduced by the pontifices—there were to be no more intercalations until AD 8. After that the Julian calendar ran smoothly. Augustus' name was, in turn, given to the month of Sextilis, which became our August. It is said that Augustus chose the month of Sextilis for renaming in commemoration of Cleopatra who had committed suicide in that month. The story goes on to suggest that August was given an extra day, taken from February, to avoid it having both an unlucky, even number of days and fewer days than Julius' month; at the same time other adjustments were made to the length of the months to avoid having three consecutive 31-day months.

Eighty generations of children have needed to learn a version of:

> *Thirty days hath September*
> *April, June and November*
> *All the rest have thirty one*
> *Excepting February alone*
> *Which hath twenty-eight days clear*
> *And twenty-nine in each leap year*
> ANON., 1555

This has been said to be the only useful poem in the English language! If you are a Quaker and need to avoid the pagan names for the months, you might memorize:

> *Fourth, eleventh, ninth and sixth,*
> *Thirty days to each affix;*
> *Every other thirty-one,*
> *Except the second month alone*
>
> CHESTER COUNTY FRIENDS, PENNSYLVANIA

There are variants on these, and other versions in Latin and most other European languages.

The honour bestowed on Julius Caesar and assumed by Augustus in having a month named after them set a precedent; the Roman senate offered the honour to several other emperors, including Tiberius (AD 14–37), the successor of Augustus. He declined it however, questioning what would happen when there were thirteen emperors in a remarkable triumph of acumen over self conceit. The next emperor was Caligula (37–41), who renamed June as Germanicus, after his father; Claudius (41–54) annexed May; and Nero (54–69) took April. Later, Domitian (81–96) renamed October after himself. September was taken in turn by Antonius (138–61) and Tacitus (275–6). Earlier, Commodus (161–9) had tried to rename all 12 months: Amazonius, Invictus, Felix, Pius, Lucius, Aelius, Aurelius, Commodus (he had to come in somewhere), Augustus, Herculeus, Romanus, and Exsuperatorius. Citizens in outlying provinces, such as Cyprus, wishing to curry favour with the emperor, went in for wholesale renamings of the months. Luckily, perhaps, none of these later renamings stuck, else we would be learning 'Thirty days hath Tacitus . . .'.

There is one curious feature of Sosigenes' recommendations. He must have been well acquainted with Hipparchus' measurement of the length of the tropical year and that $365\frac{1}{4}$ days was significantly different from its true length. There is little doubt he was capable of working out a better approximation, as did Aloysius Lilius and Clavius 16 centuries later. Maybe Caesar realized that a more complicated system was beyond the powers of the pontifices to manage; maybe he was not bothered about a discrepancy that would not show itself for several hundred years. In any event, he settled for the simple rule of one intercalation every four years.

The first of January may fall on any of the seven days of the week, and a year may or may not be a leap year. There are thus $7 \times 2 = 14$ possible Julian years, so only 14 separate calendars are needed—but you may have to wait longer before last year's becomes valid again. If you take into account the changing date of Easter, there are many more required.

FIG. 16.3 'Thirty days hath Tacitus . . .' what our children might have had to learn had the renaming of the Roman months by later emperors stayed with us.

The Julian calendar was inherited from the Roman world by Christendom and lasted well into the twentieth century; it is still used by the Greek Orthodox Church. The only adjustments that were made concerned the date of the start of the year and schemes for numbering the years.

⌛ The Christian era

The Christian era was invented by Dionysius Exiguus ('Dennis the little'—so called because of his self-demeaning manner), an Abbot from Scythia (now Moldavia). Dennis, in his writings about Easter, invented and used a new era—the Christian era—to begin on 1 January of the year following that of the Birth of Christ. This year he designated as the year of our Lord or Anno Domini (AD) 1. He used his new era to number the years because he did not wish to associate Christianity with the era of Diocletian; (used in the Alexandrian method of calculating the date of Easter, as is explained in Chapter 29).

At the start, the era was known as the Dionysian era, but was not taken up immediately by the church authorities and only became widely used in Western Europe from the eleventh century; the venerable Bede, who used it in his writings in the eighth century, was one of those responsible for its popularization. (Bede also used the notion of dates before Christ (BC)—but this did not become popular till much later.) The Christian era was not adopted in the Greek world till the fifteenth century.

Dennis had assumed from a reading of Clement of Alexandria that Christ was born in the twenty-eighth year of the reign of the Emperor Augustus. He assumed that Augustus' reign began in 727 AUC—but there he was mistaken. What he did not realize was that the reign of Augustus was always calculated from his decisive victory over his rivals for power, Anthony and Cleopatra (Cleopatra had by then switched allegiance from Caesar to Anthony), at the battle of Actium fought on 3 September 723 AUC, rather than his acceptance from the Roman people of the title of emperor on 13 January 727 AUC. Dennis may also have been confused by the uncertainty over the date of the foundation of Rome and by a numerical coincidence whereby he calculated that a Victorian cycle (see Chapter 28) began in 753 AUC. Surely, he must have felt, this must be the year of the nativity. We must conclude from this however that Christ was in fact born in 4 BC, and not 1 BC.

Several passages in the Gospels bear upon the matter. Thus, Matthew (2:1) and Luke (1:5) agree that Jesus was born during the reign of Herod the Great. The Jewish historian Flavius Josephus (37–95) establishes, by several lines of argument, that Herod died in 750 AUC. Josephus also tells us that he died shortly after a lunar eclipse, but before the Passover; astronomers calculate that such an eclipse took place on 13 March 750 AUC and the Passover would have occurred at the end of that month. This evidence draws us irresistibly to the conclusion that Jesus was born in or before 750 AUC (4 BC).

Nevertheless, some residue of doubt remains as to the correct date of the Nativity. There is an argument based upon the date of the census mentioned in Luke 2:1 and 2, which suggests that AD 6 was the correct date. From time to time astronomers come up with alternative dates from attempts to identify the Star of Bethlehem, but clues in the Gospels are too vague for much weight to be attached to these dates.

One other point remains to be discussed—the start of the year. From about 158 BC the Romans began their year on 1 January. Later, the Christian Church, wishing to dissociate itself from pagan practices, preferred to begin the year with the day of a Christian festival; a logical day might be the Feast of the Annunciation on 25 March. So, at various times and places, including England up to 1752, 25 March heralded the New Year. Some Christians preferred Christmas Day; others opted for Easter Day (which was highly inconvenient, since the actual date of Easter moved); yet others preferred 24 or 29 September.

In AD 312 Constantine had instituted a 15-year cycle of indictions (censuses of people's ability to pay taxes). These started on 1 September. A little later, in

330, he moved the capital of his empire from Rome to Byzantium, which came to be known as Constantinople. The Byzantine year started on 1 September, and this system was used by the supreme tribunal of the Holy Roman Empire till it was abolished by Napoleon in 1806.

The Julian calendar was the official calendar of the Christian Church until the end of the sixteenth century—despite its shortcomings, which had been noted since the twelfth century if not before. In 1582 it was replaced in the Catholic Church by the Gregorian calendar; the story of this reform is taken up in Chapter 19.

CHAPTER 17

The Jewish Calendar

⧗ Jewish history

The Bible is the main source of our knowledge of early Jewish history. For events reported before about 1000 BC it is difficult to disentangle fact from myth, dates are very uncertain, and corroboration from other sources rare. Thus, biblical figures such as Adam, Abraham, Moses, and Samuel remain potentially mythical. Firm dates can however be assigned to important events in Jewish history from the tenth century BC onwards and some of these are as follows:

Creation of world	3761 (!) BC
Exodus from Egypt under Moses	1235 (?) BC
King David rules Israel	1003 (?) BC
King Solomon ruled	970–925 BC
Israel split into Israel and Judah	925 BC
Domination by Assyria	722 BC
Battle of Meggido	608 BC
Fall of Jerusalem to Babylonians; exile starts	597 BC
Cyrus captures Babylon and allows Jews to return to Israel	538 BC
Jerusalem rebuilt	445 BC
Seleucid Empire founded	312 BC
Roman persecution	175 BC
Destruction of Jerusalem by Romans	70 AD
Qumran destroyed; Dead Sea scrolls hidden	73
Start of Diaspora	70
Bar Hillel reforms calendar (?)	356
Talmud finished	499
Early report of persecution of European Jews	1196
Start of Nazi holocaust	1941
Foundation of modern state of Israel	1948

Ⅺ Biblical references

References to any sort of calendar in the books of the Old Testament originating from before the Babylonian captivity are rare. The names of only four months are mentioned: two of these, Ethanum (seventh month, possibly the month of fruits) and Bul (eighth month, the month of rain) have been found in Phoenician inscriptions, but the other two, Abib (first month, month of corn or new fruits) and Ziv (second month, month of flowers) have not; on the other hand the Phoenician inscriptions also yield Marpeh, Phauloth, Mirzah, Mapha, Hir, and Zebah-shishim.

There is evidence, dating from the age of Solomon in the tenth century BC, that the calendar was lunar and the meaning of the four biblical months strongly suggests that it was tied to the seasonal year with a year that began in the autumn. It was thus probably lunisolar, with months intercalated from time to time. It is likely that the start of the months was signalled by the first sighting of the new moon.

There is some evidence that whenever it was seen that the barley would not be ripe by the 16th of Abib (or Nisan), an extra month was intercalated. In this way it was ensured that a sheaf of barley could be offered to God on the day after the Passover. Later it was noticed that this would be necessary whenever the first day of the Passover would otherwise happen before the vernal equinox.

Ⅺ The captivity

The Babylonian captivity began in 597 BC when Nebuchadnezzar II (605–562 BC), King of the Chaldean dynasty of Babylon, captured Jerusalem. During the captivity, the Jews adopted the Babylonian calendar. Except during the Seleucid period (when the Seleucid calendar was used), they continued to use this until the Diaspora in AD 70.

The Babylonian calendar was also lunisolar; each year had 12 lunar months and a 13th was intercalated as necessary. The year started in the autumn and the day at sunset; the months were deemed, as before, to begin when the new moon was first observed at Jerusalem. When a month was expected to start, observers were sent to the top of a hill near Jerusalem and required to report back to the Sanhedrim when they first saw the new moon. They were shown diagrams of the moon and when they convinced the priests that they had seen it, the priests announced the start of a month. If they had not seen it, the start was delayed till the next day. Then trumpets were blown and fires were lit which could be seen throughout the city; at the same time messengers were dispatched to outlying Jewish communities.

The Jews used a seven-day week of great antiquity, possibly of Babylonian origin, but the days were not named, just numbered—except for the seventh day, the Sabbath. At first the months too were numbered, maybe to avoid pagan connotations, but in time they acquired names clearly derived from those used by the Babylonians (as shown in TABLE 17.1) and mentioned in the Talmud; only seven of these are referred to in the Old Testament.

Alexander the Great died in 312 BC and Seleucis Necator set up his empire in 311 BC. From this date the Seleucid era was used to number the years. The era is sometimes known as the era of contracts, because the Seleucid authorities required legal documents to be dated by their era. Some Jewish communities may still date by this era, though since the fourteenth century most have used the mundane era and counted the years from the Creation (as will be discussed). Thus a year number would be designated by AM, an abbreviation for Anno Mundi—a year of the world.

ⅩⅠ The Diaspora

After the Diaspora (conventionally dated from AD 70 when the Roman general Titus burnt the temple at Jerusalem), the necessity of observing the new moon in Jerusalem to start a month led to difficulties. It was very incon-

TABLE 17.1   *The names of the Babylonian and post-captivity Jewish months*

	Babylonian		Jewish*
1	Nisanu	7	Nisan
2	Ayaru	8	Iyyar
3	Simanu	9	Sivan
4	Du'uzu	10	Tammuz
5	Abu	11	Av
6	Ululu	12	Elul
7	Tashritu	1	Tishri
8	Arahsamnu	2	Heshvan
9	Kislimu	3	Kislev
10	Tebetu	4	Tevet
11	Shabatu	5	Shevet
12	Adaru	6	Adar

*The transliterated spelling of the Jewish names follows that used in the *Encyclopaedia Judaica*.

Note that the first months of the Babylonian and Jewish year are not the same. The religious year still starts on the first of Nisan in the springtime near the vernal equinox.

venient, for instance, for a community in Spain to have to wait several days while the news travelled down the Mediterranean; it would have been impossible for those who had penetrated as far as India or China.

These inconveniences prompted some communities to celebrate the principal feasts twice—on both dates that might be dictated by the alternative starts of the months. Others experimented with methods of calculating the start of the months and of intercalation, including the 3:8 rule of intercalating three months in eight years; others settled for the more satisfactory Metonic cycle with its 7:19 rule.

In time, a definitive theory-based calendar emerged which is still in use today. Tradition attributes this to Rabbi Hillel II who lived at Tiberias on the Sea of Galilee in the fourth century. Some authorities question this, and date the final development of the Jewish calendar to the fifth century, after the Talmudic period.

Hillel II became president of the Sanhedrim, a hereditary post, and was a direct descendant of Gamaliel, the president at the time of St Paul. A large measure of the prestige of the Sanhedrim arose from its monopoly of the knowledge for regulating the calendar. It is said that Hillel, appalled by the suffering of his people, decided in AD 356 to renounce this monopoly and publish the knowledge. In this way, even the most distant communities could ascertain the correct days for their festivals. Apparently, the other members of the Sanhedrim were agreeable to this.

According to Jewish law the calendar cannot be changed except by the Sanhedrim meeting in full assembly at Jerusalem. This has only recently, since the establishment of the state of Israel, seemed even remotely likely.

Ⅺ Hillel's calendar

The day of the Jewish calendar has a conventional 24 hours, and each hour is subdivided into 1080 units called chalaks. Thus three chalakim are equal to 10 of our traditional seconds; a smaller unit, the rêga, is sometimes encountered but plays no part in the calendar. The figure 1080 ($2^3 \times 3^3 \times 5$) is said to have been chosen on account of its large number of factors (30). The day begins at 18.00 hours—six hours before the start of the conventional day to which it corresponds and which starts at midnight.

The modern Jewish calendar uses the Metonic cycle for reconciling the solar and lunar years, and employs a theoretical moon, as does the Christian calculation of the date of Easter. This theoretical moon should not be confused with the real heavenly moon. Each lunation of it is initiated by a 'molod', which is the moment of conjunction of the theoretical moon. The interval between successive molods is taken to be a constant 29 days 12 hours and 793 chalakim (29.530594 days). Sometimes the interval between successive conjunctions of the real moon is greater than this, sometimes less.

The interval between two molods may be compared with the average synodic period of 29.530589 days—a figure known to the Babylonians in 300 BC. The agreement is exact to better than half a second a month. The molods will remain in step with the average moon to within a day for more than 16 000 years.

Common years contain 12 months apiece, but embolismic years have an extra, 13th, intercalated month. In a sequence of 19 years, there are always just seven intercalated months in years 3, 6, 8, 11, 14, 17, and 19. Thus, 19 years contain 235 months (12 × 12) + (13 × 7). The total time between the first and last molod of these is 6939 days, 16 hours, 595 chalakim—so that the average number of days per year is one 19th of this or 365.24682 days. This may be compared with the length of the mean tropical year of 365.24219 days. It is accurate to about six minutes a year, and the Jewish year will slip ahead of the seasons by one day every 216 years; this will get worse as the year shortens over the millennia. The Jewish religious year starts with Nisam but Rosh Hoshanah, the Jewish civil new year, is taken to be the first day of Tishri which occurs in the autumn.

The theoretical moon, whose conjunctions are calculated to occur at precisely constant intervals, defines the first day of each month and in this way a (theoretical) astronomical calendar is specified. The civil calendar is not exactly the same, and the start of a civil year may be up to two days later than the start of the astronomical year at the molod of Tishri. These delays, called postponements, are defined by a series of rules which I will discuss later. Thus, the average length of a civil year is the same as the average length of the astronomical year; it is the point of division between two years that differs between them. Be aware then of the different levels of abstraction involved: firstly, the real moon in the heavens; secondly, the theoretical moon of constant lunation; and thirdly, the civil calendar with a molod of Tishri synchronized, on average, with the theoretical moon but subject to postponements.

⚅ The Creation of the world

In order to establish the precise day on which a molod occurs, it is necessary to specify an origin or starting point, and this has been called the molod tohu (meaning 'nothingness'—since nothing preceded it) and given the mnemonic name BeHaRD. The precise instant is stated as 4 hours 204 chalakim into Monday, 7 October 3761 BC (Julian calendar).

It is not clear exactly why this date was selected. The year was first derived by José Ben Halafta in the second century AD. It was defined by adding up the ages of various patriarchs, kings, and historical periods listed in the genealogies and histories in the Bible. In this way it is possible to estimate the number of years from Adam and the Creation as described in Genesis to a historical event whose date was known. This was the destruction of the second temple by the Romans in AD 67.

There have been many attempts at this calculation. The historian, des Vignoles, mentioned that he had collected over 200 different calculations which give dates for the creation ranging from 3483 BC to 6984 BC—but the growth of scientific knowledge in the last century has made such estimates less convincing. One of the most famous calculations was Bishop Ussher's estimate of 4004 BC, which he published in 1654.

The results vary, partly because the data itself is not consistent, partly because the three earliest manuscripts of the Bible contain different numbers, and partly due to the historical benchmark chosen to relate the date to, say, the Christian era.

Some of the components which enter the calculation are listed in TABLE 17.2. The data agrees with information in Moffat's translation of the Old Testament, which in turn is based on the Massoretic text—a manuscript dating

TABLE 17.2 *Relevant factors in estimating the year of the creation from biblical data*

	Period	Years	Biblical reference
1	Adam to birth of Noah	1056	Genesis 5: 1–32
2	Birth of Noah to birth of Abraham	892	Genesis 11: 10–29
3	Birth of Abraham to birth of Isaac	100	Genesis 21: 5
4	Birth of Isaac to birth of Jacob	60	Genesis 25: 26
5	Birth of Jacob to visit to Egypt	130	Genesis 25: 26
6	Captivity to Exodus	210	Exodus 12: 40–1
7	Exodus to dedication of first temple	480	I Kings 6: 1
8	Dedication to destruction of first temple	410	Judges, Samuel
9	Exile in Babylon	70	
10	Return to destruction of second temple	420	
		3828	

Period 2 is uncertain within two years; period 6 is given in Exodus as 430 years, but there are other manuscripts which suggest that 210 years of this represent the time spent in Canaan; period 10 is problematic and cannot be supported from the Bible alone.

The temple was destroyed by the Roman general, Titus, according to Jewish tradition, in the year corresponding to AD 67; Adam, according to these data, was created in 3761 BC.

from between the sixth and eighth centuries, the contents of which would have been established by the second century and no doubt have been known to Ben Halafta.

Once the year had been defined, it would have been possible, given the time and date of a contemporary molod, to extrapolate back by subtracting an appropriate multiple of the interval between two molods to find the time and date of the first molod of that first year. The result, 4 hours 204 chalakim into Monday, 7 October 3761 BC, is not the moment of the Creation, neither is it the moment at which either the sun or the moon were created. One might have thought that the most pleasing way of initiating the solar system would be to line up the sun, moon, and earth in a position of conjunction and then to flick them simultaneously into motion—but this was not how it was done. The sun and the moon, according to Genesis (1: 14–21), were created on the fourth day, a Wednesday (although mysteriously there was already light and darkness on the first day (1: 3–5)). Thus the sun and moon were created two days after their first fictitious molod; this seems to require some explanation and controversy surrounds the subject. The actual moment of the Creation is presumably the start of day one; that is, at 6 p.m. on Saturday, 5 October.

Before the acquisition of the appropriate scientific knowledge, it was possible to consider these cosmological calculations with perfect seriousness; nowadays, we must grope for a suitable allegorical interpretation. In particular, the ages of the first 10 patriarchs descended from Adam deserve interpretation rather than dismissal as an ignorant fancy; they are measured in hundreds of years and are stated with confident precision. Only the ages at death of Hanok (or Enoch) (365), father of Methusaleh, and Lenak (777), father of Noah, show signs of having a purely numerological significance. There is a tradition that Enoch (Hanok) was the inventor of the Jewish calendar—which would well account for the length given for his life; his son, Methusaleh lived for a thousand years, a very long time (an allegory perhaps for the calendar itself).

The use of the mundane era for counting the years was first suggested by the Jewish philosopher, Rabbi Moses ben Maimon (1135–1204), sometimes known as Maimonedes, but was not generally applied till the end of the fourteenth century. Maimonides had enormous influence on both Jewish and Christian thought. He was the personal physician to Saladin and turned down the offer of a similar post with King Richard I (the 'lion-heart').

☵ Postponements

We have seen how it is possible to calculate the date of any molod; in particular, we may calculate the date of the molod of Tishri, the first month, for any year. It might be thought that Tishri would begin at 6 p.m. of the day on which this molod fell, but in fact the Jewish calendar has an additional layer of complexity.

There are four further considerations or rules (dehiyyot) which may lead to New Year's day, the first day of Tishri, the feast of Rosh Hoshanah, being postponed for one, or in some cases two whole days. To apply these rules it is necessary first to calculate the day and time of the molod of the year in question and then that of the following year. These give the nominal, astronomical days of 1st Tishri for the two years. The difference between the two represents the nominal number of days in the year in question.

In these calculations a convention is followed whereby the days are added up and then whole multiples of seven are subtracted. This greatly simplifies the numbers involved and gives the day of the week directly.

The rules have a twofold purpose. First, they prevent the occurrence of two consecutive days on which the preparation of food and other necessary tasks such as the burial of the dead is forbidden. (It can be reasonably argued that in a hot climate, food prepared two days in advance might be contaminated by the second such day.) Secondly, they keep the lengths of all years within defined limits.

The first rule is designed to prevent the Day of Atonement (Yom Kippur), observed on 10th Tishri, from falling on a Friday or Sunday, or the Festival of Willows (Hoshanah Rabba), observed on 21st Tishri, from falling on a Saturday. One or other of these contingencies would arise if 1st Tishri were to fall on a Wednesday, Friday, or Sunday. If the astronomical calculations indicate that this would happen, the start of the year, Rosh Hoshanah, is postponed one day to a Thursday, Saturday, or Monday respectively.

The second rule is designed to allow for the fact that a new moon, which corresponds to an earlier conjunction (molod), cannot be observed till dusk. The interval of time between the true molod and the first appearance of the new crescent depends on the latitude and longitude of the point of observation. This is assumed to be six hours for observers at Jerusalem. Thus if the molod falls at noon or later (but before 6 p.m., when the next day starts), Rosh Hoshanah is postponed by a day. It may then be postponed a further day under the first rule.

The third rule is required because of the consequences of applying rules one and two to the following year, while the fourth is a consequence of applying them to the preceding year. If the molod of Tishri of a common year occurs after 3 hours 204 chalakim a.m. on a Tuesday, it can be calculated that the molod of the next year will occur after noon on a Saturday, which would cause it to be deferred one day, under rule two, to a Sunday, and a further day, under rule one. This in turn would yield a year which had 356 days, which is not acceptable. Thus, under these circumstances, Rosh Hoshanah of the current year must be postponed according to rule three by one day to a Wednesday.

Likewise, if the preceding year was embolismic (that is, it had 13 months) and the molod of the current year occurred on a Monday after 9 hours 589 chalakim, the molod of the previous year would have fallen on or after noon on a Tuesday and would have had to be postponed by two days. This would cause the previous year to have only 382 days. To counteract this, 1st Tishri of the current year has to be postponed one day to the Tuesday.

Only one of rules two, three, or four is applied. First, you test to see if rule two applies; if it does, the day is postponed; only if it does not, do you test to see if rule three applies; and only if this does not, do you test for rule four.

A consequence of this procedure is that a common year may contain 353, 354, or 355 days and an embolismic year, 383, 384, or 385 days. There are thus six types of year. Years with 353 or 383 days are called deficient; those with 354 or 384, regular; and those with 355 or 385, abundant.

In a regular common year, the 12 months contain alternately 30 and 29 days. In an embolismic regular year, an extra month of 30 days is inserted between Shevet and Adar; the original sixth month (Adar) now becomes the seventh and is renamed Adar II (or sometimes Veadar); the intercalated month takes the name Adar. The festivals which are celebrated in Adar in common years are celebrated in Adar II in embolismic years. In deficient years, the number of days in the third month, Kislev, is reduced from 30 to 29. In abundant years, the number in the second month, Heshvan, is increased to 30. The number of days in the months in the six types of year are shown in TABLE 17.3.

The Jewish calendar follows a useful convention of describing any year by its character (Keviah), which comprises three (Hebrew) letters: the first, used as a numeral, indicates the day of the week on which Rosh Hoshanah falls; the second indicates the length of the year—whether it is common or embolismic, and whether it is deficient, regular, or abundant (as indicated in TABLE 17.3); the third, used as a numeral, indicates the day of the week on which the Passover falls. It turns out that there are only 14 possible characters (on account of the postponements), so that only 14 calendars are required. Caution is required in interpreting Keviahs written in Hebrew, for that language reads from right to left, so that the extreme right character indicates the day of Rosh Hoshanah.

In ancient times, the start of the year was defined by the observation of the moon at Jerusalem. The modern calendar, which specifies the times of molods, is operated according to local time, so that molods in the Antipodes will be 12 hours later than those in Jerusalem. Since the same is true for all other modern calendars, the conversion of Jewish dates to dates in other calendars is independent of longitude.

TABLE 17.3 *The distribution of the days of the year among the months*

Month	Common year			Embolismic year		
	d	r	a	D	R	A
Tishri	30	30	30	30	30	30
Heshvan	29	29	30	29	29	30
Kislev	29	30	30	29	30	30
Tevet	29	29	29	29	29	29
Shevet	30	30	30	30	30	30
Adar	29	29	29	30	30	30
Adar II*	—	—	—	29	29	29
Nisan	30	30	30	30	30	30
Iyyar	29	29	29	29	29	29
Sivan	30	30	30	30	30	30
Tammuz	29	29	29	29	29	29
Av	30	30	30	30	30	30
Elul	29	29	29	29	29	29
	353	354	355	383	384	385

* Adar II is often called Veadar.

Key: the length of the year is indicated by the letter used in the character of the year—deficient (d, D); regular (r, R); and abundant (a, A).

⊠ The Book of Jubilees

I have described the mainstream Jewish calendar: the empirical version was used before Hillel's reform, and the reformed, calculated version, after this. Nevertheless, there is evidence of other calendars being used by sects of the Jewish faith from time to time. The apocryphal Books of Enoch and of Jubilees, dating from Maccabean times in the second century before Christ, mention a year of 364 days containing exactly 52 weeks. This year is divided into four quarters each with three months of 30, 30, and 31 days. Another passage warns of the consequences of abandoning this year in favour of a lunar year.

It has been suggested, on the basis of this, that the 364-day calendar antedated the lunar calendar and was, in essence, solar—even though there is no mention of any system of intercalation that would be required to keep this year in step with the seasons. If this supposition is correct, it would appear that the competition between an old 364-day solar year and the new lunar year was not resolved by the second century BC; indeed, the recently discovered Dead Sea scrolls suggest that the old calendar was still being used in the first century AD. The sect at Qumran that wrote them perished in 73 AD—so did their calendar.

CHAPTER 18

The Islamic and Bahá'í Calendars

▨ The Islamic calendar

Although the Islamic calendar was first used in the Arabian peninsula in the seventh century AD, its use has spread throughout the world to wherever Islam is practised. It is employed to regulate the feasts and fasts of that religion, but in many Islamic countries, the Gregorian calendar is used for civil purposes.

From the earliest times the Arabian peoples observed a strictly lunar calendar. In the twelfth month they made a pilgrimage to Mecca. There they indulged in rituals associated with the sacred meteorite given to Abraham by Gabriel and now embedded in the walls of the Ka'ba; at the end of the pilgrimage an animal was slaughtered. Since a lunar calendar does not keep step with the seasons, this sometimes led to logistic difficulties in finding food for the trip and animals for slaughter. To ensure sufficient supplies they started to intercalate a month, so as to keep the 12th in the autumn. They had learnt of this stratagem from the Jews of Yathrith, and adopted it in AD 412.

Thus, before the days of the Prophet Muhammad (c. AD 570–632), the desert Arabs used a lunisolar calendar with occasional intercalations of a month. The Kinana tribe had responsibility for the calendar, which they delegated to their calendar officials, the Nasa'a. It is not known for certain how the Nasa'a decided when to intercalate; some say there were 11 nasi or intercalations in 22 years; others say they used the same Metonic method employed by the Jews.

Certain of the months (1st, 7th, 11th, and 12th) were set aside to define two closed seasons (1st–7th, 11th–12th) in which warfare and pillaging were forbidden. The extended period of three months in between was too much for some tribes, who sometimes persuaded the Nasa'a to change the religious character of the intercalated months in such a way as to give them a welcome respite from peace. The Prophet Muhammad strongly disapproved of this ruse and its casual dispensation of the religious canons. He thus decreed, in a speech delivered in AD 632, shortly before his death, that there should be no further intercalations; to sweeten this pill he allowed that henceforth warfare against the infidel was permissible in any month.

The result of this dictum was a strictly lunar calendar, the final form of which was instituted in AD 642, 10 years after the death of the Prophet, by the Caliph Umar I. In this Islamic calendar, every year contains just 12 months. These months have alternately 30 and 29 days, excepting the last which, in kabisah (embolismic) years only, contains 30 days instead of 29. The names of the months, given in TABLE 18.1, are derived from more ancient Arabic names; some carry a seasonal connotation which is not relevant after the abolition of intercalations.

The common years thus contain 354 days, and the embolismic years, 355 days. It follows that the Islamic years begin earlier and earlier in the seasonal year, and in fact rotate through all four seasons in about 32 Islamic years. It is, in this respect, reminiscent of the Egyptian wandering year—but it rotates considerably more rapidly.

The intercalary day, the last day of the 12th month in embolismic years, is introduced 11 times in a cycle of 30 years. Numbering the years of the cycle from 1 to 30, these embolismic years are numbers 2, 5, 7, 10, 13, 16, 18, 21, 24, 26, 29. They may be seen to be spaced in such a way as to minimize the discrepancy between the start of a year and a calculated lunation. The length of the lunation, as known to Arabian astronomers of the time, was 29 days 12 hours

TABLE 18.1 *The months of the Islamic calendar*

	Name*	Days	Likely derivation†
1	Muharram	30	Sacred month; no war
2	Safar	29	Yellow (saffron), autumnal
3	Rabi' I	30	Grazing season
4	Rabi' II	29	
5	Jumada I	30	Hard, frozen, winter
6	Jumada II	29	
7	Rajab	30	Sacred month
8	Sha'bân	29	Dispersed, tribes seek water
9	Ramadân	30	Burnt, hot
10	Shawwâl	29	Camels become pregnant
11	Dhu'l-qu'da	30	Sacred month, time of truce
12	Dhu'l-hejji	29	Sacred month, time of pilgrimage

*The spelling of the transliteration of these names varies quite widely.

†The original meanings of the months lost their seasonal connotation when intercalations were abolished.

and 44 minutes (29.53055 days). It should be noted, however, that sometimes the years designated embolismic are a little different. One popular variant makes year 15 embolismic, rather than year 16.

In a 30-year cycle there are 360 months or 10631 days. The mean number of days in a month is thus 10631/360 = 29.53056. This may be compared with the mean synodic period of 29.53059 days. The Islamic months keep in step with the mean astronomical moon, gaining on it by one day in about 2500 Islamic years. If it were not for postponements (see Chapter 17), the start of the Islamic months would remain well synchronized with the Jewish months. As it is, the first day of a Jewish month falls mainly on the 28th, 29th, 30th, or 1st of an Islamic month.

The Moslems, like the Jews, observed a seven-day week. Also, like the Jews, they numbered rather than named these seven days (except that Friday is termed the Day of Assembly and is devoted to prayer).

Muhammad died in AD 632, and it was only after another 7 years that the first Caliph, Umar I, set up the calendar—17 years after the flight. Umar took 16 July AD 622 (Julian calendar) as the starting date of the Islamic era. This date is sometimes mistakenly taken to be that of the Hegira (the Prophet's flight from Mecca to Medina). The date is not that of his arrival in Medina either; the actual dates of both these events is unverified, but they probably occurred about two months later in 622.

Nevertheless, Umar chose 622 in order to commemorate the Hegira, and the era is generally known as the era of the Hegira (AH). The actual day, 16 July, was determined in a more subtle manner. Umar, like other calendar reformers, wished to set up the new calendar as seamlessly as possible. He assumed that the old calendar had been used until 632, when intercalation was first forbidden. He arranged that the start of the year in both the old and new calendars should coincide on 9 April 631. He then calculated back nine years from this date to the year 622 using the new canon with no intercalations; he arrived at the date of 16 July as the date on which the new calendar would have started had it begun in that year.

It should thus be noted that Islamic dates ascribed to the era of the Hegira before about AD 629 are proleptic and no genuine inscription will ever be found marked with such dates. In fact, the years after the Hegira, until AD 649, were given Islamic names relating to important events, rather than numbered in any way; some of these are given in TABLE 18.2. Moreover, dates before AD 622 are not adequately defined: it is not specified whether, for instance, 621 should be called −1 EH, 0 EH, 1 BH, or something else.

Over the centuries, many tables for converting Islamic dates to Gregorian dates have been published. Sometimes these are out by one day, suggesting that the era began on 15 July. The discrepancy could be explained by the fact

TABLE 18.2 *The names of the early years of the era of the Hegira*

2 AH	Permission	AD 623
3 AH	Order for fighting	AD 624
4 AH	Trial	AD 625
5 AH	Congratulation on marriage	AD 626
6 AH	Earthquake	AD 627
7 AH	Enquiring	AD 628
8 AH	Gaining victory	AD 629
9 AH	Equality	AD 630
10 AH	Exemption	AD 631
11 AH	Farewell	AD 632

that the Islamic day starts at sunset, like the Jewish day, whereas the Christian day, at any rate nowadays, begins at midnight. Another cause of discrepancy was due to some early writers failing to allow properly for the occurrence of leap days in the Gregorian calendar.

The Islamic calendar, as described, is accurate and calculable, and is used for official purposes in all Islamic countries. However, running side by side with it in some countries, is the popular calendar in which the start of each month is determined empirically from first sightings of the crescent of the actual new moon. This popular calendar runs on average in synchrony with the calculated calendar, but since the time at which a new moon could be observed depends on the latitude and longitude of the observer, the start of a month may differ from the calculated day by up to two days, and varies from one part of Islam to another—even within the same country.

Reform

The details of the Islamic calendar were not dictated to the Prophet by Allah, like the rest of the Koran. It would thus appear, without getting involved in theological disputes, that it could be changed. However, Islam does not harbour an ecclesiastical hierarchy who might be able to effect a reform. Nevertheless, from time to time, in various parts of the Islamic world, new calendars have been designed and used.

One of these was set up by Ghiyathuddin Abulfah Omar bin Ibrahim al-Khayyam, better known as Omar Khayyam (d. 1123)—the great Persian poet, astronomer, and mathematician. The name 'Kayyam' means 'tent maker', but it seems unlikely that Omar plied that trade himself. One of his school fellows was Hassan ibn Sabbah, founder of the notorious sect of the Assassins. Omar, however, was of a more academic bent and in 1074 was asked, together with a team of seven other astronomers, by Sultan Jelaledin Malik-Shah bin Alkh Ashlan Suljookee of Khorassan in Persia, to carry out astronomical observations in order to reform the Yazdegerd calendar then being used. The calendar that emerged was called the Jelali calendar in honour of the Sultan and was similar, but not identical to, the Alexandrian version of the Egyptian calendar. Unfortunately, Omar's report has not survived and there is some uncertainty about the details of the method of intercalation he proposed; it is most likely that a leap day was intercalated eight times in every 33 years to give a mean year of 365.24242 days. The Jelali era began on 15 March AD 1079. Later, the standard Islamic calendar was imposed in Persia. In 1584 the Jelali calendar was taken to India by Akber, grandson of Babur, founder of the Moghul dynasty. There, it gradually fell into disuse.

More recently, a reformed calendar was instituted in Persia (Iran) in the 1920s by Riza Shah Pahlavi. This attempted to reconcile the Islamic and Gregorian calendars. The 12 months were given ancient Persian names; the first six had 31 days, the next five had 30 days, and the 12th had 29, except in a leap year when it had 30. It is possible that it inspired the reformed Saka calendar described in Chapter 13. The new era was taken to begin in the same year as the standard Hejira era. This reform did not find favour in other parts of the Islamic world and has probably been extinct in Iran since the deposition of the Pahlavis.

Trade and the Islamic world

The Islamic calendar, with its wandering year, is not convenient for commercial transactions with Christendom. In Islamic countries on the border between the two worlds, such as Turkey, an alternative calendar, now known as the Ottoman Financial calendar, or Marti calendar, was in use from AD 1676. The years of this commenced on 1 March in the Julian Calendar. The months were numbered so that the first month corresponded with March, the second with April, and so on. The 12th month, corresponding to February, had 28 days, or 29 in a Julian leap year. For the first 200 years of its use, the different lengths of the Financial calendar and the Julian calendars led to a very confusing method of numbering the years; this difficulty was rectified in 1256 (corresponding to the Julian year 1840), when the calendar was reformed. From then on the years were numbered consecutively from 1256. This calendar was in use until 1927, when Kemal Atatürk adopted the Gregorian calendar.

⚰ The Bahá'i calendar

Every religion, it would seem, must have its own calendar so that the faithful may all perform the rites on the same day; it is something that draws them together and announces their difference from others.

One of the newest religions is Bahá'i. This is based on the teachings of Mirza Ali Muhammad of Shiraz (1819–50) who was a Sayyid, a descendant of the Prophet Muhammad, and brought up as a Shi'ite Moslem in Persia.

According to Shi'ite teaching and the strong belief of the Shaykhis sect to which he belonged, the 12th Caliph, who had disappeared 1000 years before, was prophesied to return to mediate between man and God in the manner of Moses, Jesus, and Muhammad. Ali Muhammad, by all accounts a young man of great personal charm and piety, experienced, in 1844, a divine revelation that he was the Bab, or gateway, to the reappearance of the Caliph. He announced this revelation to the Mullah Husayn who, after due consideration, accepted the claim. Before long most of the Shaykhis accepted it too. This, and his personal qualities, brought Ali Muhammad a great following. But it also aroused the emnity of the established Moslem dignitaries, and he spent most of the rest of his life in prison. Finally, he was publicly and brutally executed in 1850 at Tabriz. He left behind many manuscripts of his teaching.

After the death of the Bab, the leadership of the sect passed to Mirza Husayn Ali (1817–72), a Persian nobleman who renounced public office to follow the Bab. In due course he took the title Bahá'u'lláh (the splendour of God), which became the name of the new faith. In 1872 a deranged follower of the Bab attempted to assassinate the Shah and the other Babis were blamed. The Babis and Husayn Ali suffered atrocious punishment, but Husayn Ali escaped death and was eventually exiled. His followers were severely persecuted but his own saintly character eventually earned him great respect.

The teachings of Bahá'u'lláh emphasized the equality of men and women and of people of all races, and he is claimed to be the 12th Caliph, the latest messenger from God. The Bahá'i faith has spread throughout the entire world and is today the fastest growing religion.

Bab had 18 disciples, which together with himself made 19. This important number is commemorated in the solar Bahá'i calendar, or Badí calendar. The Badí year starts on 21 March and has 19 months each of 19 days; the 19 months are listed in TABLE 18.3 and account for 361 days. Four or five epagomenal days are inserted at the end of the 18th month to coincide with 26 February to 1 March inclusive; in leap years there are five such days. Thus the Badí year contains the same number of days as the Gregorian year (in common years: $19 \times 19 + 4 = 365$). The Bahá'i day starts at sunset as it does for Moslems and Jews.

The week plays a minor role. It starts on Saturday and the seven days are named with more of the attributes of God, but there is no day set aside specifically for religious observances.

The Badí era (BE) began in 1844, the year that Ali Muhammad declared that he was the Bab. The Badí calendar appears to keep in step with the Gregorian calendar—which implies that the extra intercalary day is inserted in the same years as in the Gregorian calendar. However, it is also claimed that the start of the year is determined by the vernal equinox, but with no specification of

TABLE 18.3 *The Bahá'i months and days*

Number	Arabic name*	Translation	First day
1	Bahá	Splendour	21 March
2	Jalál	Glory	9 April
3	Jamál	Beauty	28 April
4	Azamat	Grandeur	17 May
5	Núr	Light	5 June
6	Rahmat	Mercy	24 June
7	Kalimát	Words	13 July
8	Kamál	Perfection	1 August
9	Asmá	Names	20 August
10	'Izzat	Might	8 September
11	Mashiyyat	Will	27 September
12	'Ilm	Knowledge	16 October
13	Qudrat	Power	4 November
14	Qawl	Speech	23 November
15	Masá'il	Questions	12 December
16	Sharaf	Honour	31 December
17	Sultán	Sovereignty	19 January
18	Mulk	Dominion	7 February
	Epagomenal days: 26 February to 1 March		
19	'Alá'	Loftiness	2 March

*The months are named after the attributes of God. The same names are assigned to the 19 days of the month.

exactly where the observation is made. It would seem in practice that a nominal date of 21 March is assumed for the equinox.

It is easy enough to relate dates in the Badí calendar to Gregorian dates: count the number of days by which the Gregorian date goes beyond the nearest date in the Badí calendar (see TABLE 18.3). This immediately gives the Bahá'i day and month. To get the year, just subtract 1843 from the year of our Lord if the date is on or after 21 March, or subtract 1844 if it is before. (Bahá'i dates before 21 March 1844 are not defined because it is not specified whether they are year −1 or 0 or whatever.) As an example, consider 17 June 1995. This is the 12th day after 5 June and hence the Bahá'i date is 13th Núr 152 EB. Conversely, 9th Mulk 56 EB is 15 February 1900.

CHAPTER 19

The Gregorian Calendar

⚅ Intolerable, horrible, and derisible

The Julian calendar had several defects. The average number of days in its year was 365.25, whereas the true length of the tropical year was about 365.24219 mean solar days. This discrepancy meant that annual events such as the vernal equinox were falling earlier and earlier in the calendar year at a rate of about one day in 128 years. Furthermore, the date of Easter was falling later and later in the seasonal year.

That the year was not exactly $365\frac{1}{4}$ days long had been known to the Alexandrian astronomers and certainly to Sosigenes. The Julian calendar had the merit of simplicity, but Sosigenes must have known that it would create problems eventually.

Much more serious for the medieval Church was the observation that the dates of Easter were starting to fall on inappropriate days. It was this that was the spur to the whole reform movement. The drift of the vernal equinox was, in itself, of relatively minor importance.

Thus, the venerable Bede, writing as early as 725, notes that the full moon was ahead of its tabulated date. By the thirteenth century, alarm was spreading. Robert Grosseteste (1178–1253)—reformer, mathematician, and Bishop of Lincoln—drew attention to the problem and suggested alterations. John Holywood (d. 1244), otherwise known by his latinized name of Sacrobosco—an Englishman who was a professor at the University of Paris—suggested a scheme for rectifying the problem in his *De anni ratione*; this book was not printed till 1538, in Paris.

There were others with suggestions: Conrad of Strasburg, Campanus of Novara, and Roger Bacon (*c.* 1214–*c.* 1294). Pope Clement IV had heard of Bacon's writings and asked to be sent copies. Bacon sent these, together with a letter (which still exists in the Vatican Library). In them he quoted the earlier criticisms of Grosseteste and suggested reforms; he also described the calendar as 'intolerable, horrible, and derisible'—sentiments echoed by later writers. The matter had been drawn to the attention of the only authority who could institute reform.

Meanwhile, knowledge of classical astronomy was beginning to percolate back into Europe through the translation of Arabic works. The Alfonsine tables, giving an estimate of the length of the year, were published in 1277.

During the next century various astronomers observed the equinoxes, made complaints, and proposed remedies. At last, in 1344, Pope Clement VI invited Firmin of Belleval and John of Murs—both eminent astronomers of their day—to Avignon, to advise him on the correction of the calendar. Firmin and John proposed, in their report of 1345, that reforms should take place in 1349. Unfortunately, the Pope became preoccupied with other matters, including the Black Death in 1347/8. He died in 1352, having instituted no reform.

The complaints and proposed remedies continued. People wrote to both the Pope and to the emperor. The Church itself had problems—there were two Popes, then three contenders; it had other things to think about than the calendar. At last, in 1417, there was just one Pope again, Martin V.

In 1412 Pierre d'Ailly (c. 1350–c. 1420), recently appointed a cardinal, raised the matter of calendar reform with Pope John XXII (1410–15). (There is some confusion about the numbering of the Johns; some say he was John XXIII.) At the Council of Constance in 1415 he opined that calendrical calculation was more important to the faith than financial calculation. Pierre was not an original thinker, and his work and proposals were mainly based on those of his predecessors—Grosseteste, Sacrobosco, and Bacon—but Pope John took him seriously and issued a decree in 1412, incorporating Pierre's proposals for reform. Alas, they were never put into effect, despite discussions at the Council of Constance in 1415, for John was deposed in that year by the Council.

The next move for reform took place at the Council of Basel in 1434, which provided an opportunity for reformists to wring concessions from the Pope. A letter was read out, drawing attention to the scandal of the calendar, and the Council commissioned certain scholars to consider the matter. Nicholas of Cusa (1401–64), a cardinal, philosopher, and mathematician, presented the report of the commission in 1437. The commission was empowered by the Council to institute the reform to take effect in 1439, but again reform was thwarted; the Council fell out with the Pope, Eugenius IV, in 1438, deposed him in 1439, and appointed their own man, Felix V. Once again, there were two Popes and the Church was too busy to think about the calendar.

As the fifteenth century went by, astronomers started to ask philosophical questions about the whole basis of the calendar. What were the correct dates of the equinoxes? What was the correct longitude to detect the paschal new moon? They became less interested in the defects of the calendar as such; modern astronomy was getting under way. Even so, the complaints continued.

In 1475, Pope Sixtus IV summoned the astronomer Johannes Müller of Königsberg (1436–76; otherwise known as Monteregio or Regiomontanus, after his home town) to Rome to discuss calendar reform. Unfortunately, Johannes died the following year before anything could be done. The complaints continued.

In the Church, the Reformation was gathering strength. A council at Rome (the fifth Lateran Council held in the Church of St John Lateran) was called by Pope Leo X in 1512, in reaction to earlier schismatic conferences. At this, a new star of calendar reform arose—Paul of Middelburg (astronomer, astrologer, physician, and bishop). Paul had tried earlier to persuade Innocent VIII to reform the calendar. Leo called Paul to the Council and set up a commission, headed by him, to consider reform. The proposals put forward by Paul's commission were circulated to the kings of Christendom for consideration by their court astronomers; it was hoped that their replies could be considered by the Council. Some replied; others (including Henry VIII of England) did not. Further letters were sent, and there were further delays; those who had replied were not in agreement. The fifth Lateran Council closed in 1517 without the matter having been discussed.

Twenty years later, Pope Paul III summoned another council to meet in Trent in 1542. Plagues, hostile troop movements, difficulties with intransigent bishops, and so on, meant that further councils had to be called and it was only at the third Council of Trent, called by Pius IV in 1562, that calendar reform got on to the agenda. In the last session in 1563 the Council requested the Pope to reform the breviary; it did not say anything about calendar reform. Pius IV set out to do this but had not completed the task when he died in 1566. This was left to his successor, Pius V, who advanced the notional dates of the paschal new moons by four days, to bring them in better agreement with observation, and made a provision for the avoidance of discrepancies in the future. It was a reform—but not the one wanted by the astronomers.

Pius V died in 1572 and was succeeded by Gregory XIII; Gregory believed that the decree of 1563, issued by the Council of Trent, had not yet been fulfilled. To complete the task, he undertook to reform the calendar.

Sometime earlier Gregory had had built, in the Vatican, the Galleria della Carte Geografiche. This was equipped with maps and meteorological devices and thus resembled the Tower of the Winds, at Athens—a name it acquired itself. Later, Gregory arranged for the astronomer, Ignazio Danti (1536–86), to furnish it with astronomical instruments. Danti installed a gnomon which consisted of a hole in the roof which cast an image of the sun on a scale aligned along the meridian. With this, Gregory was able to convince himself and others that the equinoxes were occurring on the wrong dates. The Tower

FIG. 19.1 Pope Gregory XIII (Ugo Boncampagni), 1572–1585.
St Peter's Basilica, Rome.

of the Winds can still be visited in the Vatican, but you will require special permission.

⚏ Lilio's proposal

Soon after he became Pope, Gregory had been presented with a proposal. This had been written by Aluise Baldassar Lilio (1510–76; otherwise known as Aloysius Lilius, Luigi Giglio, or Luigi Lilio Ghiraldi), a physician and astronomer born in Cirò in Calabria but then living in Verona. The proposal had been presented by Lilio's brother, Antonio, also a physician, and Gregory had asked the Bishop of Sora in Calabria, Tommaso Gigli, to assess it. Tommaso consulted several experts, most of whom were well disposed to it, although some suggested alterations, and some said it was too complicated; Giovanni Carlo Ottavio Lauro kept it for many months on the excuse that he needed it to prepare his own proposals. These, when eventually forthcoming, were rejected by the other experts as being wrong and useless. Lauro then kept Lilio's manuscript for several more months while he corrected his errors. This was to no avail, however, since his proposals were again rejected, and the experts instead praised Lilio's work.

FIG. 19.2 Aluise Baldassar Lilio (1510–76), originator of the Gregorian calendar.

At this point, in 1575, the experts approached the Pope with praise for Luigi's proposal and ridicule for Giovanni Carlo's; they urged Gregory to reform the calendar along the lines suggested by Lilio. It may have been in that year (or possibly even earlier) that Gregory set up a commission, headed by Gigli, to advise him on the reform of the calendar. Among others, the commission contained Antonio Lilio and Christopher Schlüssel (1537–1612; better known as Clavius), a German Jesuit and astronomer then working at the Collegio Romano, a university at Rome. For several years the commission debated the matter, but its members could not choose between the many proposals, including Lilio's, that were available to it.

Eventually in about 1577, the president of the commission was replaced by the able Guglielmo Sileto (1514–85; born in Calabria), a cardinal, scholar, and one-time contender for the papacy. Sileto steered the commission to agree and to recommend Lilio's proposals.

An abstract or compendium of Lilio's proposals was prepared by Pedro Chacón (1526–81), a member of the commission, and in 1578 Gregory sent this, in the form of a printed book of 20 pages, to the princes of the Roman world for consideration by their court mathematicians. It would appear that they lost Lilio's original manuscript, and no copy of it survives. However, in this century, several copies of the compendium have been discovered in libraries in Italy. From the compendium it can be seen that Lilio's proposals were similar to those of Sacrobosco, written 350 years previously, and printed earlier in the century; we do not know if Lilio had read these, and we do not know the details of his proposals. Lilio died in 1576 without the satisfaction of realizing that his ideas would be accepted and his name honoured by scholars for centuries to come.

The replies to the compendium were by no means all favourable. Most of the criticisms centred on the new arrangements for calculating the date of Easter; some found these unintelligible. Nevertheless, the commission wished to push the matter forward (the Pope was nearly 80). In 1580 it wrote a report incorporating most of Lilio's proposals, for consideration by the Pope. The signatories of the 1580 report give some idea of the constitution of the commission: Cardinal Guglemo Sileto (the president), Bishop Vincenzo Lauri of Mondovi, Patriarch Ignatius (ex-patriach of Antioch who had arrived in Rome unexpectedly and uninvited from the Eastern Church; after his credentials had been checked out and somewhere found for him to stay, he was welcomed on to the commission), Leonardo Abel (who acted as interpreter for Ignatius who spoke in Arabic), Seraphinus Olivarius Rotae Auditor Gallus (a French expert in ecclesiastical law), Pedro Chacòn (a scholar and historian and author of the compendium), Antonio Lilio (brother of Luigi), and the two astronomers, Christopher Clavius and Ignazio Danti.

FIG. 19.3 Christopher Clavius (1537–1612), who wrote the definitive book on the Gregorian calendar. Source: Mary Evans Picture Library.

The Pope accepted the report and on 24 February 1582 he signed a papal bull (Pedro Chacòn wrote the text) which would put it into effect. He signed at a table still preserved in his country residence at Mondragone outside Rome. The bull, beginning with the words 'Inter gravissimas . . .' (by which it is known), was displayed on the doors of St Peter's on 1 March, and copies

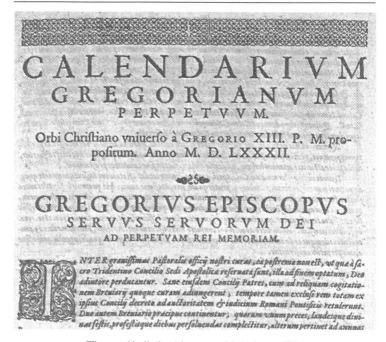

FIG. 19.4 The papal bull; the title page published on 24 February 1582.

were sent to all Catholic princes. The actual change over to the new calendar took place the same year in October. After 350 years or more the reform had at last been accomplished. Gregory struck a medal to commemorate the event.

In the preceding discussions, many had said that Lilio's proposals were, at the very least, difficult to follow and that it would be beyond the ability of a poor priest in a remote parish to work out the date of Easter by himself. To remedy this an explanatory booklet was compiled by Antonio Giglio, who finished the task in 1585. Unfortunately, Gregory died soon after, and the explanation was never published. He was succeeded by Sixtus V and then, in 1590, by a rapid succession of Popes—Urban VII, Gregory XIV, Innocent IX, and Clement VIII.

The papal bull continued to generate controversy and criticism, and the Pope delegated Clavius to publish replies. Finally, as the only survivor of the original commission, Clavius was charged with writing the final explana-

FIG. 19.5 A medal commemorating the new calendar, struck by Pope Gregory XIII.

tion. This he did and it was published in 1603—the definitive work on Gregory's reform entitled *Explicato Romani calendarii a Gregorio XIII P.M. restituti*. This was reprinted in 1612 and is now available on microfilm.

The bull was translated into many languages and sent to many churches. The Catholic countries responded by adjusting their calendars in 1582 or 1583, but the Protestant countries demurred. The main reason for this was an unwillingness to take instructions from the Pope; this is understandable considering the role Gregory XIII had played in the massacre of the French Protestants, the Huguenots, on St Bartholomew's day. Some of these countries only accepted the reform in the twentieth century; the Greek Orthodox Church still celebrates Easter according to the Julian canon.

However, commerce and diplomacy between the countries of Europe became difficult with two calendars in operation and, at the instigation of the philosopher Leibnitz, most fell into line in 1700. Nevertheless, a few, including England, her colonies, and the Eastern Orthodox Churches, resisted. The dates on which many of the countries of the world adopted the Gregorian calendar are listed in TABLE 19.1.

The failure of some countries to adopt the new calendar sometimes gave them problems. In 1908 the Imperial Russian Olympic team arrived in London 12 days too late for the games.

⌛ The Gregorian reform

I have emphasized that the primary complaint about the Julian calendar was not the fact that it was slipping with respect to the seasons and that the vernal

TABLE 19.1 *Dates on which some countries adopted the Gregorian calendar*

Country	Year	Dates omitted
Rome	1582	5–14 October
Italy	1582	5–14 October
Portugal	1582	5–14 October
Spain	1582	5–14 October
Poland	1582	5–14 October
Luxembourg	1582	15–24 December
France	1582	10–19 December
Belgium	1582[1]	
Austria	1583[1]	
Holland (Catholic)	1582[2]	
Holland (Protestant)	1700[2]	
Germany (Catholic)	1583[1]	
Czechoslovakia	1584	7–16 January
Switzerland (Catholic)	1584[1]	11–21 January
Hungary	1587	22–31 October
Germany (Protestant)	1700[1]	19–28 February
Denmark	1700	19–28 February
Norway	1700	19–28 February
Switzerland (Protestant)	1701	1–11 January
England and colonies	1752	3–13 September
Ireland	1752	3–13 September
Scotland	1752	3–13 September
Sweden	1753[2]	
Finland	1753[2]	
Alaska	1867[3]	
Japan	1873[2,4]	
Egypt	1875[4]	
Albania	1912	
China	1912[2,4]	
Bulgaria	1915[2]	
Lithuania	1915	
Latvia	1915/18[2]	
Soviet Union	1918	1–13 Febuary
Estonia	1918	
Yugoslavia	1919[2]	

TABLE 19.1 *Continued*

Country	Year	Dates omitted
Romania	1919	1–13 April
Greece	1924	10–22 March
Persia	1925[4]	
Turkey	1926	19–31 December

1. The days which were omitted varied from one part of the country to another.
2. The date of the change is problematic; in some cases because different authorities give different dates.
3. Alaska changed when it was bought by the United States of America from Russia.
4. These countries had not used the Julian calendar before their adoption of the Gregorian in the year given.

Some countries changed their New Year's Day to 1 January before abandoning the Julian calendar.

---

equinox was occurring earlier and earlier before 21 March, but that the calculation of the date of Easter was going awry. The greater part of the compendium and Clavius' explanatory work are taken up with discussions of the adjustment to the calculation of the date of Easter. Here, I am only concerned with the consequences of the fact that the average Julian year contained slightly too many days; I take up the matter of Easter in Part IV.

Clavius did not have modern measurements of the synodic period or the tropical year to help him design the new calendar; he did not even have the benefit of observations made with a telescope—even though astronomical knowledge was approaching the point of explosive growth that began in the seventeenth century. Copernicus (1473–1543) had laid the ground work for the revolution; Tycho Brahe (1546–1601) was still collecting astronomical data; Kepler (1571–1630) was a nine-year-old boy; and Galileo (1564–1642), a young man. However, there was available to Clavius and Lilio, the work of Petrus Pitatus, who had written extensively on calendar reform in the 1560s. Petrus had himself proposed that three leap years (from the Julian calendar) should be omitted every 400 years.

Controversy still surrounds the origin of this proposal. Several estimates of the length of the year were available as shown in TABLE 19.2. It may be seen that the values agree to the second sexagesimal fraction. Noel Swerdlow has pointed out that the fractional part of this common sexagesimal value—14,

TABLE 19.2 *Estimates of the length of the year available in the late sixteenth century*

Source	Date	Author	Length in days
Alfonsine tables	1252	Alfonso X	365, 14, 33, 9, 57
de Revolutionibus	1543	Copernicus	365, 14, 33, 11, 12
Prutenic tables	1551	Reinhold	365, 14, 33, 9, 24

The values are expressed as sexagesimal fractions of a day, for the ancient Babylonian sexagesimal system was still in use at the time.

The Alfonsine tables were based on Arabic sources, in turn compiled from the work of classical authors. The Prutenic tables were based on the work of Copernicus, who himself gives a number of values which differ by several minutes and was confused by the ancient theory of trepidation.

33—is exactly equal to 97/400, and believes that this accounts for Petrus' and hence Lilio's proposal.

The strategy proposed by Lilio to bring the calendar year into line with the shorter astronomical year was to remove a few of the leap days. If P leap days are suppressed from the Julian calendar in every interval of Q years, the mean length of the calendar year is reduced to $365.25 - P/Q$ days. By a judicious choice of P and Q, the calendar could be brought into precise agreement. However, precision was not the only criterion. Clavius makes it clear that the rule for deciding which leap days to drop must be simple and capable of being understood by any parish priest.

Lilio had proposed that leap years should continue to occur every four years (that is, years of the Christian era which are divisible by four, as in the Julian calendar), except that years divisible by 100 but not by 400 should be common (that is, 1600 would continue to be a leap year; 1700, 1800, 1900 would be common years; 2000 would be a leap year; and so on).

This rule gives the average length of a calendar year as $365 + 97/400$ days (365.2425) or 365 days 5 hours 49 minutes and 12 seconds. This differs from the Alfonsine value (365.242546) by about four seconds. According to this reckoning, the reformed calendar would gain a day on the sun in about 20000 years. No other simple rule would be anywhere near as accurate.

In the discussion, several points of view had been expressed as to the proper calendar date on which the vernal equinox should fall—or even whether it mattered. A special plea for 21 March (the traditional day on which it fell in AD 325, at the time of the Council of Nicaea) was made by the King of Spain and by Patriarch Ignatius, and this date was adopted. Since the equinox was currently falling on 11 March, it was necessary to omit 10 days from the calendar. Accordingly, the papal bull directed that 4 October of 1582 should be immediately followed by 15 October. These days were selected for omission because no important feasts were held on them.

Since 1582, more accurate values for the length of the mean tropical year have become available, and it has even been found that the length of the year is slowly decreasing. In 1582, the true value was 365.24222 days; in 2000 it will have fallen to 365.24219 days. This means that the length of a Gregorian year was less well matched to reality in 1582 than Lilio might have thought, and that the mismatch will get greater as time goes on.

It should be noted that since the leap days are inserted a day at a time, the equinox will not always fall on 21 March; it may be a day earlier or later. Taking into account the decrease with time of the length of the year and assuming the leap days are added 'smoothly', the Gregorian calendar will be one day out by about the year 3719. This is perhaps too far in the future to cause concern. Nevertheless, various proposals have been made to remove further leap years to make the agreement more accurate. Delambre, an eighteenth-century French astronomer, suggested that years which are multiples of 3600 should be common. In the last century, Sir John Herschel, the astronomer (whose father discovered Uranus), suggested the more elegant rule of making years divisible by 4000 common—a rule proposed for the French Revolutionary calendar. The Soviet Union and the Eastern Orthodox Churches both decided, in 1923, to run their calendar so as to make centurial years leap years only if they yield a remainder of two or six when divided by nine. This calendar will remain in step with the Gregorian until the year 2800—a leap year in the Gregorian calendar but not in the Soviet version.

In fact, any improvement resulting from Herschel's rule or the Soviet Union amendment is small; it only delays the time when the discrepancy will amount to one day to about the year 5267. The reason for this is that the discrepancy becomes dominated by the slow decrease in the length of the astronomical year. The excess time accumulates with the square of the number of years, and currently amounts to a whole day in about 3450 years. This can only be mitigated by removing a day permanently from the calendar year every so often. It will have to be left to the inhabitants of the globe alive on this date (if there are any) to decide what to do about it.

It is possible to extrapolate the Gregorian calendar backwards and assign Gregorian dates to the days before 1582 (termed 'proleptic' dates). It is then found that the day called Monday, 1 January AD 1 in the Gregorian calendar is called 3 January AD 1 in the Julian calendar (the Gregorian date is two days behind the Julian date); 28 February 100 was the first day on which the

Gregorian date became only one day behind; on 1 March 200 it became identical to the Julian date; and on 1 March 300 it was one day in advance. Thereafter, it advanced by one day on each 1 March in centurial years not divisible by 400. This knowledge provides a ready means of converting a Gregorian date to the corresponding Julian date, or vice versa.

Many countries, including Scotland and Sweden, who continued to use the Julian calendar, decided to begin their years on 1 January before adopting Gregory's reform; but whenever the Gregorian calendar has been used, the year has begun without exception on 1 January.

⧗ The English experience

The United Kingdom and her colonies (which included the American colonies) converted to the Gregorian calendar in September 1752 during the reign of George II: Wednesday, 2 September was followed by Thursday, 14 September, and the intervening 11 days just did not exist in that year. At the same time, New Year's Day in England and her colonies and Ireland was moved from its old position on 25 March—the Feast of the Annunciation and Lady Day—to its new position on 1 January. This meant that 31 December 1751 was the day before 1 January 1752 and that no days bore the dates 1 January to 24 March 1751. Scotland had done the same thing earlier in 1600.

There had been an earlier attempt to introduce the new calendar in 1584, in the reign of Elizabeth I. When she received the bull of 1582, she passed it to Dr Dee (1527–1608), her astrologer and confidant, who had pretensions as a magician and astronomer (but who was not, in fact, a doctor; he was a contemporary of Nostradamus who had studied medicine, but it is not recorded whether they ever met). Dee considered the bull and suggested alteration to some of the details. However, in his report, he advised Elizabeth to adopt the reform. Dee's report was passed, by Lord Burghley, to a committee consisting of Thomas Digges (an astronomer), Henry Savile (1549–1622; a classical scholar), and a Mr Chambers. The committee endorsed Dee's advice and prepared an act entitled 'An act to give Her Majesty authority to alter and make a new calendar, according to the calendar used in other countries'. This bill passed two readings in the House of Lords.

Meanwhile Sir Francis Walsingham wrote to the Archbishop of Canterbury urging the approval of the bishops. The Protestant bishops were not eager to adopt a proposal from the Pope, whom they considered to be the Antichrist. Accordingly, they replied with delaying tactics, saying that they needed time for widespread consultations (which would have taken years to complete). They also used other arguments including the notion that the world was about to end so there was not much point in reforming the calendar.

This and popular anticatholic opinion ensured that the bill was quietly ignored. There were further sporadic attempts at reform during the next 150 years, such as that of John Greaves, Professor of Astronomy at Oxford in 1645. He attempted to resurrect Dr Dee's proposals.

In 1699, the Archbishop of Canterbury discussed, with the Savilian Professor of Geometry at Oxford, John Wallis (1616–1703), a proposal to drop all leap days for 40 years till the calendar had caught up with the sun. Wallis, who is credited with the invention of the sign for infinity (∞), perceived 'the hand of Joab' in the proposal, and opined that it had originated from papists; reform again came to naught.

At last, in the eighteenth century, the flames of religious bigotry were subsiding. H. Wilson wrote a pamphlet in 1735 arguing convincingly for reform; this was taken up by the Earl of Macclesfield and, in 1751, an Act of Parliament (24 Geo. II, ch. 23) was passed, under the guidance of Lord Chesterfield. This was drawn up by Peter Davall who was secretary of the Royal Society at the time and called 'An act for regulating the commencement of the year, and for correcting the calendar now in use'. It was presented to Parliament on 25 February 1751 by Lord Chesterfield and passed its second reading on 18 March. It finally received royal assent to become law on 22 May.

Dr Bradley, the Astronomer Royal, prepared tables for the Government. The new act ordained that 1751, which had begun on 25 March (old style), should end on the last day of December and that the next day (1 January) should be the first day of 1752; it also stipulated that the day following 2 September 1752 should be 14 September, and arranged for the appropriate changes in the reckoning of leap years and for the use of the Gregorian method of calculating the date of Easter. Thus, the English legal year of 1751 had 282 days, while 1752 had 355.

The act also specified the correct date on which various annual events (either those that recur after an interval of a full year or those that are tied to specific dates) should be observed. Among the former kind were fairs, the opening and closing of common land, the payment of debts and interest, the day a person reached his majority, and the start of the financial year; among the latter, the election of officials and ecclesiastical feasts and fasts.

The English financial year used to start on Lady Day—the first day of the year, 25 March; it did so in 1752. The next financial year began 365 days later in 1753 on 5 April. It continued to start on this date until 1800 when it was moved one day on to 6 April. It has stayed there ever since.

Christmas Day 1752 was observed 355 days after Christmas Day 1751; this presented a source of anxiety to some. In certain places Christmas was celebrated on 5 January in 1752—365 days after the 1751 festival. It is said that in several parts of England the common people observed Christmas on 5 January

FIG. 19.6 Philip Stanhope, the Earl of Chesterfield (1694–1773). He guided the bill
to reform the English calendar through Parliament. Photo: AKG London.

until as late as 1889. This should not be confused with any celebration of the
Feast of Epiphany on 6 January—the day following Twelfth Night.

   In some parts of England it was held that the moment when Christmas
Day began the cattle went down on their knees in their stalls; it was said that
they knelt on 5 January. On the other hand, the famous Glastonbury thorn,

FIG. 19.7 'Humours of an election entertainment' published by Hogarth in February 1755. The banner on the floor carries the inscription 'Give us back our eleven days'. Source: Ancient Art & Architecture Collection.

which was believed to flower on Christmas Day, was reported by the Vicar of Glastonbury to have done so on 25 December, despite newspaper reports to the contrary. Similar uncertainties were absent in Roman Catholic countries in 1582; the blood of St Januarius (martyred at Pozzuoli, near Naples, in 305) acknowledged papal infallibility and liquified in its vial at Naples in 1583 on its usual date of 19 September.

To many the act was unpopular; a common cry was 'Give us back our eleven days'. The matter even became an issue in the 1754 election. Hogarth satirized the sentiment in an engraving in 1755 entitled 'Humours of an election entertainment'. There were said to have been riots in Bristol but recent attempts to find reports of these in contemporary newspapers have drawn a blank. It was, of course, perfectly reasonable to be upset by the loss of 11 days of your life if you believed that your days were numbered and it was preordained that you would die on a certain calendar date. People who had been destined to die on one of the missing eleven days might have been expected to have been overjoyed, but, of course, they did not know the allotted day. Others, however,

would question the validity of that belief. It is perhaps odd that nobody
objected to losing 1 January to 24 March, 1751—but maybe that was seen in a
different light.

Apart from these problems the act created many real difficulties, not all of
which could have been anticipated. Many of these problems were left to the
people concerned to sort out as best they could, but in Chester, in particular,
they were acute. There, a fair, which lasted for several days, had been held from
time immemorial on Michaelmas Day, 29 September; the citizens would
indulge in all manner of jollity, feasting, and drinking. They would, however,
have had ample time to return to a suitable state of sobriety by the Friday after
St Denis' Day (9 October) when they elected the mayor and other civic offi-
cers. The 1751 act was clear: the fair would be delayed 11 days and held on
10 October in 1752, but the mayor would continue to be elected on the Friday,
which was 13 October in that year. This was felt to be a very difficult situation;
there was a grave danger that the mayor would be elected by a citizenry still
dazed from its revels. An act was petitioned and duly granted (26 Geo II *c.* 34);
the election could be postponed in 1752, and thereafter, till the Friday after 20
October.

The Gregorian calendar was later taken to the four corners of the globe on
the back of the British Empire. It is now all but universally used.

CHAPTER 20

The French Republican Calendar

▓ The Revolution and the calendar

An aim of the Revolution in France of 1789 was to produce a brand new calendar; instead, it succeeded in resurrecting the oldest calendar of all. This was introduced not to correct some astronomical discrepancy but to mark a break with the past. The Church—the guardian of the calendar with its feasts and fasts and saints' days—was one of the sources of power of the hated ancient order; it had to be done away with.

On 6 November 1792, the two deputies, Louis Pierre Manuel and Antoine Joseph Garsas, asked the Committee of Public Instruction to reform the calendar. Other advocates included François de Neufchâtel, who suggested a competition and a prize for the best suggestion. Pierre Sylvain Maréchal (1750–1803) had previously published a reformed calendar based on that of ancient Egypt and in which the days, instead of being dedicated to the various Catholic saints and martyrs, were dedicated to various benefactors of mankind (including himself, on his birthday, 15 August). He now submitted an amended version to the Committee.

The Committee of Public Instruction set up a subcommittee under the chairmanship of the mathematician, Charles Gilbert Romme (1750–95). Perhaps the most eminent of its members were Louis Lagrange (1736–1813) and Gaspard Monge (1746–1818), who were members of the Academy of Science and on the committee of the body that was designing the new metric system. Lagrange, who had made important contributions to the theory of planetary dynamics, was one of the most noted mathematicians and astronomers of his time, if not of all time. The committee also included Antoine François Fourcroy and Guyton de Morveau (both chemists), Joseph Jérôme le François de Lelande (astronomer), Joseph Lakanal (1762–1845; educationalist), and Marie-Joseph de Chénier (1764–1811) and Philip François Nazaire Fabre d'Eglantine (1750–94; both poets and dramatists). The poet Fabre bore the title 'd'Eglantine' to commemorate his winning of the award of the Golden Eglantine at a festival, the Floral Games (Jeux Floraux), at Toulouse. Maréchal himself, who had been in prison before the Revolution for

FIG. 20.1 Pierre Sylvain Maréchal (1750–1803).
© Photothéque des Musées de la Ville de Paris.

publishing his reformed calendar, was not on the Committee; he had become dismayed by the Terror of the Revolution and now had a wise policy of keeping clear of politics.

Romme reported back to the Committee of Public Instruction on 17–18 September 1793, and on the 20th he presented his proposals to the National Convention, where they were accepted in principle. Eventually, on 5 October 1793, a decree was promulgated by the Convention; this set forth the new calendar.

The new era was deemed to have started on 22 September 1792. This date also marked the foundation of the Republic and also, by coincidence or not, the autumnal equinox. Much was made of this by proclaiming it as the day when all Frenchmen became equal, and the days and nights were also equal, and so forth.

Each year was to begin at the midnight that initiated the day on which the true autumnal equinox was observed at the Paris Observatory. There were to be, in each year, 12 months of 30 days each. A common year of 365 days was made up by five 'jours complémentaires', originally called the 'sanculotides',

FIG. 20.2 Philip François Nazaire Fabre d'Eglantine (1750–94), author of the names of the months and the days of the year in the Republican calendar.

and a leap year was to have an extra intercalary day, a sixth sanculotide, placed after the other five. Leap years were supposed to occur normally every four years and the four-year period was to be known as a 'franciade'. The leap year was described as a sextile year and the first one was year III.

This scheme, like Maréchal's original proposal, was almost identical to the

Alexandrian version of the ancient Egyptian calendar, currently used by the Coptic and Ethiopian Churches and described in Chapter 11. No doubt the eminent scholars on Romme's Committee knew this, but did the Revolutionaries? The curious name, sanculotide, is based on a contemptuous phrase used by the old aristocrats to refer to the common men who wore trousers instead of the aristocratic breeches or 'culottes'; in this way the 'sans-culottes' became heroes of the Revolution. (Note that modern French dictionaries spell the word 'sansculottide'; I have retained the original spelling.)

The months were divided into three equal decades of 10 days each, and the day was to be divided into 10 decimal hours; each hour was subdivided into 100 decimal minutes and each decimal minute into 100 decimal seconds. There were thus 1000 decimal minutes and 100 000 decimal seconds in the day, rather than 1440 old minutes or 86 400 old seconds. Watchmakers were instructed to manufacture decimal clocks and watches.

FIG. 20.3 A decimal watch (1795) manufactured in Condom. Separate hands show the time according to the 10-hour republican day and the 24-hour traditional day. Photograph Courtesy of The Time Museum, Rockford, Illinois.

The Committee decided, after some argument and false starts, that each month and day should be given a name, and they asked Fabre d'Eglantine to devise these. This he succeeded in doing by 25 October. The full details of the new calendar, including the new names, were made public in a decree dated 24 November. The names of the months in order were Vendémiaire, Brumaire, Frimaire, Nivose, Pluviose, Ventose, Germinal, Floréal, Prairial, Messidor, Thermidor, Fructidor.

Fabre may have been influenced in his invention of these names by the old names in use in Germany. The names were supposed to reflect the character of the month. This they do for France, but they are hardly appropriate for universal use in Canada or equatorial Africa, let alone south of the equator. Carlyle, in his history of the French Revolution, written 35 years later, offered his own translation Vintagearious, Fogarious, Frostarious, Snowous, Rainous, Windous, Buddal, Floweral, Meadowal, Reapidor, Heatidor, Fruitidor. The names also invited caricature and in England there appeared Wheezy, Sneezy, Freezy, Slippy, Drippy, Nippy, Showery, Flowery, Bowery, Wheaty, Heaty, Sweety.

The ten days of each decade were prosaically called Primedi, Duodi, Tridi, Quartidi, Quintidi, Sextidi, Septidi, Octidi, Nonidi, Decadi. (The Republican decade is discussed at greater length in Chapter 21.)

Fabre did better with the 366 days of the year. The five sanculotides were named. Fête du Vertu, Fête du Génie, Fête du Travail, Fête de l'Opinion, and Fête des Récompenses.

The sixth, the leap day, was called Fête de la Révolution. The sanculotides were to be public holidays dedicated to celebrating virtue, genius, work, opinion, and reward. Games would be held on the Fête de la Révolution and citizens would come from all parts of France.

The names of the other 360 days defined the *Almanac of shepherds* and are set out in Appendix III. It is not recorded whether anybody succeeded in memorizing them all. A calendar published in Avignon for the third year of the Revolution seems to have been recalled from an imperfect memory; both 'Orange' and 'Olive' occur twice, and the items are sometimes different or in a different order.

It is recorded that some ardent Revolutionaries named their children from the *Almanac of shepherds*, instead of by the name of the saint of their day of birth. It is thus possible that some baby was appropriately called Pissenlit (26th Ventose); but pity the child born on 5th Frimaire (Cochon). Taken as a whole, the names represent a snapshot of French rural life at the end of the eighteenth century. The calendar did not allow for the Industrial Revolution. There was no room left for a day of la voiture, le métro, le sandwich, let alone la bombe-atomique and similar twentieth-century delights; several of the words themselves have become archaic and are not found in modern dictionaries.

Alas for poor Fabre d'Eglantine, he ran foul of the Terror, was tried on a trumped-up charge, and went to the guillotine on 5 April 1794. On his way to the 'iron maiden' he distributed manuscript copies of his poems to the crowd.

After Napoleon had fought and conquered most of Europe, the names were translated into Italian, German, and Dutch, so that the Republican calendar could be used throughout the Empire. The translations of the months are given in TABLE 20.1.

FIG. 20.4 A five-franc piece struck by Napoleon in 13 ER.
© The British Museum.

TABLE 20.1 *The names of the Republican months translated into other languages*

French	Italian	German	Dutch
Vendémiaire	Venemmiaio	Herbstmonat	Wijnoogstmaand
Brumaire	Brumaio	Nebelmonat	Mistmaand
Frimaire	Frimaio	Kältemonat	Rijpmaand
Nivose	Nevoso	Schnemonat	Sneeuwmaand
Pluvoise	Piovoso	Regenmonat	Regenmaand
Ventose	Ventoso	Windmonat	Windmaand
Germinal	Germinale	Sprossenmonat	Kiemmaand
Floréal	Fiorile	Blütemonat	Bloemmaand
Prairial	Pratile	Wiesenmonat	Grasmaand
Messidor	Messidoro	Erntemonat	Oogstmaand
Thermidor	Termidoro	Hissemonat	Hittemaand
Fructidor	Fruttidoro	Obstmonat	Vruchtmaand

⌧ The start of the year

According to article III of the decree of 4 Frimaire, the original intention was to begin the year at the midnight which preceded the instant of the autumnal equinox. It was pointed out, shortly after the publication of the decree, by the astronomer Jean Baptiste Joseph Delambre (1749–1822), that this policy would have two unwelcome results. First, the intervals between two sextile years would sometimes be five years rather than the usual four; this would happen irregularly three times every century and it would first occur in year XX (1811) when the leap year expected in year XIX (four years after the previous one in year XV) would be delayed a year. Secondly, from time to time, the equinox would fall very close to midnight so that a small error of timing might cause an intercalary day to be inadvertently inserted or omitted. This first would happen, according to Delambre, in the year CXLIV.

Delambre brought the matter to the attention of Gilbert Romme. He in turn informed the Committee of Public Instruction, which asked him to head a subcommittee to formulate proposals for the perfection of the calendar (rather than refer to the dangerous possibility of correcting errors!). Romme duly proposed a decree which would annul article III about the start of the Republican year and the true equinox. In its place it was suggested to institute intercalary days in a manner very similar to that used in the Gregorian calendar; the only difference was an additional provision to suppress leap years every 4000 years. This provision, it was thought, would bring the average length of the year even closer to the mean tropical year. As these matters were discussed, one member of the Committee was heard to remark: 'You see now we make decrees for eternity'; 4000 years hence was not far short of eternity.

Unfortunately, shortly after this, Romme was seen to give support to the wrong political faction. He was arrested and condemned to die; rather than face the guillotine, he stabbed himself to death. The proposal was never considered; the calendar was never perfected.

⌧ The decline of the calendar

Despite its revolutionary symbolism, the new calendar turned out to be unpopular and inconvenient. The French workman objected to working a 10-day week; most people objected to working on Sundays; the decimal hours caused intolerable confusion; business with the rest of Europe (still adhering to the Gregorian calendar) became difficult.

As revolutionary fervour ebbed away, the provisions of the calendar were gradually repealed. The first things to go (in 1795) were the decimal division of the day and the sanculotides (which were renamed 'jours supplementaires'). In 1799, after Napoleon Bonaparte had became First Consul, he repealed the imposition of the Republican calendar on Rome; he did this as part of his agreement with the Pope. In 1805, after he had been crowned emperor, he restored France to Catholicism and prepared to restore the old calendar.

In 1805, Regnault de Saint-Jean d'Angely (lawyer and old friend of Napoleon) and Jean Joseph Mounier (a statesman) criticized the calendar and proposed to the senate that it restore the old calendar. A senate commission, headed by the mathematician Pierre Simon Laplace, took up this suggestion. In September 1805, Sunday was restored as the day of rest, and from 1 January 1806, the Gregorian calendar was reinstated.

The Republican calendar may be seen to have had a symbolic rather than a practical value; this was confirmed in 1871 when it was revived for a brief period by the Communards in Paris. That was the last that was heard of it—though several decimal watches still exist.

CHAPTER 21

The Week

⧗ Introduction

You and I may have to refer to a calendar to find out today's date, but a person who does not know what is the day of the week is sometimes thought to be on the verge of mental disintegration. However, in the absence of newspapers, the radio, and other cultural clues, keeping track of the day of the week is difficult—as Robinson Crusoe discovered. In fact, his bout of fever caused him to miss a day from his tally, as he realized on his eventual rescue: Man Friday should indeed have been named Man Saturday.

Our week of seven days may well be among the oldest surviving human institutions, with a history of continuous observation over 3000 years or more; the seven-day cycle, it would seem, has continued mostly without interruption for most of recorded time. There have been, however, two interruptions. Alaska used to be part of the Russian Empire; known as Russian America, it was considered to be even further east than Siberia; its local time was thus in advance of that of Eastern Russia. In 1867, Alaska was purchased by the USA (a shrewd investment; it cost them $7.2 million). It immediately became further west than California and this necessitated winding the calendar back a day. The inhabitants thus suffered a one-off eight-day week. This was not the least of their problems; Russia was still observing the Julian calendar at the time so that 12 days were omitted from October in 1867. I would imagine that the Eskimos and Aleuts who formed the bulk of the population were not too put out by this, but it is not recorded whether there were any Jewish people there at the time. The reverse happened in the Philippines, which had originally been colonized from the east and America, rather than from Europe and the west. In 1884, people there found themselves on the wrong side of the new international date-line, and had to endure a six-day week.

Most settled societies organized their affairs so that meetings, markets, religious rites, and days of rest could take place at regular intervals. There are different categories of 'week' which we may name according to their primary function: the market week, the religious week, the working week, the astrological week.

If you wish to meet at regular intervals, one possibility is to meet at every full moon and enjoy the convenience of the light it provides; another is to meet when the new moon first becomes visible. Such arrangements are fine provided the weather is good and the moon is visible, but without some knowledge of the number of days between one full or new moon and another, there is no way of predicting in advance when the next meeting is due; unfortunately, this period is variable. Another possibility, which many societies adopted, is to meet regularly after a fixed number of days. If you can count, there is a reasonable prospect of getting to the meeting on time. Such a period is the week; most societies had one, but the number of days varied from society to society—5, 7, 8, 9, 10 were some of the more common numbers of days in their weeks. The seven-day week was observed by the Babylonians, Jews, and several nations in West Africa.

A seven-day week is the norm in Europe and in many European languages the word for 'week' is either the word for 'seven' or a word derived from it; the English speaking world has been known to use the adjective 'hebdomadal'.

There is a Hebrew word for 'week'—'shavu'a'; this is etymologically connected with the number seven, and the number seven is also connected to the word 'sabbath' itself. This accounts for the curious extension of the word 'sabbath' in relation to 'sabbatical years'. During these, university teachers and other academics may be relieved of their formal duties for one year in seven so that they may travel, study, or skulk.

ⅩⅡ The seven-day week of the Jews

The word 'sabbath' in Hebrew is often used interchangeably to mean either the seven-day week or the day of rest; it is probably of Babylonian origin. The Babylonians were interested in both astrology and astronomy; perhaps because there were seven known planets, the number seven had great religious and cosmological importance in Babylonian art and ritual. The Assyrians considered certain days of the lunar cycle—the 7th, 14th, 19th, 21st, and 28th—to be unlucky and cooking, travelling, and other activities were forbidden on these days. Similar beliefs were held by the Persians, and even by the Buddhists further to the east. Such practices may have been widespread and shared by the early ancestors of the Jewish people. The Jewish month to this day is scheduled to start at a new moon.

It is nevertheless necessary to distinguish between these practices, tied to the phases of the moon, and a regular and strict seven-day cycle—though the latter could be, in origin, a formalized version of the former. Certainly, a regular seven-day period is more convenient and easier to use than a variable one of seven or eight days. Inevitably it leads to a more regular life. It has been suggested that the dissociation of the week from the lunar cycle was a step in the transition from primitive moon worship to monotheism. Thus, the two great Jewish contributions to Western civilization—monotheism and the week—may be inherently linked.

The Jewish seven-day period is mentioned in the story of the Creation in the Book of Genesis (Genesis 2: 2–3):

> *And on the seventh day, God ended his work which he had made; and he rested on the seventh day from all his work which he had made.*
>
> *And God blessed the seventh day, and sanctified it: because that in it he had rested from all his work which God created and made.*

It is also referred to in the Decalogue in the Fourth Commandment (Exodus 20: 8–11) and again in Deuteronomy (5: 12–15). In Exodus we read:

> *Remember the Sabbath day to keep it holy.*
>
> *Six days shalt thou labour and do all thy work.*
>
> *But the seventh day is the Sabbath of the Lord thy God: in it thou shalt not do any work, thou, nor thy son, nor thy daughter, thy manservant, nor thy maidservant, nor thy cattle, nor thy stranger that is within thy gates:*
>
> *For in six days the Lord made Heaven and Earth, the sea, and all that in them is, and rested the seventh day: whereby the Lord blessed the Sabbath day and hallowed it.*

These books—Genesis, Exodus, and Deuteronomy—form part of the Pentateuch and are traditionally attributed to Moses. If this attribution is correct, they date back, via an oral tradition, to the flight from Egypt, which may have taken place in the thirteenth century BC, or even before. They were, most likely, first written down in around the ninth century BC. Certainly by the eighth century BC, the tradition was well established.

In 597 BC, Nebuchadnezzar II (605–562 BC), King of the Chaldean dynasty of Babylon, captured Jerusalem (the Jews had unwisely changed sides in the war between Babylon and Egypt) and many Jewish people were transported to Babylon in Mesopotamia (now Iraq). Jerusalem was later sacked in 587 BC and more Jews were transported in that year, and again later in about 568 BC. These events are mentioned both in surviving cuneiform documents and in the Bible, in the Second Book of Kings. The Jewish people remained in captivity in Babylon for about 60 years (70 was predicted by the prophet Jeremiah) and there they learnt much of Babylonian culture including their calendar and, conceivably, the seven-day week.

It is possible that the stories told in Genesis had been altered during the Captivity and were only later written down in their present form. In the last analysis, it is unclear whether the Jews invented the week all by themselves or acquired at least the idea from the Babylonians. There is a suggestion that the Babylonians ended their week on a Friday, as the Moslems do today, and that after their return from captivity in Babylon, the Jews moved the end of their week to Saturday—because, it is said, of their hatred of all things Babylonian. Be this as it may, the Jews hold their Sabbath on this day according to the Fourth Commandment.

In classical and even medieval times, there were occasional attempts to link the observation of the Jewish Sabbath to Saturn worship, since Saturn was the regent of Saturday in the astrological week (as discussed later). The only thing to be said in favour of this theory (and it is not much) is that Saturn has a reputation as a bringer of bad luck and it might have been circumspect to refrain from important activities on his day (that is, to rest). The idea will not seem strange if you believe Friday, Satan's birthday, is an unlucky day. In passing, I note that the likely origin of the unlucky nature of Friday is that it was on this day that Christ was crucified; Friday the 13th is deemed to be particularly unlucky.

One point is of interest—by the time of Christ the observance of the Jewish Sabbath as a day of rest had become prevalent over much of the Eastern Mediterranean. This is attested to by several Classical authors—Horace, Ovid, Seneca, Juvenal, and others. It may perhaps be the result of the general attractiveness of a day of rest in a life of otherwise continuous toil; on the other hand, it might stem from the unlucky character of Saturn's day and its identification with the Sabbath.

♊ The astrological week

At about the time of the birth of Christ, an astrological week of seven days was becoming popular in the Roman world—though it was not recognized officially till much later. In this week, each of the seven days was allocated to one of the seven planets, according to a well-defined scheme. The origin of all this is obscure. The idea behind these so-called 'chronocratories' was that periods of time, hours in this case, were ruled by the corresponding planets.

To peoples with an astrological bent, seven is, as I have said, an important (if not lucky) number. Seven moving celestial objects were known to the ancients long before the telescope: the sun, the moon, and the five planets (Mercury, Venus, Mars, Jupiter, and Saturn)—all visible to the naked eye. They were referred to as 'wandering stars' or 'planets' (from a Greek word meaning wanderer), and I shall follow suit.

The Greeks, and later the Romans, called the astrologers of Babylon, 'Chaldeans'. These certainly believed that the planets exerted an influence on human affairs. However, there is no evidence that they associated the planets with a cycle of seven days, even though they apparently recognized a variable week of seven or eight days.

Definite knowledge of this astrological week comes from the writings of the Roman historian, Dion Cassius, (AD 150–235) and there are many references in surviving classical literature which suggest that such knowledge was well known among the populace. Dion tells us that it originated not long before, in Egypt. Plutarch (AD 46–120), in his *Moralia*, included an essay entitled 'Why are the days named after the planets reckoned in a different order from the actual order?'. Alas, the text is lost, and only the title survives.

The astrological scheme depends on a unique ordering of the planets. Fashions changed but perhaps the most popular order was that of their assumed distance from the earth. This order happens to coincide with that of their orbital periods—though the ancients had no conception of these. The ordering and the orbital periods are as follows:

Saturn	29 years
Jupiter	12 years
Mars	687 days
Sun	365 days
Venus	224 days
Mercury	88 days
Moon	29 days

Saturn takes years to complete his orbit (the Sanskrit name for Saturn means 'slow moving'; its metal is lead); Mercury darts about rapidly (like quicksilver?); and the moon traverses the whole sphere every month.

Following Dion Cassius, first consider the 24 hours of the first day: assign the first to Saturn (the slowest planet), the second to Jupiter, and so on, assigning the seventh hour to the moon (the fastest). Repeat this several times, assigning the eighth hour to Saturn, the ninth to Jupiter, and so on. The 24th hour is thus assigned to Mars. Continue through the second day, assigning the next planet, the sun, to its first hour and the 24th hour to Mercury. Continue in this way through the remaining days. The 24th hour of the seventh day is thus assigned to the moon, so that the first hour of the eighth day is back to Saturn and the whole cycle of assignments is repeated for the second week, and so on, for ever. These two interlocking cycles of seven planets and 24 hours repeat after 168 (24 × 7) hours or seven days. The full set of assignments is shown in TABLE 21.1.

This process of assigning in succession the planets to the 168 hours of a seven-day week is mentioned in many later astrological works. The planet assigned to the first hour of a day presides over the whole day and is termed the regent of the day. You can see from TABLE 21.1 that the first day of the cycle is the day of Saturn, the next the day of the sun, and so on. The planets, arranged in the order of their assignment to the first hour of successive days (according to the theory of chronocratories) are as follows:

1	Saturn
2	Sun
3	Moon
4	Mars
5	Mercury
6	Jupiter
7	Venus

Indirect evidence for knowledge of the astrological week is provided by finding the planets, or the gods representing these planets, listed in this particular order. There are thus hints that the astrological week was known in the first century BC (for example, in the writings of the Roman poet Tibullus (54–19 BC)), but none survive from earlier than that. The first certain evidence comes from two wall inscriptions and a fresco which were unearthed in Pompeii;

FIG. 21.1 Roman graffiti from Pompeii. This shows, on the left, abbreviations for the names of the planets in the order of the days of the week. The inscriptions suggest that they were part of a school lesson on the Roman calendar.

these show the gods in the astrological order. They must date back at the latest to AD 79, when Vesuvius erupted and buried the city.

A familiarity with three elements is required to create the astrological week: a day divided into 24 hours, the seven planets, and a predilection for astrology and the doctrine of chronocratories. These were all present in Egypt after the excursions of Alexander the Great into Western Asia in about 327 BC. It is thus likely, though direct evidence is lacking, that Dion Cassius was right, and the astrological week was invented by the Greeks in Egypt sometime after the death of Alexander in 323 BC. It is interesting that the word 'horoscope' itself betrays an interest in hours.

TABLE 21.1 *The assignment of planets to the hours of the day*

Hour	Day 1 (Saturday)	Day 2 (Sunday)	Day 3 (Monday)	Day 4 (Tuesday)	Day 5 (Wedensday)	Day 6 (Thursday)	Day 7 (Friday)
1	Saturn	Sun	Moon	Mars	Mercury	Jupiter	Venus
2	Jupiter	Venus	Saturn	Sun	Moon	Mars	Mercury
3	Mars	Mercury	Jupiter	Venus	Saturn	Sun	Moon
4	Sun	Moon	Mars	Mercury	Jupiter	Venus	Saturn
5	Venus	Saturn	Sun	Moon	Mars	Mercury	Jupiter
6	Mercury	Jupiter	Venus	Saturn	Sun	Moon	Mars
7	Moon	Mars	Mercury	Jupiter	Venus	Saturn	Sun
8	Saturn	Sun	Moon	Mars	Mercury	Jupiter	Venus
9	Jupiter	Venus	Saturn	Sun	Moon	Mars	Mercury
10	Mars	Mercury	Jupiter	Venus	Saturn	Sun	Moon
11	Sun	Moon	Mars	Mercury	Jupiter	Venus	Saturn
12	Venus	Saturn	Sun	Moon	Mars	Mercury	Jupiter
13	Mercury	Jupiter	Venus	Saturn	Sun	Moon	Mars
14	Moon	Mars	Mercury	Jupiter	Venus	Saturn	Sun
15	Saturn	Sun	Moon	Mars	Mercury	Jupiter	Venus
16	Jupiter	Venus	Saturn	Sun	Moon	Mars	Mercury
17	Mars	Mercury	Jupiter	Venus	Saturn	Sun	Moon
18	Sun	Moon	Mars	Mercury	Jupiter	Venus	Saturn
19	Venus	Saturn	Sun	Moon	Mars	Mercury	Jupiter
20	Mercury	Jupiter	Venus	Saturn	Sun	Moon	Mars
21	Moon	Mars	Mercury	Jupiter	Venus	Saturn	Sun
22	Saturn	Sun	Moon	Mars	Mercury	Jupiter	Venus
23	Jupiter	Venus	Saturn	Sun	Moon	Mars	Mercury
24	Mars	Mercury	Jupiter	Venus	Saturn	Sun	Moon

The regent of the day is the planet assigned to the first hour (6 a.m.). The day takes its name from the regent.

Most of us would recognize several potent influences in human life: political power and order, sex and procreation, food and agriculture, trade and communications, force and war. It is not difficult to see how these came to be associated with gods, but it was the Babylonians, given their interest in astrology, who first associated them with the planets. This association passed to the Greeks, thence to the Romans, and later to the tribes of northern Europe and elsewhere. It was easy to connect the local god associated with each influence, with the corresponding planet, and hence with the day of the week. The associations are shown in detail in TABLE 21.2 (see page 280).

The hour lore lived on until the Middle Ages and was known to Chaucer (as in 'The knight's tale'). For all I know, modern astrologers may still use it. Which hour is which depends on conventions concerning the hour at which the day starts; in Roman and medieval times this was sunrise, for astrological purposes.

An interesting question arises as to how the first day of the astrological week, Saturday, was selected to be the Jewish Sabbath day. It has been suggested that it was invented by astrologers conversant with the Old Testament and Jewish tradition. Certainly the Jews, from the first century, came to call the planet Saturn by the name Shabtai—an old name for the Sabbath. Others have confused Saturn worship with the Sabbath.

In Genesis, it is said that the stars and planets were created on the fourth day. This began, according to Jewish reckoning, at sunset on the first Tuesday. Before this moment, the planets just did not exist and could not therefore exert any influence on human affairs or anything else; it is an appropriate place to start the rigmarole of assigning planets to the hours as described by Dion Cassius. Accordingly, it is suggested, the hour starting at 6 p.m. on what we now call Tuesday was assigned to Saturn, and the process continued from there. We then find that 6 a.m. on Wednesday is assigned to Mercury. According to the Roman tradition this is the hour assigned to the regent of the day. All else follows.

This is either a coincidence or it points to a Jewish origin of the astrological week—or at least to its invention by people conversant with the Creation as described in Genesis. There are several Hebrew documents dating from the first few centuries of the Christian era which mention or describe the astrological week; nevertheless, there is no mention of it in the Bible.

In the first years of the Christian era, the astrological week spread through the Roman Empire. In some cases, Christianity followed it; it may be that its attraction was enhanced by the popularity of Mithraism, with its worship of the sun. In other cases, Christianity and the week—with Sunday and Saturday already changed to Dies Dominica and Dies Sabati—travelled together. By the first century AD the astrological week had reached India, where the names of the seven days in Sanskrit follow the names of the same seven planets. There is a story related by Philostratus in his biography of the philosopher and vegetarian, Apollonius of Tyana (born about 4 BC), that the latter, on a visit to India, was given seven rings, marked with the names of the seven planets, by a Brahmin. Philostratus goes on to say that Apollonius wore them 'according to the names of the days'. Apollonius is said to have visited India in the middle of the first century.

From India, the week diffused to Tibet, Burma, Nepal, Thailand, and Ceylon; by the Song dynasty, at the end of the first millennium, it had reached China. More recently, it has spread throughout the world in the wake of the great European Christian empires. Likewise, the seven-day week, adopted by Islam, spread to other places of the world on the back of that religion. In more recent times, virtually every country in the world has come to use it, if only for commercial reasons.

In the Eastern Church, represented by the Greek and Slavic countries, the astrological week either never took hold or was abolished. The days of the week are mostly referred to by number or position in the week, except for the Lord's day and the Sabbath.

⚅ The Sabbath and the Lord's day

The early Christians took over the Old Testament and the seven-day week from the Jews. They set aside the first day of the week, Sunday—the day of the Resurrection—as a day of prayer, while those of Jewish origin continued to observe the Jewish Sabbath (though St Paul did not support this). Sunday became known as 'The Lord's day' and Saturday remained the Sabbath.

Another explanation of the assignment of Sunday to the Lord has been suggested. In the early days of the Christian era, there were two new religions contending for the souls of Roman citizens: Christianity and Mithraism. Mithraism derives from the ancient Persian cult of Zoroastrianism, which today survives in the beliefs of the Parsees of India. Mitra was a sun god and member of the Zoroastrian pantheon and a central feature of Mithraism was the worship of the sun. Possibly the establishment of the astrological week, in the minds of the devotees, was helped by its ready-made provision of a day of the sun, Sunday. But eventually, Christianity won, and maybe it was convenient to worship the new god on the same day as the old—Sunday.

Later, after Christianity had become established, the need to observe the Fourth Commandment gave rise to confusion and controversy, since the Sabbath, or day of rest, was the Saturday (the seventh day) and the Lord's day was the Sunday. Only in AD 321 did an edict of the Emperor Constantine establish Sunday, the Lord's day (Dies Dominica), as a day of both rest and prayer—at any rate for the city dwellers. Constantine also abolished the old nundinae and decreed that markets should be held on the first day of the seven-day week, that is, Sunday. Later, in AD 400, the day of rest was extended to country dwellers. Sunday, as a day of rest, was later incorporated into the Justinian code in AD 789.

Sunday is now, in a great part of the Christian world, a day of rest and recreation, in which all members of a family can be together. For many years the Lord's Day Observance Society and the trade unions have sought to preserve this in England against the onslaughts of modern commercialism. There are signs that they are losing the battle.

In the non-Christian world, the day of prayer and rest is not Sunday but Friday or Saturday:

Friday (The day of assembly)	Islam
Saturday (The Sabbath)	Judaism
Sunday (The Lord's day)	Christianity

It has been suggested that different days were selected by the prophets, first of Christianity and then of Islam, to mark and perpetuate the difference between their faith and that of their rivals. The Seventh Day Adventists have reverted to the earlier practice of observing the Sabbath on the seventh day, taking this to be the Saturday—perhaps for the same sort of reason. In Jerusalem, where Christianity, Islam, and Judaism uneasily meet, those who prefer talk to bombs have difficulty in arranging meetings between interested parties: only four days in the week are free of religious affiliations.

An interesting point concerns modern conventions about the day which starts the week. In Britain and America, most people would now say that the week begins with Monday—although this is not the traditional view. A British Standard Specification, rigorously followed by the compilers of diaries, tells us that the week begins on Mondays; this makes Sunday (our rest day) the seventh day, enabling us to abide by the Fourth Commandment. On the other hand, printed calendars and almanacs invariably have Sunday as the first day—following the old tradition.

⌛ Other weeks

I shall now enlarge the definition of the term 'week' to include periods of time with a fixed and integral number of days greater than one day and less than a month. These, like the seven-day week, are necessarily conventional, since there is no natural period with which they may be identified.

It is not surprising that many societies had weeks consisting of periods other than seven days. The Romans had a market week known as the 'internundinum tempus', which they inherited from the earlier Etruscans; 'internundinum tempus' translates as the 'period between the ninth'. (You may remember the Roman custom of inclusive numbering—they would say that Saturday was the eighth day after the previous Saturday.) The nundinae, or market days, recurred every eight days. They also had a system of dividing their months into irregular periods marked by the kalendae (first day), the nonae (fifth or seventh), and the idus (13th or 15th). These divisions are understood to have their origin in the phases of the moon. Later, at the beginning of the Christian era, the official nundinae were replaced by the popular astrological week—which has persisted ever since.

Market weeks allow an interesting variation. Several co-operating villages may form a ring; each village holds a market at regular intervals. The timing of this market, for each member of the ring, is staggered, so that on each day of the market week there is a market at one or other of the villages. Market weeks of 4, 8, or 16 days—with and without such rings—are popular in Africa.

The ancient Egyptians had 10-day periods and each one was associated with a constellation called a decan. The start of each period was marked by the heliacal rising of its decan. The French Revolutionary calendar copied this system and had 10-day periods they called decades. The Chinese, with a predilection for decimal arithmetic, also recognized a 10-day period, but they also had a 12-day cycle; these two cycles resonated with a period of 60 days. The days of both periods are named; the days of the 10-day period are named after the 10 heavenly stems and those of the 12-day period, after the 12 earthly branches (which are the names of animals).

In the New World, people in Central America adopted a 13-day period (associated with the 13 lords of the heavens), a nine-day period (referring to the nine lords of the night), and a 20-day period (which derives from their vigesimal (20) counting system); the Maya also believed in chronocracy. The Baha'i religion observes a 19-day period; 19 is an important number in this faith. In the 20-day period of the Mayans and the 19-day period of the Baha'i week, the days are named; the Mayan and Baha'i calendars also share the feature of not observing lunar months.

One of the more elaborate systems of 'weeks' is the Waku system of the Indonesians in Java, developed in the tenth century; it consists of no fewer than nine separate cycles: the two-day duwiwara, the three-day triwara, the four-day tjaturwara, the five-day pantjawara, the six-day sadwara, the seven-day saptwara, the eight-day astawara, the nine-day sangawara, and, finally, the 10-day dasawara. The days in each of these cycles are named and the simultaneous presence of these nine temporal rhythms gives ample scope for divination—the purpose of the system; the sequence of days in these inter-locking cycles repeats in 1260 days. The seven-day cycle derives from the astrological week which arrived in Java from India at that time; eight is sacred to Hinduism and is related to the points of the compass; six is associated with a sacred geometrical figure with six sides.

⚅ The Revolutionary week

In more recent times, the protagonists of the two great anticlerical revolu-tions—the French Revolution of 1792 and the Russian Revolution of 1917—abolished the seven-day week, which they saw as a reprehensible Christian institution, and replaced it with another. This is perhaps understandable in that, among other things, revolutions tend to be directed against the estab-lished religion—one of the perceived sources of power. Both attempts eventu-ally failed.

The French Revolutionaries abolished the traditional week despite a prophetic remark by Bishop Henri Grégoire: 'Sunday existed before you, and it will survive you'. The National Convention abolished the week in a decree dated 24 November 1793. In its place they set up a 10-day week in which the days were referred to, prosaically, by number; the days of rest, it was decreed, should occur at 10-day intervals on the day named 'Decadi'. Although the new period of 10 days was in harmony with the longer-lasting reform of weights and measures and the introduction of the metric system, there is no doubting the anticlerical intentions of its prime architect, Gilbert Romme; when the latter was asked about the purpose of the new calendar, his reply was 'to abolish Sunday'. Popular cartoons of the day portray the struggle between M. Sunday and Citizen Decadi.

During the reign of Terror which followed, churches were forbidden to open and shopkeepers forbidden to close their shop on Sunday. In 1794, Robe-spierre came to power; he had romantic ideas and instituted a new religious cult which included 'Decadic festivals' to be observed every 10 days. The festi-vals were to be dedicated to patriotism, filial piety, and so forth, and patriotic hymns such as 'L'hymne des Marseillais' had to be sung at these occasions. They were inaugurated on 8 June 1794 with a festival dedicated to the 'supreme being' and held in the garden of the Tuileries; Robespierre himself officiated. Alas, on 27 July of the same year, Robespierre fell from power and was exe-cuted. The reign of Terror ended, and the festivals were ignored.

Three years later, on 4 September 1797, after an attempted coup d'état, the Directory proceeded to reaffirm the festivals and Robespierre's cult. A decree, introduced by Merlin de Douai on 3 April 1798, made the observance of the decades compulsory and, from 1798, heavy fines and imprisonment were imposed on recalcitrant shopkeepers, priests, and others who dared to open their shops on Decadis or go to church on Sundays. It was also forbidden to date documents in the old manner. On 30 August 1798, a decree reintroducing Robespierre's festivals required that births, deaths, and divorces should be announced at them and magistrates should deliver sermons. Offices, schools, shops, and tribunals were all required to close on Decadis and on Quintidi afternoons; the people thus had one and a half days off in ten (15 per cent of their lives to rest in), whereas under the old system they had had one in seven days off (14.3 per cent); they were slightly better off under the new system.

Nevertheless, the French people, unlike Robinson Crusoe, never lost track of the old seven-day week. People would go through a civil marriage service on Decadi and a Christian one on Sunday; some took both Decadi and Sunday off from work. In the country districts particularly, the decrees abolishing Sunday could not be implemented.

Eventually, revolutionary zeal began to wane and the festivals had all but disappeared by June of 1799 when Merlin de Douai resigned from the Directory. The Revolutionary calendar and the decades lingered on until after Napoleon's coup d'état of November 1799, when Laplace annulled the decree of April 1798. A decree of 26 July 1800 restored to the French people the right to worship or open a shop whenever they wanted. Finally, the Gregorian calendar and the Sunday day of rest were reinstituted by a decree dated 9 September 1805.

The next attempt to abolish the Christian week came shortly after the Russian Revolution in 1917, when use of the Julian calendar had been abolished, and the Gregorian calendar instituted. There were attempts to replace the seven-day week by one of 10 days, following the model of the French Revolution. These came to nothing at the time.

It was only later, in May of 1929, at the fifth Congress of the Soviets of the Union, that it was proposed to alter the existing work week in order to utilize expensive machinery in the factories more effectively and so raise productivity. The proposer, Larin, managed to interest Stalin in the project and on 26 August of the same year, the Council of People's Commissaries announced that the old work week would be abolished and be replaced by a continuous production week known as 'nepreryvka'. A decree of 24 September modified the proposal by replacing the seven-day cycle with a five-day one. The new arrangements were to come into effect on 1 October.

An increase in productivity of about 60 parts in 300 (20 per cent) was optimistically anticipated. The idea was that one fifth of the workforce would work for four days and have the fifth day off; the next fifth would start their four-day stint on the second day; the third fifth would start on the third day; and so forth. In this way everybody would have one day off in five, but not all on the same day; four-fifths of the workforce would always be on duty to man the machinery.

The five days of this week were to retain their original names (which in Russian carried no religious connotation); the other two, Saturday and Sunday, were abolished. Emigré bourgeois cartoons showed Saturday and Sunday being shot.

The workers were issued with coloured slips of paper to tell them which day to take off: yellow, orange, red, purple, and green (presumably white would have been too counter-revolutionary). The days of the five-day week came to be named by their colour. Two people might arrange to meet a nepreryvka next purple day—at least they might if they were both 'purple day' people.

It is obvious that the reform had two components: the introduction of a staggered work week and the replacement of the seven-day cycle by a five-day cycle. The motive for introducing the second feature was to 'combat the religious spirit' by abolishing Saturday and Sunday.

Once again, however, revolutionary enthusiasm inhibited detailed consideration of the consequences of the reforms and how the people might respond to them. There were two main disadvantages to nepreryvka. First, it had a profound, disruptive, and unwelcome effect on family and social life; members of the same family, friends, or marriage partners, who had been given different colours, were rarely able to spend time together. Although the Soviet workers got more time off than their Western counterparts, they complained 'What is there for us to do at home if our wives are in the factory, our children at school, and nobody can visit us'. Maybe this was deliberate; Lenin's widow, in good Marxist fashion, regarded Sunday family reunions as a good enough reason to abolish that day. Nevertheless, the authorities bowed to popular demand and, on 16 March 1930, allowed requests from families to synchronize their days off. A curious by-product was that general workers' meetings, considered necessary by the Marxists, were effectively prevented. The second problem with nepreryvka was in maintaining a continuity in the administration and responsibility for the machinery; in effect, responsibility was so divided that no-one took it. Stalin himself recognized this problem in a speech bemoaning the lack of responsibility in Soviet society.

Although it was hard to admit that nepreryvka was a mistake it was abolished by decree on 23 November 1931. The staggered work week was dropped, but the seven-day week was not restored. Antireligious opinion still prevailed and a six-day week, the 'chestidnevki' was instituted, starting on 1 December 1931; every sixth day was a common day of rest. Even this reform still left Russian society divided. The rural people, as in France a century earlier, continued to use the seven-day week clandestinely, while the urban population abided by the chestidnevki. This became an administrative nightmare and on 26 June 1940, in the interests of industrial production, the chestidnevki was abolished, the seven-day week (with less time off) restored, and Sunday reinstated as a day of rest after an absence of 11 years.

Even with the decay of religious sentiment, which has occurred over the last 30 years or more, it is likely that any future attempts to abolish the seven-day week will similarly founder—so many aspects of our lives are built around the week and the habit of regulating our affairs according to a seven-day cycle. It would be scarcely more difficult to force us all to speak Esperanto than to alter the number of days in the week. Although a week of seven days is a convention, and there is little reason to believe that it is in any way rooted in our biorhythms, it is too embedded in the lives of all to be changed.

The names of the days of the week

A comparison of the names of the days of the week in a range of different languages is not only interesting in itself, but also sheds light on their history. The names in some 69 languages, arranged in most cases according to their linguistic grouping, are shown in Appendix II. The selection is a tiny fraction of the several thousand languages currently in use, let alone the extinct ones. Many more, particularly from Africa and South-East Asia, could have been added.

The names used are derived in various ways and have different meanings within the languages. Important distinctions may be made between those that are derived from the names of the planets (planetary names); those derived from the names of gods (deistic names); those that merely give their numerical position in the week (ordinal names); and names describing some conventional feature of the day (descriptive names).

The languages of Western Europe and India use planetary or deistic names, whereas those of Eastern Europe and the Middle East are ordinal. Thus, the Slavic languages and Greek, of Eastern Christianity, are predominantly ordinal; Portuguese is an exception, and for some unexplained reason is ordinal.

A further subdivision separates the planetary names into two types. First, there are those that are phonetic copies from the Latin (for instance, 'de Lun' in Cornish or 'Lundi' in French from the Latin 'dies Lunae'); secondly, there are those that have substituted the local name of a planet (Somavare in Sanskrit or Zamigmar in Tibetan for the moon's day). In the North, the Germanic and Scandinavian languages retain the day of the sun and of the moon, but translate them into the local idiom, while the planetary names of the other days are replaced by the names of local gods equivalent to the Roman ones as indicated in TABLE 21.3. Icelandic is interesting because the old deistic Scandinavian names were mostly replaced, by a reforming Bishop in 1121, by ordinal names.

To the Romans, the planets were closely identified with a planetary god; Northern people had their own gods who in some cases patronized the same activities as a Roman god. Thus, Freya or Frigg was the goddess of love, as was the Roman Venus; it was thus appropriate to identify Venus with Frigg, and Dies Veneris became Friday. Likewise, Mars and Tiw both supported war, so Dies Martis became Tuesday; and both Thor and Jupiter had control of thunder and lightening, so that Dies Iovis became Thursday or, more literally, Donnerstag in German. Both Woden or Odin and Mercury were responsible for conducting the souls of fallen warriors to their new dwellings and led to Dies Mercurii being called Wednesday. Apparently, there was no deity equivalent to Saturn, though some nineteenth-century scholars supposed there to be a Norse god, Saeterne.

In more recent times a similar process has taken place in the Caribbean. Slaves taken from Benin in West Africa and transported to Haiti took with them their spirit cults. These spirits in turn became identified with Catholic saints and even associated with the days of the week. Among these is the notorious Baron Samedi, head of the spirits of the dead and identified with Saint Expedit.

Some of the African people had indigenous seven-day weeks. The names of the days in their languages (such as Ewe or Yoruba) tend to be descriptive.

TABLE 21.2 *Planets and gods and their equivalents*

Influence	Planet	Babylonian	Greek	Roman	Teutonic
War	Mars	Nergal	Ares	Mars	Tiw, Tyr
Trade*	Mercury	Nebo		Mercurius	Woden, Odin
Power†	Jupiter	Marduk		Iupiter	Thunar, Thor
Fecundity	Venus	Ishtar	Aphrodite	Venus	Freya, Frigg
Agriculture	Saturn	Ninib	Kronos	Saturnus	

* Although Mercurius was the patron of merchants, Woden (or Odin) played a more bloodthirsty role, requiring human sacrifices for the promise of victory in battle, and was the chief god in the Teutonic pantheon. The Romans identified Woden with Mercurius.

† Iupiter was the chief god in the Roman pantheon, and Woden/Odin in the Teutonic pantheon.

Some ordinal sets start the week on Monday and call Monday, day 1; others start with Sunday and call Monday, day 2; and yet others start with Saturday, so that Monday is day 3. The Eastern Christians start with Monday, whereas in the Middle East and the Semitic languages, the week starts with Saturday. (Note that the Arabic name for Saturday derives from seven rather than Sabbath, and that the counting starts with Saturday.) In many languages spoken where Islam is practised, most days are ordinal, but the name of Friday (Juma), the holy day, means 'The day of assembly'. Swahili is the odd man out; the names of both Wednesday and Thursday are related to the number five; the Swahili week starts on Saturday, the day after the Islamic holy day, so that Wednesday is the fifth day. The name for Thursday is believed to derive from an older, pre-Islamic week which began, like the Jewish week, with Sunday.

Planetary weekday names in early Christian countries are frequently modified by replacing the day of Saturn (dies Saturn) by the Sabbath, and the day of the sun (dies sol) by the day of the Lord. The Greeks call Sunday the 'Lord's day' and Friday, following a Jewish custom, the 'day of preparation'. The name

FIG. 21.2 Baron Samedi, head spirit of the Gédé and counterpart of St Expedit.
Painting by Edward Duval Carrié.

of Saturday in the Eastern Church is the Sabbath; there are two names for Sunday in the Slavic languages—the first, used in Russian, means 'Resurrection', and the second, used in other Slavic languages means 'rest'. Monday in the Slavic languages means 'day after the rest day'.

The names in ordinal, planetary, and deistic weeks may be modified in other ways. Wednesday frequently becomes 'the middle of the week' particularly in the Scandinavian and Germanic languages, as in modern Icelandic, German, and Russian. Friday is sometimes called 'the day of preparation'.

A curious feature of the Scandinavian languages is the use of 'bath day' for Saturday. There is a report by John of Wallingford (d. 1214) that Ethelred overcame and massacred the Danes in 1002 by attacking on St Bride's day, 13 November, which, he says, was their bath day. Alas for the story, 13 November 1002 was a Friday or a Thursday—the ambiguity arises from uncertainty concerning the day John supposed the year to have started on.

The Celtic names are of interest and, whilst closely similar, do fall into two groups. In Welsh, Cornish, and Breton, the names are taken phonetically from the Latin with no attempt to translate them into the local language. This might suggest that the astrological week arrived before Christianity and the Lord's Day had become established in Rome. On the other hand, the name for Sunday in Irish and Gaelic is a variant, of 'dies Dominica', suggesting that the week arrived later (after the rise of Christianity) to these more distant Celtic lands. Other Gaelic and Irish names for the days refer to fast days.

Sunday in Turkish is market day; Turkish and Kurdish are similar—both are mainly ordinal, as is Luganda (which took Sunday from the British as the Welsh took it from the Romans). Several of the Hungarian names are phonetic copies of the Slavic names.

The Yoruba names used in Nigeria are neither planetary nor ordinal but, in order, starting with Sunday, mean: immortality, profit, victory, confusion, creation, failure, three meetings. In the related Ewe language, spoken in Ghana, the names mean: work begins, second work day, muddle, lucky, unlucky, feast of the god Amiy, rest day—a sort of epitome of human existence. These two weeks are indigenous and owe nothing to either the Jewish or astrological week.

Euskera—the language spoken by the Basque communities in the Pyrenees and the Alps—is of uncertain origin; there are several dialects in which the names of the days of the week differ considerably.

In Sanskrit there were two names for each day, as shown in TABLE 21.3; one or the other is still used in a differing form throughout India. The names are derived from the names of the planetary gods or their epithets and probably

TABLE 21.3 *The names of the days in Sanskrit*

Day	Sanskrit names	Planetary god	Meaning
Sunday	Revivara	Sun	
Monday	Somavara		Cool, moist, soma[*]
Tuesday	Mangalavara	Mars	Happiness, bliss
Wednesday	Saumayavara	Mercury	Auspicious, gentle
Thursday	Brhaspativara	Jupiter	Great master[†]
Friday	Sukravara	Venus	Bright, resplendent[§]
Saturday	Sanivara	Saturn	Slow moving
Sunday	Adityavasara		Of sun
Monday	Induvara	Moon[*]	
Tuesday	Bhaumavasara		Of Mangala
Wednesday	Budhuvasara		Awakening
Thursday	Guruvara		Teacher
Friday			
Saturday	Mandavara		

[*] Soma was a yellow beverage extracted from an unidentified plant. It was drunk by the Brahmins and associated with the moon god, Indra.

[†] Brhaspati is connected with the god Brahma, who is associated with the Hittite god of thunder (as was Iuppiter in the Roman pantheon).

[§] Venus is the bright morning star.

date from the first century AD. The attributes of the Indian planetary gods and the corresponding Roman planetary gods are different, with the exception of Iuppiter (Jupiter) and Barhaspati, who are both responsible for thunder. Some of the names used in Ceylon and Burma are also derived from the Sanskrit; the Tibetan names are direct translations of the names of the planets. The Chinese names, recognized throughout China, are ordinal but mean 'The day of the first planet', and so on. (Planet is used in the Greek sense.)

The Romany names are unexpectedly ordinal. The gypsies are said to have come from India and their language is closely allied to Sanskrit. Either they left India before the arrival of the planetary week or, more probably, were influenced by Eastern Christianity. It is known that they spent some time in Asia Minor before wandering further to arrive in Western Europe in the tenth century. Their word, 'Koóroki', for Sunday, is phonetically the same as the Greek word 'Kiriaki', as is the Armenian word 'Giragi'.

⧗ Epilogue

> *Solomon Grundy*
> *Born on a Monday*
> *Christened on Tuesday*
> *Married on Wednesday*
> *Took ill on Thursday*
> *Worse on Friday*
> *Died on Saturday*
> *Buried on Sunday*
> *This is the end*
> *Of Solomon Grundy*
>
> TRADITIONAL

. . . and of Part II.

PART III

CALENDAR CONVERSIONS

CHAPTER 22

Calendar Conversions

⌘ Introduction

The problem of converting dates from one calendar to another has often been solved by using tables. This is still the only method if the calendar is so ill-defined that recourse must be made to contemporary records, as is the case for strictly empirical calendars in which the start of a month or a year, or the decision of whether an intercalation is to be made, is determined by observation of the heavens, the seasons, or nature itself.

Some astronomical calendars (see Chapter 6), such as the Chinese and many Indian calendars, depend on a detailed theory of the behaviour of heavenly bodies, and some, like the Chinese calendar, have suffered innumerable modifications. These theories and modifications are usually not available, and it is beyond the scope of this book to provide algorithms for conversions of these: recourse must again be made to conversion tables.

Nevertheless, there are sufficiently precise, unambiguous, and stable rules for a number of arithmetic calendars for the conversion of these to be done using only arithmetic procedures. This obviates the need for tables. It is these calendars and their conversions that I consider in Part III of this book.

In this chapter I describe algorithms or methods for converting the date of a day expressed in one arithmetic calendar to the date expressed in another. If we are interested in n calendars, there are $n(n + 1)$ possible conversions to consider, and if n is greater than a few, a large number of routines would be required. This tiresome difficulty can be circumvented by first converting the date in calendar A into a day number, J, and then converting J into a date in calendar B; in this way only $2n$ routines are required. This limits us to calendars for which an exact conversion is possible, and which are defined by stable and rigid rules—but there are plenty of these. It is useful to categorize them as regular or irregular: regular calendars can be converted using a generalized algorithm; irregular calendars, such as the Jewish calendar, are more complicated and require special treatment.

For the purposes of this section, I shall assume that a calendar date or day is specified by the day of the month, D, the month number, M, and the year of the era, Y—written as D/M/Y (using the European convention). The names of the month are of no relevance.

The calendars considered count their years from some fixed point which defines their era. Such calendars have an epoch which is the first day of the first month of the first year of the era, the date expressed as 1/1/1. Days before the epoch can still be specified in the calendar, but they are generally termed proleptic dates.

Different conventions exist for numbering the years of proleptic dates. Christendom has for many centuries referred to proleptic dates before 1 January AD 1 as dates 'before Christ' or 'BC'. In this convention, the year before AD 1 is 1 BC; there is no year 0. This is inconvenient for many purposes and an alternative convention used by astronomers is to refer to 1 BC as the year 0, 2 BC as the year −1, or generally n BC as the year $1 - n$. Zero or negative years are thus BC and positive years are AD. I shall generally use this convention here for all calendars.

Day numbers

The notion of assigning a unique number to each day was originally invented by Joseph Scaliger (though some say he took it from the Greeks of Constantinople, as mentioned in Chapter 6); it is universally used by astronomers. Julian day numbers, J, are positive integers which are assigned in sequence to consecutive days and which so define each day unambiguously.

It is necessary to specify the epoch of the Julian day numbers—that is to say, the day at which the counting starts; I will call this 'day zero'. Day zero is 1 January of the year −4712 (1 January 4713 BC) in the Julian calendar, and its Julian day number is 0. Days before this have negative Julian day numbers, so that 31 December 4714 BC has a day number of −1, and this is preceded by day −2, and so forth.

The start of the day

I mentioned in Chapter 3 that caution needs to be exercised when a day in one calendar is to be identified with a day in another if the times at which the day is reckoned to start are different. Some start their day at dusk and some at dawn; some (like ourselves) opt for midnight; while astronomers choose noon. FIG. 22.1 indicates the conventional identification of days which start at different times. You can see that a day that starts at dusk begins some six hours before the corresponding day which starts at midnight; the day which starts at dawn, some six hours after; and the Julian day, which starts at noon, some 12 hours after.

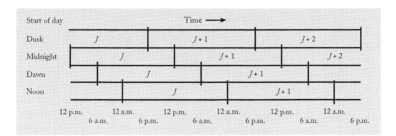

<small>FIG. 22.1</small> The identification of days.

Intercalation cycles

Many calendars employ intercalation and insert an extra day—a leap day—from time to time. A crucial idea is that of an 'intercalation cycle'. This defines a sequence of dates that is repeated endlessly; each sequence is identical to every other, and each cycle contains the same number of years and of days, and zero or more leap days. All common years in an intercalation cycle contain the same number of days, but an extra leap day (if any) is intercalated in leap years—and always at exactly the same point.

Some calendars, such as the Egyptian, do not have leap years and every year is identical to every other; in this case the intercalation cycle contains just one year of constant length. The Julian calendar, with a leap year every four years, has an intercalation cycle of four years.

Regular calendar

A regular calendar must contain intercalation cycles, but there need not be any leap days. A further requirement is that it must be possible to calculate the month from the 'day of the year', and vice versa. This can always be done in one way or another, but is particularly easy in regular calendars. If we assign 1 to the first day of the year, 2 to the second, and so on, the 'day of the year' is the number of the day under consideration.

A list of some regular calendars is given in TABLE 22.1, which also gives the epoch of each calendar as a Julian day number and as a date in the Julian calendar.

Not all calendars can be considered to be regular; notable irregular calendars are the Jewish calendar, the Roman calendar, and the Mayan calendar. Special algorithms must be devised to convert these. Regnal calendars and calendars like the Christian ecclesiastical calendar defy conversion in a simple manner. Some calendars, such as the Gregorian, Republican, and Saka calendars need special treatment if they are to be considered regular.

TABLE 22.1 *Regular calendars*

Number	Calendar name	Julian date	Epoch	Day number	Era
Egyptian calendars					
1	Egyptian	26-02-0747	BC	1448 638	Nabonassar
2	Armenian	11-07-0552	AD	1922 868	Armenian
3	Khwarizmian	21-06-0632	AD	1952 068	Yazdegerd
4	Persian	16-06-0632	AD	1952 063	Yazdegerd
Alexandrian calendars					
5	Ethiopic	29-08-0008	AD	1724 221	Incarnation
6	Coptic	29-08-0284	AD	1825 030	Diocletian
7	Republican	11-09-1792	AD	2375 840	Republican
Julian calendars					
8	Macedonian	01-09-0312	BC	1607 709	Alexander
9	Syrian	01-10-0312	BC	1607 739	Alexander
10	Julian Roman	01-01-0001	AD	1721 424	Christian
11	Gregorian	03-01-0001	AD	1721 426	Christian
Islamic calendars					
12	Islamic A	15-07-0622	AD	1948 439	Hegira
13	Islamic B	16-07-0622	AD	1948 440	Hegira
Modern calendars					
14	Bahá'í	09-03-1844	AD	2394 647	Bahá'í
15	Saka	24-03-0079	AD	1749 995	Saka

1. The calendars based on a common system are arranged in groups.

2. The two versions of the Islamic calendar are both in use: A begins on a Thursday, and B on the next Friday.

3. The Gregorian, Republican, and Saka calendars are not, strictly speaking, regular calendars, but they can be forced into the mould with minimal difficulty.

🔢 The Algorithms

The word algorithm itself is derived from the name of Abu Jafar Muhammad ibn Musa al-Khwarizmi (as noted in Chapter 5) and an algorithm can be looked upon as a recipe for calculating something of interest.

The algorithms for converting dates in this section are presented in a form suitable for use with a pocket calculator or, more expeditiously, on a personal computer using a language such as BASIC, C, or FORTRAN.

Algorithms which use decimal numbers are best avoided when working with computers or calculators since they may not record with sufficient accuracy; for example, when the result of a calculation should be exactly 2, some computers might record it as 1.999 . . . ; if we ignore the decimal, we are wrongly left with 1 rather than 2. As a bonus, integer arithmetic is performed faster than decimal arithmetic in many computer systems.

An algorithm consists of a set of instructions written in the imperative mood. Since these look like equations (which are in the indicative mood), we distinguish them by writing them in the form: $A \Leftarrow B$ (which is a command to store or denote by A, the value denoted by B). The necessity for this seemingly pedantic point is highlighted by the instruction $X \Leftarrow X + 1$: this tells us to increase the value of X by 1; it is not true in general that $X = X + 1$.

Some of the algorithms presented could be improved by algebraic simplification; they are presented as they are for the sake of clarity. Some are compressed into a form which uses fewer multiplications and divisions; I make no claim that they are in their most efficient form.

The algorithms described in Part III have all been implemented in FORTRAN 77 and tested by converting every day number in 1000 years (starting at the epoch of each calendar) into the corresponding calendar date, and then back to the day number. Invariably, the same day number was obtained as was started with; this test cannot detect errors of the same nature present in both algorithms. Accordingly, the algorithms have been subjected to numerous spot checks against equivalents taken from published tables, or, when possible, by comparing the results for an extended period with dates obtained using other programs believed to deliver correct conversions.

CHAPTER 23

Mathematical Notes

⧗ Conventions

The descriptions of the algorithms presented in this section are unavoidably mathematical in character. Nevertheless, I have aimed to keep the algebra and arithmetic as simple as possible. All the numbers involved are whole numbers or integers (with one exception in algorithm H); these may be combined in various ways to yield further integers by addition, subtraction, multiplication, division, and remaindering. The evaluation of the equations requires some comment. First you should evaluate expressions within brackets starting with the innermost bracket. To avoid an excessive number of these, a rule of precedence (familiar to computer programmers) has been used: first evaluate multiplications and divisions; only when this has been done should you evaluate additions and subtractions. Multiplication is usually implied in the usual way by writing the symbols for two or more quantities next to one another (but note that in deference to tradition I have used SOL and LUN to stand for single quantities). Invariably in the algorithms, and sometimes elsewhere, multiplication is denoted by $*$.

Division denoted by /, is invariably integral division; this is slightly different but simpler than the division we are used to. So, A/B denotes the result of dividing the dividend, A, by the divisor, B, and discarding any remainder; this quotient is thus an integer. Some writers on calendrical matters have used $(A/B)_w$ to denote the same thing.

If we divide A by B, there is often a remainder. This is denoted by $\text{MOD}(A,B)$; mathematicians would write instead 'A modulo B'; and other writers have used $(A/B)_r$—the MOD function is available in many computer languages. This remainder lies in the range 0 to $B-1$. If A is exactly divisible by B, the remainder $\text{MOD}(A,B)$ is zero. We always assume that A and B are integers ($A \geq 0$ & $B > 0$). Some examples are, in order:

$$3 + 5 = 8 \qquad 6/3 \quad\;\; = 2$$
$$5 - 3 = 2 \qquad \text{MOD}(5,3) = 2$$
$$5/3 \;\; = 1 \qquad \text{MOD}(3,5) = 3$$
$$3/5 \;\; = 0 \qquad \text{MOD}(6,3) = 0$$

Sometimes it is necessary to limit a quantity to some maximum value. We use the MIN function to denote this so that MIN(A,B) represents the smaller of A and B; this function is available in many computer languages.

Quotients and remainders

I have mentioned integral division and remaindering; the results of these two operations are related:

$$\text{MOD}(A,B) = A - (A/B)*B \tag{23.1}$$

or $\quad A = (A/B)*B + \text{MOD}(A,B) \tag{23.2}$

or $\quad A/B = (A - \text{MOD}(A,B))/B \tag{23.3}$

We may add any positive multiple, n, of B to A, in the MOD function, without changing the result, so that:

$$\text{MOD}(A + n*B, B) = \text{MOD}(A,B) \tag{23.4}$$

This is sometimes useful as A might be negative; we may add some multiple of B to A so as to make $A + n*B$ positive without altering the result. Likewise we find:

$$(A + n*B)/B = n + A/B \tag{23.5}$$

as long as A, B, and n are positive.

It is also useful to note that MOD$(P + Q, B)$ where $Q = $ MOD$(R + S, B)$ is equivalent to MOD$(P + R + S, B)$ and that MOD$(A + C, B)$ is equivalent to MOD(MOD(A,B) + MOD$(C,B), B)$.

A useful feature of the MOD function is that if A is exactly divisible by B, then there is no remainder when A is divided by B and MOD$(A,B) = 0$. In particular, MOD$(A,2) = 0$ if A is even and $= 1$ if A is odd.

Finally, we may note that:

$$0 \leq \text{MOD}(A,B) \leq B - 1$$

or $\quad 1 \leq 1 + \text{MOD}(A,B) \leq B$

where '$\leq$' means 'less than or equal to'. It is convenient to say that the 'range' of MOD(A,B) is 0 to $B - 1$.

Further uses of the MOD function and integer division

A useful technique involves the use of cyclic permutations; one example suffices. The months of the year are usually numbered from January (1) to December (12). Sometimes it is useful to assign a new number, M', to each month of number M; so March (month (M) = 3) is assigned the number M' = 0, April (M = 4) is given M' = 1, May is given 2, and so forth. December is thus given 9 and the numbering continues into the new year with January and February getting 10 and 11. This has the advantage that any leap day (29 February) is now the last day in the new sequence. This amounts to renumbering the months from 0 to 11 instead of 1 to 12, and then arranging them in a new cyclic permutation. The numbers M and M' are related in a simple manner:

$$M' = \text{MOD}(M + 9, 12) \tag{23.6}$$

$$M = \text{MOD}(M' + 1, 12) + 2 \tag{23.7}$$

Sometimes we may want to add B to A, but only if some other quantity, C, is greater than or equal to D. This can all be done using a succinct formula based upon the notion that C/D = 0 if C is less than D and C/D = 1 if C is greater or equal to D but less than $2D$. We thus find that:

$$A + B*(C/D) = A \qquad \text{if} \quad 0 \leq C < D \tag{23.8}$$

$$A + B*(C/D) = A + B \quad \text{if} \quad D \leq C < 2D \tag{23.9}$$

It is relatively easy to devise a formula in which B is added to A only if it falls in a certain range of values.

⚅ The number of days in the months

The number of days in the months of our Gregorian calendar would seem to be an arbitrary sequence of 30, 31, with 28 or 29 thrown in for February. It is unexpected that a simple, albeit somewhat ponderous, formula can deliver the number of days in month M.

First rearrange the months so that January and February follow December, as already explained. I list the months in TABLE 23.1, together with the number of days, L, in each and the 'excess number of days', X, which is the number of days beyond 28 (that is, $L - 28$).

Note that when we start with March, the sequence of Xs forms a repeated pattern—3,2,3,2,3. The third repetition stops uncompleted at February, the last of the renumbered months. The pattern itself consists of 3,2 repeated three times with the third repetition truncated.

It is thus clear that, for the first five renumbered months, X = 3 for months for which M' is even and X = 2 if M' is odd. Thus, we may write for these:

$$X = 3 - \text{MOD}(M', 2) \tag{23.10}$$

This holds for the other months as well if we replace M' by $\text{MOD}(M', 5)$. We then obtain, after adding in the 28 which was subtracted in calculating X:

TABLE 23.1 *Days in the months*

	Jan	Feb	Mar	Apr	May	Jun	Jul	Aug	Sep	Oct	Nov	Dec	Jan	Feb
M	1	2	3	4	5	6	7	8	9	10	11	12	1	2
M'	10	11	0	1	2	3	4	5	6	7	8	9	10	11
L	31	28	31	30	31	30	31	31	30	31	30	31	31	28
X	3	0	3	2	3	2	3	3	2	3	2	3	3	0
Z	—	—	0	3	5	8	10	13	16	18	21	23	26	29

Key

M Conventional month number

M' Rearranged month number

L Number of days in month

X Excess $= L - 28$

Z Total excess days by first of month starting from 1 March

$$L = 31 - \mathrm{MOD}(\mathrm{MOD}(M',5),2) \tag{23.11}$$

or in terms of the conventional month number, M:

$$L = 31 - \mathrm{MOD}(\mathrm{MOD}(\mathrm{MOD}(M + 9,12),5),2) \tag{23.12}$$

which is valid for all months except February. This ponderous formula must be evaluated from the inside out, starting with $\mathrm{MOD}(M+9,12)$.

This result has its uses, but perhaps of greater interest is the day of the year, T, of the day defined by D and M. The day of the year is clearly equal to D plus the sum of the days in all the preceding months. If the year were to begin on 1 March, T would thus be given by:

$$T = 28M' + Z + D \tag{23.13}$$

where Z is the sum of the Xs of the preceding months as shown in TABLE 23.1. We may note that the excess, X, for February (which we know is wrong) is never included in Z.

A simple formula for Z was discovered by Christian Zeller (1824–99), a director of a women's teacher training college and a school inspector in Germany. He delivered a short account in Latin at a meeting of the Mathematical Society of France on 16 March 1883, and it is reprinted in the journal of that society. Zeller discovered that:

$$Z = (13M' + 2)/5 \tag{23.14}$$

so that:

$$T = 28M' + (13M' + 2)/5 + D$$
$$= (153M' + 2)/5 + D$$
$$= (2 + 153\text{MOD}(M + 9,12))/5 + D \qquad (23.15)$$

However, we have tacitly assumed that the year starts on 1 March. We need to adjust this formula for T if we take the year to start properly on 1 January.

If the month is January or February ($M = 1$ or 2 with $M' = 10$ or 11) we must subtract the number of days from 1 March to 31 December inclusive (306). For the other months, we must add the number of days from 1 January to the last day of February inclusive; this is 59 or 60 in a leap year. This results in the formula:

$$T = (153M' + 2)/5 + 59 - 365(M'/10) + P + D \qquad (23.16)$$

This stratagem depends on the fact that $M'/10 = 1$ for January and February, but is otherwise zero. P must be set equal to 1 in a leap year if the month is not January or February; otherwise set it to zero. Those adept at simple algebra might care to compress this expression further.

The lengths of the months in other calendars often follow a much simpler pattern. The months of the Islamic calendar alternate between 29 and 30; the first six months of the Saka calendar contain 31 days (except in a leap year) and the remaining months contain 30. In Chapter 24 I present similar formulae for these calendars.

⌛ Congruence relations

Zellor's congruence is an example of a type of expression that is very useful in calendar calculations. Many such expressions were used by Gauss in his work in this field. He did not, however, describe how they might be discovered. Burnaby went some way to remedy this omission and the following discussion is based on his work.

The problem is to find a formula which takes values from 1 to N as n increases from 1 to B, but not in a regular way. Here n is an integer variable. As an example, consider the number of common years, C, which have occurred by year n of a Metonic cycle as employed by the Jewish calendar. We may tabulate values of C for all 19 values of n in TABLE 23.2.

The problem is to derive a formula for C in terms of n. If the embolismic years were equally spaced, we might suppose that the relation required would be $C = (12n)/19$—but this formula is plainly wrong. We therefore try the relation $C = (12n + x)/19$ and the problem then is to determine x. It must be emphasized that it is not necessarily true that any value of x can be found or that, if there is a value, that it is unique.

Burnaby proceeds as follows. First, we may see from TABLE 23.2 that when $n = 1$, $C = 1$, and that $12n + x = 12 + x$. Now $(12 + x)/19$ only gives the right value of 1 for C when $12 + x$ lies between 19 and 37. This gives an upper and lower limit to $12 + x$, and hence to x: $7 \leq x \leq 25$. Likewise, when $n = 2$, $C = 2$, and the limits are $14 \leq x \leq 32$. In this way we can construct tables of the minimum and maximum values of x for each value of n as shown in TABLE 23.2. Any value of x which delivers the correct value of C for all n must, simultaneously, lie in all 19 of these ranges. That can only happen if the largest minimum is less than or equal to the smallest maximum. The largest minimum is seen to be 17 and the smallest maximum is also 17, so 17 is a valid value of x which gives a correct relation:

$$C = \frac{(12n + 17)}{19} \tag{23.17}$$

In general we would try first a relation of the form:

$$C = \frac{(An + x)}{B} \tag{23.18}$$

and note that for each value of n, with a corresponding value of C:

$$BC \leq An + x \leq BC + B - 1 \tag{23.19}$$

TABLE 23.2 *Embolismic years in the Metonic cycle*

n	1	2	3	4	5	6	7	8	9	10	11	12	13	14	15	16	17	18	19
C	1	2	2̇	3	4	4̇	5	5̇	6	7	7̇	8	9	9̇	10	11	11̇	12	12̇
min. x	7	14	2	9	16	4	11	−1	6	13	1	8	15	3	10	17	5	12	0
max. x	25	32	20	27	34	22	29	17	24	31	19	26	33	21	28	25	23	30	18

Key	
n	Year number in the Metonic cycle
.	Embolismic year
C	Aggregate number of common years by year n
min. x	$(19 * C) - (12 * n)$ (See page 297.)
max. x	$19 * (C + 1) - (12 * n)$ (See page 297.)

so that:

$$BC - An \leq x \leq BC + B - An - 1 \qquad (23.20)$$

It may occur that no value of x can be found which satisfies all of the relationships; in this case one could try an expression of the form:

$$C = (AnK + x)/(BK) \qquad (23.21)$$

using several values of K in turn. If this fails one might try more complicated expressions involving n.

A particularly simple congruence is the formula $G = 3C/4$, which transforms the number, C, of a century in the Gregorian calendar into the difference between the day numbers of a date in the Julian and Gregorian calendars; this difference is valid throughout the century. The relation between C and G is illustrated below:

$$C = 1, 2, 3, 4, 5, 6, 7, 8, 9, \ldots$$
$$G = 0, 1, 2, 3, 3, 4, 5, 6, 6, \ldots$$

Note how G skips an increment every fourth century. I discuss the use of this formula in a later chapter.

CHAPTER 24

To Calculate the Day of the Week

⊞ Introduction

There are several ways in which you can work out the day of the week of any date in either the Julian or Gregorian calendar. These are generally arithmetic and require us to specify the day of the week by a number. We shall use the traditional numbering as shown in TABLE 24.1 (although other numberings have been used from time to time).

Many methods result in a number, W, which may be much larger than 7; it is then necessary to subtract multiples of 7 till this is no longer so. This can be expressed succinctly by the formula $1 + \text{MOD}(W - 1,7)$.

The traditional method involves the use of tables which are printed in the Book of Common Prayer of the Anglican Church (amongst other places). I shall start however with a method described by Lewis Carroll.

⊞ Lewis Carroll's method

Lewis Carroll, author of *Alice in Wonderland*, was, in real life, Charles Dodgson, a don at Oxford and a mathematician and logician. He described his method in a letter to *Nature* in 1887.

To use Lewis Carroll's method, follow the directions given, which are presented exactly as he wrote them, punctuation and all. Carroll claimed he could do the calculation in 20 seconds:

> *Take the date in four portions, viz. the number of centuries, the number of years over, the month, the day of the month.*
>
> *Compute the following 4 items, adding each, when found, to the total of the previous items. When an item or a total exceeds 7, divide by 7, and keep the remainder only.*
>
> *The Century-Item.—For Old Style (which ended September 2, 1752) subtract from 18. For New Style (which began September 14) divide by 4, take overplus from 3, multiply remainder by 2.*

TABLE 24.1 *The numbers and letters of the days of the week*

Day	Number	Letter
Sunday	1	A
Monday	2	B
Tuesday	3	C
Wednesday	4	D
Thursday	5	E
Friday	6	F
Saturday	7	G

The Year-Item.—*Add together the number of dozens, the overplus, and the number of 4's in the overplus.*

The Month-Item.—*If it begins or ends with a vowel, subtract the number denoting its place in the year, from 10. This, plus its number of days, gives the item for the following month. The item for January is '0'; for February or March (the 3rd month), '3'; for December (the 12th month), '12.'*

The Day-Item is the day of the month.

The total, thus reached, must be corrected, by deducting '1' (first adding 7, if the total be '0'), if the date be January or February in a Leap Year: remembering that every year, divisible by 4, is a Leap Year, excepting only century-years, in New Style, when the number of centuries is not so divisible (e.g. 1800).

EXAMPLES
1783, September 18

17, divided by 4 leaves '1' over; 1 from 3 gives '2'; twice 2 is '4.'

83 is six dozen and 11, giving 17; plus 2 gives 19, i.e. (dividing by 7) '5.' total 9, i.e. '2.'

The Item for August is '8 from 10,' i.e. '2'; so, for September, it is '2 plus 3,' i.e. '5.' Total 7, i.e. '0' which goes out.

18 gives '4.' Answer, 'Thursday.'

1676, February 23

16 from 18 gives '2.'

76 is 6 dozen and 4, giving 10; plus 1 gives 11, i.e. '4.' Total '6.'

The item for February is '3.' Total 9, i.e. '2.'

23 gives '2.' Total '4.'

Correction for Leap Year gives '3.' Answer, 'Wednesday.'

The reader is urged to try this out; 'overplus' is used to signify a remainder.

⌘ Traditional methods

The traditional methods employ the notions of calendar letter and dominical number and make extensive use of tables. I shall discuss these briefly, and then go on to develop algebraic methods that may be readily used with computers.

Begin by considering the calendar letter of the day in question. Imagine that the seven letters, A to G, are allocated sequentially to every day of a common year: A is allocated to 1 January, B to 2 January, and so on, with 7 January receiving G. Then, begin again, giving A to 8 January and so on to the end of the year. Such letters are called calendar letters. In a leap year the procedure is slightly different: 29 February is allocated the same letter, D, as 1 March.

We can readily see that the last day of the year, 31 December, receives the same letter, A, as 1 January, and that every day in the year has a specific calendar letter. The following couplet makes it easier to memorize the calendar letters of the first day of every month:

At Dover Dwell George Brown Esquire,
Good Christopher Finch And David Friar.

This tells us, for instance, that the calendar letter of April, the fourth month, is G(eorge)—the fourth word of the ditty. If we wish to find the calendar letter of a given day in a given month, begin with the letter of the first day of the month, count on from there till we reach the day required, returning to A each time we pass G. For 13 April, we start with G and count on 12 letters; in this way we arrive at E as the calendar letter of this day.

You may find it easier to convert the calendar letter of the first of the month to a calendar number by setting A to 1, B to 2, and so on, up to G to 7. To this calendar number add the day of the month and subtract 2; 13 April thus yields $7 + 13 - 2 = 18$. Now divide the result by 7; keep the remainder (in this case 4); and add one to it. If we convert the result, 5, back to a letter we obtain E as before.

You may prefer to write this algebraically in terms of the MOD function. If we denote the calendar number of a day by C, the calendar number of the first day of the month by F, and the day of the month by D we have:

$$C = 1 + \text{MOD}(F + D - 2, 7) \tag{24.1}$$

We may go on to derive an algebraic expression for the calendar letter (C) of any day (D) of month M. To do this we make use of Zeller's congruence (discussed in Chapter 23). This results in the following expressions:

$$M' \Leftarrow \text{MOD}(9 + M, 12) \tag{24.2}$$

$$Z = (13M' + 2)/5 \tag{24.3}$$

$$T = 28M' + Z + D - 365(M'/10) + 59 \tag{24.4}$$

$$C = 1 + \text{MOD}(T - 1, 7) \tag{24.5}$$

The expressions for M' and Z are further explained in Chapter 23, but T requires further comment. Equation 23.16 (see page 296) contains an extra term, P, to allow for the extra day in February in leap years. It does not appear in equation 24.4 because the calendar letter or number of 29 February is taken to be the same as that of 1 March (4 or D).

The point of calendar letters is that every Sunday (or any other day of the week) in a normal year, has the same calendar letter. In a leap year this rule is slightly different: every Sunday (or any other day) before 1 March has the same letter, and every Sunday (or other day) on or after 1 March also has the same letter, but the two letters differ; the letter of the later Sundays (or other day) is one letter back from that for the earlier Sundays (or other day).

⌘ Dominical numbers

The letter corresponding to Sundays is called the dominical or Sunday letter. These vary from year to year and, as just explained, there are two dominical letters for leap years. Corresponding to each dominical letter is a number—the dominical number. Numbers can be assigned to the letters in a straightforward manner as indicated in TABLE 24.1.

The next step is to find the dominical letter (or number) for the year in question. First note that every common year ends (31 December) on the same day of the week as the one it began on (1 January). This is a consequence of there being 52 weeks and one day in a common year. In leap years, the year ends one day of the week later than the one it began with.

This implies that the dominical letters of succeeding years are displaced back one place (G being before A)—except after leap years, when they are displaced two places. We may say that the dominical letter regresses one or two places for each succeeding year. In a leap year, the second of the two letters has regressed by two places. For instance, the dominical letter in the Gregorian calendar for 1990 is G; for 1991 it is F; 1992 is a leap year, so for that year the dominical letters are E and D; for 1993 it is C; for 1994, B; for 1995, A; for 1996, another leap year, it is G and F; and so forth.

If there were no leap years, the sequence of dominical letters would repeat every seven years. However, in the Julian calendar, leap years recur every four years. The lowest common multiple (LCM) of 4 and 7 is 28 so that, in this calendar, the sequence of dominical letters repeats every 28 years. In the Gregorian calendar the cycle is much longer—400 years.

The 28-year cycle is sometimes known as the Sunday letter cycle, but more often as the solar cycle. It was somewhat arbitrarily assumed that the first year of a solar cycle should be a leap year, and that 1 January should be a Monday, so that the dominical letters are G, F. This convention allows us to construct TABLE 24.2 which shows the relation between the position of a year in the solar cycle and its dominical letter.

The old computists devised methods for calculating the position of any year in the solar cycle. It was then a simple matter to look up the dominical letter in the table. This method is too unwieldy for the Gregorian calendar.

It is possible, however, to dispense with any table and instead to calculate the dominical letter of any year, Julian or Gregorian. We have seen that, if we were to ignore leap years, the dominical letter regresses by one each year. Thus the number of regressions that have occurred by the end of year Y, starting

TABLE 24.2 *Relation between the position of a year in the solar cycle and its dominical letter*

Year of the cycle	Dominical letter	Year of the cycle	Dominical letter
1	G,F	15	C
2	E	16	B
3	D	17	A,G
4	C	18	F
5	B,A	19	E
6	G	20	D
7	F	21	C,B
8	E	22	A
9	D,C	23	G
10	B	24	F
11	A	25	E,D
12	G	26	C
13	F,E	27	B
14	D	28	A

at year AD 1 is $Y - 1$. However, each leap year ensures one extra regression, so we must add to this the number of leap years. In the Y years of the Julian calendar, there are $Y/4$ leap years, so the total number of regressions by year Y is $Y + Y/4 - 1$.

In the Gregorian calendar there is no leap day in the centurial years (1800, 1900, and so on) unless they are divisible by 400. This implies that we must subtract $Y/100$ from the Julian formula and add back $Y/400$. We thus find that the total number of regressions, R, is given by:

Julian calendar: $R = Y + Y/4 - 1$ (24.6)

Gregorian calendar: $R = Y + Y/4 - Y/100 + Y/400 - 1$ (24.7)

It turns out that the dominical number for the year 1996 is 7 in the Julian calendar and 1 in the Gregorian calendar. In the Julian calendar, $\text{MOD}(R,7) = 2$ in 1996, so that whenever $\text{MOD}(R,7) = 2$ the dominical number $(N) = 7$; when $\text{MOD}(R,7) = 3$, $N = 6$; and so forth. Similar considerations apply for the Gregorian calendar.

It is then easy to verify that for year Y in the Julian calendar:

$$N = 7 - \text{MOD}(R + 5, 7)$$ (24.8)

or:

$$N = 7 - \text{MOD}(Y + Y/4 + 4, 7)$$ (24.9)

and for the Gregorian calendar:

$$N = 7 - \text{MOD}(R, 7)$$ (24.10)

or:

$$N = 7 - \text{MOD}(Y + Y/4 - Y/100 + Y/400 - 1, 7)$$ (24.11)

It is useful to define a quantity, P, as:
for the Julian calendar:

$$P = Y + Y/4 + 4$$ (24.12)

and for the Gregorian calendar:

$$P = Y + Y/4 - Y/100 + Y/400 - 1$$ (24.13)

so that we may write succinctly for either calendar:

$$N = 7 - \text{MOD}(P, 7)$$ (24.14)

As an example, if $Y = 1066$, we find for the Julian calendar that $P = 1066 + 1066/4 + 4 = 1336$ and the dominical number of the year AD 1066 is $7 - \text{MOD}(1336,7) = 1$.

If the year in question is a leap year, the dominical number (or letter) given by this formula is the second of the two, since we have counted its leap day in R. Thus, if Y is a leap year and the month is January or February we must increase N by one. This can be done by subtracting a quantity V from R; V takes the value 0, except when $M' \geq 10$ (January or February) and Y is a leap year. It is easily verified that V is correctly given by:

for the Julian calendar:

$$V = (Y/4 - Y'/4) \qquad\qquad (24.15)$$

and for the Gregorian calendar

$$V = (Y/4 - Y'/4) - (Y/100 - Y'/100) + (Y/400 - Y'/400) \qquad (24.16)$$

where:

$$Y' = Y - M'/10 \qquad\qquad (24.17)$$

As an example we may calculate N for a date in January in 1992 in the Gregorian calendar. We find:

$$M' = 10$$
$$M'/10 = 1$$
$$Y' = 1991$$
$$V = (1992/4 - 1991/4) - (1992/100 - 1991/100)$$
$$\quad + (1992/400 - 1991/400)$$
$$\ = 1$$
$$P = 1992 + 1992/4 - 1992/100 + 1992/400 - 1$$
$$\ = 2474$$
$$N = 7 - \text{MOD}(2474 - 1, 7)$$
$$\ = 5 \quad \text{or } E \text{ as found previously}$$

If the calendar number of the date in question happens to be the same as the dominical number, the day is clearly a Sunday. If not, a further step is required. If we suppose that the calendar number is C, the difference $C - N$ is the number of days beyond Sunday that the date falls on: that is, if $C - N = 1$, the day is a Monday; if $C - N = 2$, the day is Tuesday; and so on. It may however happen that $C - N$ is zero or less. This can be allowed for by adding

7 and removing any excess with the MOD function. This gives, for the number of the day of the week, W:

$$W = 1 + \text{MOD}(7 + C - N, 7) \tag{24.18}$$

For instance, for 10 January 1992 in the Gregorian calendar, we find that $C = 3$ and $N = 5$. Thus $W = 1 + \text{MOD}(7 + 3 - 5,7) = 6$, so that this day is a Friday. For 29 February 1992 we have $C = 1 + \text{MOD}(3 + 29,7) = 4$ (the same as for 1 March), and $N = 5$; we then have $W = 1 + \text{MOD}(7 + 4 - 5,7) = 7$, so 29 February 1992 is a Saturday.

Equations 24.12 and 24.13 are valid for all dates in the Christian era when Y is positive; a minor adjustment will render them valid for all dates. For this it is necessary to do two things. First, if the date is before the epoch of the era (1 January AD 1) the year, Y, should be replaced by $1 - X$ for a year X BC. Secondly, it is necessary to augment Y' by a constant. For the Julian calendar the constant should be of the form $28n$, and for the Gregorian calendar, $400n$. In either case n should be chosen in such a way that Y' is positive for the dates of interest.

In fact, the value of W is unchanged if we augment Y by any multiple of 28 (Julian case) or 400 (Gregorian case). If we augment a Gregorian year Y by $400n$, we find that $Y + Y/4 - Y/100 + Y/400$ is increased by $497n$; since this is

TABLE 24.3 *Regulars and concurrents*

Month	Regular	Dominical		Concurrent
		Letter	Number	
January	1	A	1	6
February	5	B	2	5
March	5	C	3	4
April	1	D	4	3
May	3	E	5	2
June	6	F	6	1
July	1	G	7	7
August	4			
September	7			
October	2			
November	5			
December	7			

a multiple of 7, W remains unchanged. A similar argument holds for the Julian case if Y is increased by $28n$—$Y + Y/4$ increases by $35n$.

⬚ Regulars and concurrents

The early computists sidestepped the use of equation 24.18 by using tables of regulars and concurrents. Each month has a number called its 'regular' which indicates the calendar letter of its first day. Likewise, each year has a number related to its dominical number and called its 'concurrent'. The regulars of the months and the relation between concurrents and dominical numbers is shown in TABLE 24.3.

The day of the week of the first day of any month of any year is obtained by adding the regular of the month to the concurrent of the year. It is then a simple matter to derive the day of the week for any day of that month. For instance, the regular of February is 5 and the dominical letter of the first part of 1992 is E (or number 5). Thus the concurrent of 1992 is 2. Then the first day of February in 1992 is $5 + 2 = 7$ or Saturday. We may express this by the formula:

$$W = 1 + \mathrm{MOD}(R + U, 7) \tag{24.19}$$

where U is the concurrent and R the regular. It is readily seen that concurrents and dominical numbers are related by:

$$U = 1 + \mathrm{MOD}(13 - N, 7) \tag{24.20}$$

Equation 24.19 follows immediately from 24.18 and 24.20 when R is set equal to C for the first day of the month.

⬚ The algorithms

Here I collect the steps necessary for calculating the day of the week, W, from the date $D/M/Y$ in either the Julian or Gregorian calendar using formulae based on the traditional method.

JULIAN CALENDAR

$\mathcal{A}\,l\,g\,o\,r\,i\,t\,h\,m\ \ \mathcal{A}$

1 $M' \Leftarrow \mathrm{MOD}(9 + M, 12)$ (24.2)

2 $Q \Leftarrow M'/10$

3 $Z \Leftarrow (13*M' + 2)/5$ (24.3)

4 $T \Leftarrow 28*M' + Z + D - 365*Q + 59$ (24.4)

$$5 \quad C \Leftarrow 1 + \mathrm{MOD}(T - 1, 7) \tag{24.5}$$

$$6 \quad Y' \Leftarrow Y - Q \tag{24.17}$$

$$7 \quad V \Leftarrow Y/4 - Y'/4 \tag{24.15}$$

$$8 \quad P \Leftarrow Y + Y/4 + 4 - V \tag{24.12}$$

$$9 \quad N \Leftarrow 7 - \mathrm{MOD}(P, 7) \tag{24.14}$$

$$10 \quad W \Leftarrow 1 + \mathrm{MOD}(7 + C - N, 7) \tag{24.18}$$

GREGORIAN CALENDAR

$\mathcal{A} l g o r i t h m \quad \mathcal{B}$

$$1 \quad M' \Leftarrow \mathrm{MOD}(9 + M, 12) \tag{24.2}$$

$$2 \quad Q \Leftarrow M'/10$$

$$3 \quad Z \Leftarrow (13 * M' + 2)/5 \tag{24.3}$$

$$4 \quad T \Leftarrow 28 * M' + Z + D - 365 * Q + 59 \tag{24.4}$$

$$5 \quad C \Leftarrow 1 + \mathrm{MOD}(T - 1, 7) \tag{24.5}$$

$$6 \quad Y' \Leftarrow Y - Q \tag{24.17}$$

$$7 \quad V \Leftarrow (Y/4 - Y'/4) - (Y/100 - Y'/100) \\ + (Y/400 - Y'/400) \tag{24.16}$$

$$8 \quad P \Leftarrow Y + Y/4 - Y/100 + Y/400 - 1 - V \tag{24.13}$$

$$9 \quad N \Leftarrow 7 - \mathrm{MOD}(P, 7) \tag{24.14}$$

$$10 \quad W \Leftarrow 1 + \mathrm{MOD}(7 + C - N, 7) \tag{24.18}$$

These algorithms are valid for years for which Y is positive, that is, AD 1 or later.

Careful inspection of these algorithms will show that several quantities are calculated more than once. This is wasteful of time and effort and the algorithms may be greatly compressed and simplified to the following:

JULIAN CALENDAR

$\mathcal{A}lgorithm$ C

1 $M' \Leftarrow \mathrm{MOD}(9 + M, 12)$ (24.2)

2 $Y' \Leftarrow Y - M'/10 + 28 * n$

3 $W \Leftarrow 1 + \mathrm{MOD}(D + (13 * M' + 2)/5 + Y' + Y'/4, 7)$

GREGORIAN CALENDAR

$\mathcal{A}lgorithm$ D

1 $M' \Leftarrow \mathrm{MOD}(9 + M, 12)$ (24.2)

2 $Y' \Leftarrow Y - M'/10 + 400 * n$

3 $W \Leftarrow 1 + \mathrm{MOD}(2 + D + (13 * M' + 2)/5 + Y' + Y'/4$
$- Y'/100 + Y'/400, 7)$

These algorithms are valid for any year if a year X BC is entered as $Y = 1 - X$ and n is chosen so that $Y' \geq 0$.

The day of the week from the Julian day number

In succeeding chapters I show how the Julian day number, which uniquely defines any day, can be calculated from the date in a variety of calendars. It is then very easy to calculate the day of the week from the day number. The day with Julian day number zero is a Monday ($W = 2$). It follows that for any day, J:

$$W = 1 + \mathrm{MOD}(J + 1, 7)$$ (24.21)

⌛ An alternative

If you really do not like arithmetic, you could try and obtain what was described as a 'toy' in the pages of *Nature* and invented by the Revd J. P. Wiles in 1906. At the time it cost 2/- and gave not only the day of the week of any date but also the date of Easter for any year.

CHAPTER 25

The Conversion of Regular Calendars

⌛ The algorithm

The same algorithm may be used to convert a date in any regular calendar into its corresponding Julian day number—or vice versa. The algorithm, which I describe here, is based on the work of D.A. Hatcher published in 1986.

Twelve constants are required for each calendar. These are all integers or whole numbers and are listed in TABLE 25.1 for the calendars mentioned in TABLE 22.1 (see page 290). (Three additional constants are required for Gregorian-type calendars and these are given in TABLE 25.4.)

All the calendars, except for those based on the Egyptian wandering year, require the intercalation of a leap day at regular intervals; this defines a cycle of intercalation, generally of four years. These intercalated days are inserted at the end of some particular month, which need not necessarily be the last. For instance, in the Julian calendar they are inserted at the end of February, the second month.

A key feature of the algorithm is that the date ($D/M/Y$) in the given calendar is converted first into a corresponding date in another calendar, the computational calendar. The day of the month, the month, and year in this computational calendar are denoted by D', M', and Y'. Thus any day, unambiguously defined by its Julian day number, J, corresponds to a date $D/M/Y$ in the given calendar and $D'/M'/Y'$ in the computational calendar. It is convenient to define also a day of the year which we denote by T in the given calendar and T' in the computational calendar.

The crucial idea is to arrange the start of the year of the computational calendar so that the leap day (if any) falls at the very end of it. This means that the start of the computational year falls on the day which immediately follows the leap day. This stratagem greatly simplifies the conversion formulae and any difficulties due to the leap days evaporate.

TABLE 25.1 *Parameters for the conversion of dates in regular calendars*

Number	Calendar	y	j	m	n	r	p	q	v	u	s	t	w
1	Egyptian	3968	47	1	13	1	365	0	0	1	30	0	0
2	Armenian	5268	317	1	13	1	365	0	0	1	30	0	0
3	Khwarizmian	5348	317	1	13	1	365	0	0	1	30	0	0
4	Persian	5348	77	10	13	1	365	0	0	1	30	0	0
5	Ethiopian	4720	124	1	13	4	1 461	0	3	1	30	0	0
6	Coptic	4996	124	1	13	4	1 461	0	3	1	30	0	0
7	Republican	6504	111+g	1	13	4	1 461	0	3	1	30	0	0
8	Macedonian	4405	1401	7	12	4	1 461	0	3	5	153	2	2
9	Syrian	4405	1401	6	12	4	1 461	0	3	5	153	2	2
10	Julian Roman	4716	1401	3	12	4	1 461	0	3	5	153	2	2
11	Gregorian	4716	1401+g	3	12	4	1 461	0	3	5	153	2	2
12	Islamic A	5519	7665	1	12	30	10 631	14	15	100	2951	51	10
13	Islamic B	5519	7664	1	12	30	10 631	14	15	100	2951	51	10
14	Bahá'í	6560	1412+g	20	20	4	1 461	0	3	1	19	0	0
15	Saka	4794	1348+g	2	12	4	1 461	0	3	1	*	*	*

1. The epagomenal days of the Egyptian (1–4) and Alexandrian (5–7) calendars are counted as belonging to month 13.

2. *g* for calendars 7, 11, 14, and 15 represents a term required to take into account the rate of Gregorian intercalation.

3. Except for the Islamic, Bahá'í, and Saka calendars we have:

   $q + v = r - 1$

   $t + w = u - 1$

4. The Saka calendar requires special treatment and the quantities marked * depend on T' or M.

The steps in the process of converting dates into day number and vice versa are summarized as:

$$D/M/Y \leftrightarrow D'/M/Y' \leftrightarrow J$$

The lengths of the months in the given and computational calendars are identical, as are the days on which each starts. The points of intercalation are also the same. The two calendars differ only in their epochs and in the days selected to be the first days of their years.

In most calendars (as mentioned in Chapter 3) the years, months, and days are specified using current rather than elapsed time. The computational calendar, however, uses elapsed time. This means that the first year of the era is called year 0 (not year 1); the first month of the computational year is called month 0 (not month 1); the first day of each month is called day 0 (not day 1); and the first day of the year in the given calendar is day 1 and in the computational calendar, day 0. The advantage of this scheme is that the number of years, for instance, that precede the year labelled Y', is equal to Y'. In the more conventional scheme (starting the labelling with 1), the number of years preceding year Y would be $Y - 1$.

We denote the number of the month in the given calendar which corresponds to month 0 in the computational calendar by m. Thus, in the Gregorian calendar m = 3 because the leap day is inserted between 28 February and 1 March so that the computational year starts on the first day of March, the third month. March thus becomes month 0 in the computational calendar, April becomes month 1, and so forth; while January and February become months 10 and 11.

The various symbols used in the conversions are defined in TABLE 25.2.

The conversion of given dates to computational dates

The following algebraic equations allow the calculation of Y', M', and D' given Y, M, and D:

$$Y' = Y + y - (n + m - 1 - M) /n \tag{25.1}$$

$$M' = \mathrm{MOD}(M - m + n, n) \tag{25.2}$$

$$D' = D - 1 \tag{25.3}$$

Equation 25.3 ensures that the days of the month in the computational calendar start with 0. Equation 25.2 ensures that M' is in the range 0 to $n - 1$ inclusive, and that $M' = 0$ when $M = m$. Equation 25.1 and the term y are discussed further below; in this equation, the third term, $(n + m - 1 - M)/n$, is zero if $M \geq$ m, but is otherwise 1; this ensures that the computational year starts with the first day of the month M', for example, 1 March in the case of the Julian and Gregorian calendar.

The conversion of computational dates to given dates

The converse operations of calculating Y, M, and D, given Y', M', and D' can be performed using:

$$D = D' + 1 \tag{25.4}$$

$$M = \mathrm{MOD}(M' + m - 1, n) + 1 \tag{25.5}$$

$$Y = Y' - y + (n + m - 1 - M) /n \tag{25.6}$$

TABLE 25.2 *Definition of terms*

Symbol	Definition	Range
Y	Year number in given calendar	
M	Month number in given calendar	$1 \leq M \leq n$
D	Day of month in given calendar	$1 \leq D$
T	Day of year in given calendar	$1 \leq T$
Y'	Year number in computational calendar	$0 \leq Y'$
M'	Month number in computational calendar	$0 \leq M' \leq n - 1$
D'	Day of month in computational calendar	$0 \leq D'$
T'	Day of year in computational calendar	$0 \leq T'$
n	Number of months in year	
m	Month number in given calendar for which $M' = 0$	$1 \leq m \leq n$
r	Number of years in a cycle of intercalation	
p	Number of days in a cycle of intercalation	
C	Number of days in a common year $= p/r$	
q,v	Parameters required in calculating years	
u,s,t,w	Parameters required in calculating months	
J_g	Day number of epoch of given calendar	
J_1	Day number of first day of first cycle of intercalation which starts after J_g	
J_c	Day number of epoch of computational calendar	
y	Computational year in which J_1 falls	
j	Number of days that J_c falls before day zero, that is, $j = -J_c$	
x	Number of days that J_1 falls after J_g, that is, $x = J_1 - J_g$	

Equations 25.4 and 25.6 follow immediately from equations 25.3 and 25.1, and it is readily verified that 25.5 yields $M = m$ when $M' = 0$.

The conversion of dates into day numbers

Once we have converted our given date into a corresponding date in the computational calendar, the next step is to convert the computational date into a day number, J. We may write J as:

$$J = c + d + D' - j \tag{25.7}$$

Here, c is the number of days in the computational calendar which precede year Y'; d is the number of days in year Y' which precede month M'; D' is the number of days in the current month, M', which precede the day of interest;

and j is the number of days in the computational calendar which precede the epoch of the computational calendar (that is, day 0 of this calendar).

c is given by the generalized congruence relation:

$$c = (pY' + q)/r \tag{25.8}$$

Here p is the number of days in an intercalation cycle and r the number of years, so that the ratio p/r is the average number of days in a year. The constant q is equal to zero except for the Islamic calendars. For these, equation 25.8 correctly allows for the 11 intercalary days in the 30-year cycle of the Islamic calendar. The reader may care to verify by direct calculation that the relation is correct. If such a relation cannot be devised, the particular calendar cannot be considered to be regular.

Likewise, d is given by another congruence relation which, again, the reader may care to verify:

$$d = (sM' + t)/u \tag{25.9}$$

Here u is related to a periodicity in the lengths of the months of the year. Except for the Islamic and Saka calendars, we find $t + w = u - 1$. For the Julian and Gregorian calendars however, the values of s, t, and u follow immediately from Zellor's congruence described in Chapter 23.

For some calendars, such as the reformed Saka calendar, it may not be possible to find appropriate values of s, t, and u. Nevertheless, the method can still be used if it is allowed for that these parameters are not constant but depend on M' (see next section).

The conversion of day numbers into dates

The first step of the converse conversion of day numbers into dates is to convert the day number into a computational date, and then to convert the latter into a date in the given calendar using equations 25.4, 25.5, and 25.6.

We first calculate the number of days in the computational calendar, J', which precede the day number, J, of interest:

$$J' = J + j \tag{25.10}$$

We next find the number of complete years, Y', contained in this period and the number of days in the current year, T', which remain when these years are removed. This is done with the aid of two congruence relations which the reader may again care to verify for each calendar:

$$Y' = (rJ' + a)/p \tag{25.11}$$

Note that $q + v = r - 1$, except for the Islamic calendar. Any days not included in these Y' years belong to the current intercalation cycle and it may be verified that the day of the year in the computational calendar is given by:

$$T' = \text{MOD}(rJ' + v, p) / r \qquad (25.12)$$

Given T', we may now calculate the number of months, M', which precede the current day and the number of days remaining, which gives the day of the month, D':

$$M' = (uT' + v) / s \qquad (25.13)$$

and

$$D' = \text{MOD}(uT' + w,) / u \qquad (25.14)$$

The computational date, $D'/M'/Y'$ may now be converted to the corresponding date in the given calendar as already explained.

⧖ The start of the computational calendar

Some of the formulae given previously involve the quantities y and j. These define the epoch of the computational calendar, and require further detailed explanation.

I have already stated that the first year of the computational calendar ($Y' = 0$) is the first year of a cycle of p days (or r years).

Consider the first day, J_1, of the first cycle which starts after the epoch of the given calendar which we suppose is day J_g. We suppose that J_1 is x days after J_g so that:

$$x = J_1 - J_g \qquad (25.15)$$

A little thought will now convince you that the number of complete cycles, b, each containing p days, that fall on or after day zero and before day J_1 is given by:

$$b = (J_g + x) / p \qquad (25.16)$$

The first of these cycles will start less than p days after day zero, so that the cycle which precedes this cycle must start on or before day zero. We take the first day of this cycle to be the epoch of the computational calendar and denote it as day J_c.

Then the number of days in the computational calendar which precede the day J_1 is readily seen to be $p(b + 1)$ and the number of years which precede it is $r(b + 1)$. Since the first computational year is called year 0, the computational year for which J_1 is the first day is called year $r(b + 1)$.

On the other hand, J_1 is x days after day J_g, and these x days contain the first x/C years of the given calendar, where C is the number of days in a common year. We may calculate C from:

$$C = p/r \qquad (25.17)$$

It follows that the day J_1 lies in year $1 + x/C$ of the given calendar. In other words, year $Y' = r(b + 1)$ in the computational calendar corresponds to year $Y = 1 + x/C$ in the given calendar. We may therefore substitute these values of Y and Y' into equation 25.1 to obtain:

$$r(b + 1) = 1 + x/C + y - (n + m - 1 - M)/n \qquad (25.18)$$

Day J_1 falls at the start of year Y in month $M = m$, so that $(n + m - 1 - M)/n = 0$, and we thus find a formula that enables us to calculate y for each calendar:

$$y = rb + (r - 1 - x/C) \qquad (25.19)$$

To calculate j, first note the following:

1. Day zero falls j days after J_c;
2. J_g falls J_g days after day zero;
3. J_1 fall x days after J_g;
4. J_1 falls $p(b + 1)$ days after J_c.

It immediately follows that:

$$p(b + 1) = j + J_g + x \qquad (25.20)$$

and from this it may be shown that:

$$j = p - \text{MOD}(J_g + x, p) \qquad (25.21)$$

In the arguments just deployed, I have tacitly assumed that day zero is the start of the Julian period (1 January 4713 BC). We are then assured that the conversion formulae will be valid for any date which falls on or after day zero. In fact, the adoption of the start of the Julian period as the first valid day is quite arbitrary. We can extend the validity of the formulae backwards (or forwards) in time, if we so wish, by subtracting (adding) any constant from all the J_g values given in TABLE 25.1. Having done this, we must recalculate new values of y and j from equations 25.19 and 25.21 using the new values of J_g.

The positions of the leap days and the corresponding values of x are given in TABLE 25.3.

TABLE 25.3 *The starts of the first cycles*

Number	Calendar	Day following first leap day after epoch	x	C	x/C
1	Egyptian	1/1/1	0	365	0
2	Armenian	1/1/1	0	365	0
3	Khwarizmian	1/1/1	0	365	0
4	Persian	1/10/1	245	365	0
5	Ethiopian	1/1/3	1096	365	3
6	Coptic	1/1/3	1096	365	3
7	Republican	1/1/3	1096	365	3
8	Macedonian	1/7/3	912	365	2
9	Syrian	1/6/3	882	365	2
10	Julian Roman	1/3/4	1155	365	3
11	Gregorian	1/3/4	1155	365	3
12	Islamic A	—	0	354	0
13	Islamic B	—	0	354	0
14	Baha'i	1/20/1	347	365	0
15	Saka	1/2/2	396	365	1

1. C is the number of days in a common year.

2. The day after the leap day is specified using the normal numbering of the months in which the first month of the year is number 1.

3. There is no intercalation in calendars 1 to 4, and their years start on Thoth 1st or its equivalent. In calendars 1 to 3, the epagomenal days are placed after the twelfth month to form a thirteenth, but in the Persian calendar (4) they are placed after the eighth month.

4. The Macedonian and Syrian calendars follow the Julian pattern. Their leap days all fall on 29 February, but the Macedonian year starts on 1 September and the Syrian on 1 October. The Julian and Gregorian years start on 1 January.

5. There are 11 intercalary days in each 30-year cycle of the Islamic calendars (which must be treated as special cases).

⊞ Gregorian type calendars

The Gregorian calendar is not a regular calendar in the proper sense since the leap days are not intercalated at strictly regular intervals and there is no well-defined intercalation cycle.

Nevertheless, Gregorian dates can be converted into day numbers by first assuming them to be Julian dates, and then adjusting the day number so found for the Gregorian mode of intercalation. Day numbers can likewise be converted into Gregorian dates. The steps in the conversion are:

$$D/M/Y \leftrightarrow \quad D'/M'/Y' \leftrightarrow \quad J_J \quad \quad J_G$$

Here J_J is the day number corresponding to the date $D/M/Y$ interpreted as being in the Julian calendar, and J_G is the day number of the same date interpreted as being in the Gregorian calendar.

This stratagem requires a method of interconverting J_J and J_G. To do this, suppose that $g = J_J - J_G$. We must then replace equation 25.7 by:

$$J = c + d + D' - j - g \qquad\qquad (25.22)$$

when converting Gregorian dates into day numbers. To use this we must be able to calculate g in terms of Y'. Likewise, we must replace equation 25.10 by:

$$J' = J + j + g \qquad\qquad (25.23)$$

when converting day numbers into Gregorian dates, and to use this we must be able to calculate g in terms of J.

The problem then is to calculate g in terms of Y' if we are converting dates to numbers or in terms of J if we are converting numbers to dates. At this point we may note that the problem is not confined to the Gregorian calendar as such, but is relevant to any calendar which employs a Gregorian mode of intercalation.

In a calendar with a Gregorian mode of intercalation there is primarily a leap day every four years, but the leap day is suppressed three times in every 400 years at 100-year intervals. This is in contradistinction to the Julian mode in which there is a regular leap day every four years. Calendars that use the Gregorian mode include the Republican and Saka calendars and a version of the Bahá'í calendar. I will therefore approach the problem in general terms and assume that day numbers and dates can be interconverted in a hypothetical Julian-type version of these calendars. I next explain how to calculate g.

Firstly, note that the Gregorian calendars have a cycle of 400 years containing exactly 146 097 days ($365 \times 400 + 400/4 - 3$), and suppose that this cycle ends with the one leap day at the end of a century that is not suppressed. A new cycle thus starts the following day, which is the start of a new year (and century and intercalation cycle) in our computational calendar. Let the day number of this day (at the start of the new year) be J_o, the year of the Gregorian-type calendar in which it falls be Y_o, and the difference between J_J and J_G on this day be g_o. Note also that this day is the first in the computational year Y_o' with $Y_o' = Y_o + y$. Y_o and J_o mark the start of a 400-year period which may be chosen arbitrarily. The selection of a value for J_o implies a value of Y_o.

We now observe that every time a leap year is suppressed thereafter, g increases by 1. Thus, one century after J_0, g increases by one; likewise, after two and three centuries, but not after four centuries (when a leap year occurs under Gregorian rules). So:

In centuries	0 1 2 3 4 5 6 7 8 9 . . .
the increase is	0 1 2 3 3 4 5 6 6 7 . . .

The number of centuries elapsed after J_0 is given by $(Y' - Y_0')/100$ and the reader may verify that the required increase in g is given by the expression: $3(1 + (Y' - Y_0')/100)/4$. The factor $\frac{3}{4}$ arises from a congruence relation discussed in Chapter 23. We thus have:

$$g = g_0 + 3(1 + (Y' - Y_0)' /100 /4 \tag{25.24}$$

It is easy enough to see that this is mathematically equivalent to:

$$g = g_0 + 3(1 + 4\alpha + (Y' - Y_0)' /100 /4 - 3\alpha \tag{25.25}$$

where α is any arbitrary positive integer. The effect of using different values of α is the selection of different 400-year cycles.

It then follows that:

$$g = 3((Y' + A /100 /4 + G \tag{25.26}$$

where:

$$A = 400\alpha + 100 - Y_0' \tag{25.27}$$

$$\text{and} \quad G = g_0 - 3\alpha \tag{25.28}$$

It is essential that the quantity that is divided by 100 in equation 25.26 be positive, and this can always be arranged by choosing a suitable value of α. We should thus chose α so that $Y' + A \geq 0$ or:

$$\alpha \geq (Y_0 + y - 100 /400 \tag{25.29}$$

Equation 25.26 then allows us to calculate g for any year. A similar procedure results in a formula for g in terms of J.

If we define I as $J - J_0$ we may see, after careful thought, that:

$$(h-)1 \; K(+ \; h) - 1 /4 + 1 \leq I + 1 \leq hK + h/4 - 1 \tag{25.30}$$

where h is the century in which I is located, starting with century 1 for $I = 0$; $K = 36524 \; (365 * 100 + 100/4 - 1)$ being the number of days separating the position at which two consecutive leap days were suppressed. It follows that:

$$h - 1 \geq 4(I +)(/ 4K +) 1 \geq h - 1 \tag{25.31}$$

so that:

$$h = 4(I +)(/ 4K +) 1 + 1 \tag{25.32}$$

Since leap days are suppressed in centuries: $h = 1,2,3,3,4,5,6,6, \ldots$, we use the stratagem already described and obtain the analogues of equations 25.26, 25.27, and 25.29:

$$g = g_0 + 3((4J + B)(/ 4K +) 1) /4 - 3\beta \tag{25.33}$$

$$= 3((4J + B)(/ 4K +) 1) /4 + G \tag{25.34}$$

where G is given by $g_0 - 3\beta$ (see the following). In equation 25.33 we have:

$$B = 4(\beta(4K +) +(K +) 1 - J_0) + 1 \tag{25.35}$$

and we must choose β so that $4J + B \geq 0$ so that:

$$\beta \geq (4(J + j - (K +)) -) /((4K +) 1) \tag{25.36}$$

Equation 25.34 allows us to calculate g in terms of J. Values of A and B, based upon suitable values of α for four Gregorian-type calendars are given in TABLE 25.4.

TABLE 25.4 *Parameters for calculating the Gregorian correction*

Number	Calendar	J_0	Y_0	g_0	A	B	G
7	Republican*	2375475	0	0	396	578797	−51
11	Gregorian	2305508	1600	10	184	274277	−38
14	Baháʾíʈ	2451606	156	1	184	274273	−50
15	Saka§	1867268	322	3	184	274073	−36

*It is assumed that leap days were intercalated in the French Republican calendar in the autumn preceding their intercalation into the Gregorian calendar. In fact, the calendar was repealed before any leap days were suppressed. I have also ignored the proposal, never made legal, to retain the leap day in years divisible by 4000.

ʈThe Baháʾí calendar is, strictly speaking, an astronomical calendar. I have assumed here, for illustrative purposes only, that leap days were intercalated one day after their insertion in the Gregorian calendar.

§The Saka calendar was designed to keep in step with the Gregorian calendar, and leap days are intercalated in the same years as in the Gregorian calendar.

It turns out that the same values of α and β satisfy equations 25.29 and 25.36, so that we are justified in putting $G = g_0 - 3\alpha = g_0 - 3\beta$.

⌛ Notes on individual calendars

Several categories of regular calendar are listed in TABLE 25.1; each requires some additional explanation.

The first three calendars are based on the Egyptian calendar with its vague year and its five epagomenal days located after the twelfth month. We suppose that each intercalation cycle starts on the first day of the first month. The Persian calendar (4) is anomalous in that the epagomenal days are placed immediately after the eighth month, so that the cycle is taken to start straight after these.

The next two calendars (5 and 6) are based on the Alexandrian calendar, which is a modified form of the Egyptian calendar with a leap day every four years, the first being in year three. The leap day is placed after the epagomenal days, which in turn are placed at the end of the year. Each four-year cycle starts on the first day of the first month.

The Republican calendar (7) is another Alexandrian calendar, similar to the previous three, but I have incorporated Romme's proposal to intercalate days in a manner similar to that in the Gregorian calendar. In fact, his proposal included the reinstatement of a leap day every 4000 years—but I have ignored this feature which would complicate the algorithm even further. The calendar was abolished by Napoleon before it had run for a century so, in practice, no leap days were ever suppressed.

Calendars 8, 9, and 10 are based on the Julian calendar, with a leap day inserted every four years on 29 February. They differ in their epochs and the position of the start of the year: the Macedonian (8) calendar starts on 1 September; the Syrian (9) on 1 October; and the Julian and Gregorian (10 and 11) are taken to start on 1 January. The Gregorian calendar requires j to be modified by the addition of g (as described) to account for the modified scheme for introducing leap years of leap days.

The two Islamic calendars (12 and 13) only differ in their epochs. They are anomalous in that each cycle of 30 years contains 11 leap years. This complication requires the introduction of the parameter, q, which is otherwise zero for the other calendars.

The Bahá'í calendar (14) is in some respects analogous to the Persian calendar. Four epagomenal days are placed at the end of month 18, and these are followed by a leap day in leap years. These are assumed to occur with the same frequency and in the same year as those in the Gregorian calendar, but this may not be strictly true, since leap days are said to be introduced on an empirical basis.

The Saka calendar

The reformed Saka calendar (15) presents a major problem and is not strictly regular. The first five months of the computational calendar contain 31 days, the next six contain 30 days, and the last contains 30 (or 31 in a leap year). No simple congruence relations equivalent to equations 25.9 and 25.13 can be found. Saka dates can only be processed as special cases in which s, t, and w depend on M' or J.

In the conversion of Saka dates to day numbers we must redefine the constants s and t as s = 31 − Z and t = 5Z, where $Z = M'/6$.

Likewise, in the conversion of day numbers to Saka dates we must redefine s and w as s = 31 − Z and w = −5Z, where $Z = T'/185 − T'/365$. Moreover, we must replace equation 25.13 by:

$$D' = (\text{MOD}(uT' + w,) + X)/u \tag{25.37}$$

where $X = 0$, except when T' (the maximum value) = 365 (when we put $X = 6$).

These alterations render correct conversions; the reader may care to verify that correct values are obtained with their use and, perhaps, discover a simpler scheme.

The Saka calendar also introduces a leap day every four years at the end of the second month. They are introduced in the same years as in the Gregorian calendar, and the usual correction, g, must be applied to allow for the Gregorian character of the calendar.

The Roman calendar

The Romans used a more elaborate system than the mere counting of days. The first step in calculating a Roman date from a day number must be to calculate a Julian date. Only then can the Roman date be calculated. I know of no satisfactory algorithm for doing this, and the best method of calculation employs the use of a table such as TABLE 16.2 (see page 213). Likewise, to calculate the day number of a Roman date, first obtain the Julian date and then convert that.

It may nevertheless be of interest to calculate the day of the month, D, on which the nones, ides, and kalends of a month fall. This can be done by looking up the category of the month, Q, in TABLE 25.5. The nones then fall on day N and the ides on day I as given by:

TABLE 25.5 *The Roman months*

	Ian	Feb	Mar	Apr	Mai	Iun	Iul	Aug	Sep	Oct	Nov	Dec
M	1	2	3	4	5	6	7	8	9	10	11	12
Length (days)	31	28/9	31	30	31	30	31	31	30	31	30	31
Category (Q)	1	1	2	1	2	1	2	1	1	2	1	1
Nones (N)	5	5	7	5	7	5	7	5	5	7	5	5
Ides (I)	13	13	15	13	15	13	15	13	13	15	13	13
Kalends (K)	1	1	1	1	1	1	1	1	1	1	1	1

$$N = 2Q + 3 \tag{25.38}$$

$$I = N + 8 \tag{25.39}$$

The kalends always falls on the first day of the succeeding month ($K = 1$). Given N, I, or K it is then easy to count back to get the Roman name of the day, as TABLE 16.2 (see page 213) will show.

⊠ The algorithm

Finally, the components of the two conversion algorithms are collected together for convenience. Their efficiency can only be slightly improved, and I have not attempted this. To convert a date $D/M/Y$ into a day number J:

Algorithm ℰ

---

1. $Y' \Leftarrow Y + y - (n + m - 1 - M)/n$ $\tag{25.1}$

2. $M' \Leftarrow \mathrm{MOD}(M - m + n, n)$ $\tag{25.2}$

3. $D' \Leftarrow D - 1$ $\tag{25.3}$

4. $c \Leftarrow (p * Y' + q)/r$ $\tag{25.8}$

5. $d \Leftarrow (s * M' + t)/u$ $\tag{25.9}$

6. $J \Leftarrow c + d + D' - j$ $\tag{25.7}$

For Gregorian-type calendars, replace step 6 by:

6a $g \Leftarrow 3 * ((Y' + A)/100)/4 + G$ $\tag{25.26}$

6b $J \Leftarrow c + d + D' - j - g$ (25.22)

For the Saka calendar, insert before step 5:

4a $Z \Leftarrow M'/6$

4b $s \Leftarrow 31 - Z$

4c $t \Leftarrow 5 * Z$

To convert a day number, J, into a date, $D/M/Y$:

Algorithm F

1   $J' \Leftarrow J + j$ (25.10)

2   $Y' \Leftarrow (r * J' + v)/p$ (25.11)

3   $T' \Leftarrow \text{MOD}(r * J' + v, p)/r$ (25.12)

4   $M' \Leftarrow (u * T' + w)/s$ (25.13)

5   $D' \Leftarrow \text{MOD}(u * T' + w, s)/u$ (25.14)

6   $D \Leftarrow D' + 1$ (25.4)

7   $M \Leftarrow \text{MOD}(M' + m - 1, n) + 1$ (25.5)

8   $Y \Leftarrow Y' - y + (n + m - 1 - M)/n$ (25.6)

For Gregorian-type calendars replace step 1 by:

1a $g \Leftarrow 3 * ((4 * J + B)/146097)/4 + G$ (25.34)

1b $J' \Leftarrow J + j + g$ (25.23)

For the Saka calendar, insert before step 4:

3a $X \Leftarrow T'/365$

3a $Z \Leftarrow T'/185 - X$

3b $s \Leftarrow 31 - Z$

3c $w \Leftarrow -5 * Z$

and replace step 5 by:

$$\textbf{5a} \quad D' \Leftarrow \left(6 * X + \text{MOD}(u * T' + w,) \right) / u \qquad (\textbf{25.37})$$

Relevant constants are listed in the tables of this chapter. It is hoped that the reader will try his or her hand at adapting the method to other calendars of personal interest.

CHAPTER 26

The Jewish Calendar

⌛ Introduction

The Jewish calendar has an undeserved reputation of being hard to understand. In reality it is not so difficult and is much more comprehensible than the rules for the calculation of the date of Easter or the intricacies of the Mayan calendar round. The important thing to grasp is the difference between the highly regular astronomical calendar and the civil calendar with its postponements.

An essential requirement of any method of converting dates in the Jewish calendar is the calculation of the molod of Tishri which defines the start of each year. We must first calculate the astronomical instant of this event and then apply postponements to obtain the civil date of the first day of Tishri.

⌛ The molod of Tishri

The molods of Tishri occur at regular intervals, and the interval between two consecutive molods being a constant 29 days 12 hours and 793 chalakim. Here, I express the subdivision of the hour in terms of the traditional unit of time, the chalakim (ch.). There are 1080 chalakim in an hour and hence 25 920 in a day. I shall describe all time intervals in terms of chalakim; so the interval between two molods is 765 433 ch. (For brevity, I shall denote this interval by m and the number of chalakim in a day by k.)

The decision to calculate in terms of chalakim only, has the advantage of simplicity in the resulting formula. It has the disadvantage of setting a trap for the unwary computer programmer in that the number of chalakim can be too great to store in a 32-bit word. It also departs from the traditional method of calculating in terms of days, hours, and chalakim using the sort of procedure familiar before the onset of decimalization, but now less often used.

The Jewish era (AM) began with the Creation, reckoned to have taken place at 18.00 hrs on the day whose date in the Julian calendar was 6/10/3761 BC—that is, six hours in advance of the start of 7/10/3761 BC; the Julian day number of this day, a Monday, was 347 998. It is generally assumed that a Jewish day corresponds to the conventional day that starts six hours later. Thus, day 1 AM corresponds to Julian day number 347 998.

We shall calculate the time elapsed (in chalakim) from the moment of the Creation to the molod of Tishri of the year Y specified in the mundane era. This supposes a first Tishri, and this is taken to have occurred a short time after the Creation on day 2 at 5 h, 204 ch. This moment, known as BeHaRD (Beth = 2, He = 5, Res = 20, Daleth = 4) thus took place 1 day, 5 hrs. and 204 ch. (or 31524 ch.) after the moment of the Creation. We shall denote this period by b.

A potent source of confusion is the traditional habit of specifying BeHaRD as 2 d, 5 h, 204 ch. This should be interpreted to mean that the molod occurred on day 2 at 5 h, 204 ch., rather than after 2 days, 5 hours, and 204 chalakim; that is, the day denotes current time but the hour and chalakim denote elapsed time. Most people in fact use this convention.

I will now explain how to calculate the number of months that have elapsed since the Creation and the molod of Tishri of the year Y. This is done by calculating the number of months contained in the $y (= Y - 1)$ complete years that have elapsed before the molod.

The Jewish calendar is based on the Metonic cycle of 19 years which contains 235 months. The number of complete cycles in a period of y years is readily seen to be $y/19$ and the number of years remaining in the last cycle is:

$$y' = \mathrm{MOD}(y, 19) \tag{26.1}$$

There are 12 months in a common year and 13 in an embolismic year. In terms of the 19-year cycle, the 13th month is intercalated in years 3, 6, 8, 11, 14, 17, 19. A useful relation, $(7y + 1)/19$, gives the number of embolismic years in a cycle up to and including year y, as may be readily verified. It then follows that the number of months up to the molod of Tishri is:

$$\mu = 235y'/19 + 12y' + (7y' + 1)/19 \tag{26.2}$$

This can be readily simplified by algebraic manipulations to give the more concise formula:

$$\mu = (235y' + 1)/19 \tag{26.3}$$

It then follows that the time elapsed from the moment of the Creation to the molod of Tishri of year Y is

$$t = b + m\mu = b + m((235Y - 234)/19) \tag{26.4}$$

The number of complete days elapsed is given by:

$$d = t/k \tag{26.5}$$

and the precise time in chalakim of the molod on day $d + 1$ is given by:

$$t' = \text{MOD}(t, k) \tag{26.6}$$

The day of the week of the molod (Sunday = 1, and so on) is given, bearing in mind that the Creation (on 1/1/1 AM) took place on a Monday, by:

$$w = 1 + \text{MOD}(d, 7) \tag{26.7}$$

We must now consider whether the year Y is embolismic or not. It is easily verified that Y is embolismic if $\text{MOD}(7Y + 13, 19) \geq 11$. It is sometimes convenient to define a parameter, E, which is zero for common years and one for embolismic years. It may be seen that:

$$E = \text{MOD}(7Y + 13, 19)/12 \tag{26.8}$$

We also need to know whether the previous year, $Y - 1$ was embolismic or not; we therefore define:

$$E' = \text{MOD}(7(Y - 1) + 13, 19)/12$$
$$= \text{MOD}(7Y + 6, 19)/12 \tag{26.9}$$

It remains now to consider the postponements which may have to be applied to the day of the molod to give the date of the first day of Tishri. The astronomical postponements which depend on t', w, E, and E' are applied first. Then the political postponements are applied; these depend only on the day of the week, recalculated after the application of the astronomical postponements.

The rules for the astronomical postponements tell us that we must add one day to d if:

1	FacH		$t' \geq 19\,940$	
2	GaTRad	or	$t' \geq 9924$	and $w = 3$ and $E = 0$
3	BaTu	or	$t' \geq 16\,788$	and $w = 2$ and $E = 0$
				and $E' = 1$

$$\tag{26.10}$$

Although it is possible to derive a formula for a quantity, P, which is one, if a postponement is required, or otherwise zero, it is somewhat ponderous. I therefore leave the decision of whether to postpone or not to the reader. If the conditions for an astronomical postponement are met, d must be incremented by 1 and w re-evaluated using equation 26.7.

Next we must consider the political postponement; this depends on the new w, obtained after the application of the astronomical postponements. We must add 1 to d again if:

$$4 \quad \text{Adu} \qquad\qquad w = 1 \quad \text{or} \quad w = 4 \quad \text{or} \quad w = 6 \qquad\qquad (26.11)$$

It is readily verified that the postponement should be applied if MOD $(d + 4,7)$ is an odd number. It follows that whatever the value of w, the postponement may be applied by adding P to d where:

$$P = \text{MOD}\left(\text{MOD}\left(d + 5,7\right),2\right) \qquad\qquad (26.12)$$

The final value of d now represents the number of complete days between the moment of the Creation and the start of the first day of Tishri in the civil calendar. The last day before the moment of the Creation had a Julian day number of 347 997 so that the Julian day number, J, of the day corresponding to Tishri 1st is given by:

$$J = d + 347997 \qquad\qquad (26.13)$$

The converse calculation of the year, Y, in which a Julian day number falls, proceeds in a similar manner. First, we calculate the number of months that were completed before day J, by dividing the number of chalakim elapsed by day J from the moment of the Creation by the number of chalakim in a lunation, m. The current month (in which J falls) is thus given by:

$$M' = \left(k(J - 347996)\right)/m + 1 \qquad\qquad (26.14)$$

Next, we estimate the number of years in these M' months. The estimate is given by Y' in:

$$Y' = 19(M'/235) + \left(19(\text{MOD}\left(M',235\right) - 2)\right)/235 + 1 \qquad\qquad (26.15)$$

In this expression the first term represents the number of years in completed cycles, while the second represents the number of completed years in the current cycle, as the reader may verify. The final 1 ensures that Y' represents the current year. This year is in fact only an estimate and may be one too high if the start of the year was postponed. It may be necessary to correct Y' by subtracting 1 if the day J', calculated for the molod of Tishri of year Y', is greater than J.

⧖ An algorithm for calculating the day of Tishri 1st for year Y

For convenience, the relevant formulae can be collected together and presented as an algorithm for calculating the Julian day number of the 1st of Tishri in year Y of the mundane era.

$\mathcal{Al}\,g\,o\,r\,i\,t\,h\,m\ \ \mathcal{G}$

---

1   $t \Leftarrow 31524 + 765433((235 * Y - 234)/19)$ (26.4)

2   $d \Leftarrow t/25920$ (26.5)

3   $t' \Leftarrow \text{MOD}(t, 25920)$ (26.6)

4   $w \Leftarrow 1 + \text{MOD}(d, 7)$ (26.7)

5   $E \Leftarrow \text{MOD}(7 * Y + 13, 19)/12$ (26.8)

6   $E' \Leftarrow \text{MOD}(7 * Y + 6, 19)/12$ (26.9)

7   If $t' \geq 19940$ (26.10)

    or $t' \geq 9924$ and $w = 3$ and $E = 0$

    or $t' \geq 16788$ and $w = 2$ and $E = 0$ and $E' = 1$

    then $d \Leftarrow d + 1$

8   $J \Leftarrow d + \text{MOD}(\text{MOD}(d + 5, 7), 2) + 347997$ (26.12) and (26.13)

It is very likely that the value of t calculated in step 1 will exceed the capacity of a 32-bit computer word. If this is not acceptable, steps 1, 2, and 3 should be replaced by:

a $\mu \Leftarrow (235 * Y - 234)/19$

b $t_c \Leftarrow 204 + 793 * \mu$

c $t_h \Leftarrow 5 + 12 * \mu + t_c/1080$

d $d \Leftarrow 1 + 29 * \mu + t_h/24$

e $t' \Leftarrow \text{MOD}(t_c, 1080) + 1080 * \text{MOD}(t_h, 24)$

An algorithm for calculating the year Y in which day J falls

Algorithm H

1 $M \Leftarrow (25920 * (J - 347\,996))/765\,433 + 1$ **(26.14)**

2 $Y \Leftarrow 19 * (M/235) + (19 * (\mathrm{MOD}\,(M, 235) - 2))/235 + 1$ **(26.15)**

3 Calculate the day number of Tishi 1st, J', for year Y using the previous algorithm.

4 If $J' > J$

   $Y \Leftarrow Y - 1$

It is possible that the value of $k(J - 347\,996)$ calculated in step 1 will exceed the capacity of a 32-bit computer word. If this is not acceptable, step 1 may be replaced by an instruction that involves decimals. The following is adequate for the first 16 000 years after the Creation:

1a $M \Leftarrow [(J - 347\,996) * 0.03386318] + 1$

Here, the square brackets [. . .] indicate that the integral part only of the product should be retained. The decimal constant is calculated as the ratio a/m; you should ensure that all eight places after the decimal point are utilized so that the product is accurate to a month.

The relation between Anno Domini and Anno Mundi

The molod of Tishri of the year Y (Anno Mundi) currently falls in the autumn (in the Northern hemisphere) of the corresponding Christian year, Y_c; this is given by AD $Y - 3761$. The mean length of the year is 365.24682 days which is slightly longer than the Gregorian year and slightly shorter than the Julian year. Thus, the molod of Tishri will occur later and later in the Gregorian year and earlier and earlier in the Julian year. However, these discrepancies are small and neither amounts to as much as a day in 200 years. The year, Y EM, will therefore start in the Christian year $Y - 3761$—not for ever, but at least for as long into the future as it has in the past. With this proviso, we may write:

$$Y = Y_c + 3761 \qquad (26.16)$$

TABLE 26.1 *Characters of years*

K	Type	Variety	Days/year
1	Common	Deficient	$353 = K + 352$
2	Common	Regular	$354 = K + 352$
3	Common	Abundant	$355 = K + 352$
4	Embolismic	Deficient	$383 = K + 379$
5	Embolismic	Regular	$384 = K + 379$
6	Embolismic	Abundant	$385 = K + 379$

or $Y_c = Y - 3761$

If the year designated by Y_c is a year before the Christian era (BC) you must ensure that it is properly represented by a negative number; that is, for y BC you must put $Y_c = 1 - y$.

⌛ The days in the months

The number of days in the months varies with the character of the year. This in turn depends on the number of days in the year as indicated in TABLE 26.1.

In this table we see that the six types of year may be identified by a single parameter, K. To find this for year Y we must first establish the number of days in the year. To do this, calculate the day numbers, J and J', of the first day of Tishri in year Y and in year $Y + 1$. We then find that:

$$K = J' - J - 352 - 27\text{MOD}(7Y + 13, 19)/12 \qquad (26.17)$$

The last term on this expression depends on the fact that if year Y is embolismic, $\text{MOD}(7Y + 13, 19) \geq 11$ (as the reader may verify).

TABLE 17.3 (see page 229) shows the number of days in every month for each of the six types of year. Of more use in the present context is TABLE 26.2 which shows the number of days in the years which precede the start of a given month.

⌛ To convert a day number into a Jewish date

We may first calculate the year, Y, in which the day, J, falls using algorithm H. Next we calculate the character of the year, K, from equation 26.17. It remains to calculate the month, M, and the day of the month, D. A general formula for

TABLE 26.2 *The aggregate number of days preceding each month*

K	Month (M)												
	1	2	3	4	5	6	7	8	9	10	11	12	13
1	0	30	59	88	117	147	176	206	235	265	294	324	—
2	0	30	59	89	118	148	177	207	236	266	295	325	—
3	0	30	60	90	119	149	178	208	237	267	296	326	—
4	0	30	59	88	117	147	177	206	236	265	295	324	354
5	0	30	59	89	118	148	178	207	237	266	296	325	355
6	0	30	60	90	119	149	179	209	238	267	297	326	356

We generally denote the number of days preceding month M in a year of character K by $A(K,M)$. The table gives values of A for every M and K.

It is possible to derive a formula which gives $A(K,M)$ in terms of K and M, but it is complicated and not given here.

calculating the month remains elusive but M can be found quite easily using TABLE 26.2.

The first step is to calculate the day of the year. To do this we calculate the day number, J_1, of Tishri 1st for year Y using algorithm G. We then find that the day of the year is given by d in:

$$d = J - J_1 + 1 \tag{26.18}$$

We must now consult the line in TABLE 26.2 which corresponds to the character, K, of the year in question. We must find the latest month whose entry, A, is not greater than d. For instance, for a year of character $K = 6$ and a day $d = 234$, we find that month 8 has an entry 209, whereas month 9 has an entry 238; 209 is less than d but 238 is greater. The month is then $M = 8$ with $A = 209$. The day of the month is then given quite simply as:

$$D = d - A \tag{26.19}$$

⊠ To convert a Jewish date to a day number

We first calculate the day number, J_1, of Tishri 1st for year Y using algorithm G. We next find the character, K, of the year using equation 26.17, and then look up the number of days, $A(K,M)$, preceding the start of month M for a year of appropriate character. Finally, we calculate J by:

$$J = J_1 + A(K,M) + D - 1 \tag{26.20}$$

⌛ An algorithm for calculating the date $D/M/Y$ from a day number, J

$\mathcal{A}l\,g\,o\,r\,i\,t\,h\,m$ I

1 Calculate the Jewish year, Y, in which J falls using algorithm H

2 Calculate the day number, J', of the Tishri 1st in year Y and the corresponding value of K, using algorithm J   **(26.17)**

3 $d \Leftarrow J - J' + 1$ **(26.18)**

4 Find from row K in table 26.2, the highest value of M such that $d > A(K, M)$

5 $D \Leftarrow d - A(K, M)$ **(26.19)**

Programmers may note that if we put $n = d/30$, then M is either $n - 1$, n, or $n + 1$.

⌛ An algorithm for calculating the day number, J, of a date $D/M/Y$

$\mathcal{A}l\,g\,o\,r\,i\,t\,h\,m$ J

1 Calculate the day number, J, of the Tishri 1st in year Y using algorithm G

2 Calculate the day number, J', of the Tishri 1st in year $Y + 1$ using algorithm G **(26.17)**

3 $K \Leftarrow J' - J - 352 - 27 * \mathrm{MOD}(7*Y + 13, 19)/12$ **(26.17)**

4 Look up $A(K, M)$ from TABLE 26.2.

5 $J \Leftarrow J + A(K, M) + D - 1$ **(26.20)**

This algorithm is readily programmed in a computer with the aid of a lookup table for $A(K, M)$.

CHAPTER 27

The Mayan Calendar

⊠ Introduction

The Mayan people reckoned dates in two ways: by the calendar round and by the long count. The former allowed them to specify a date within a period of about 52 years; the latter was a direct measure of the number of days elapsed since the start of a great cycle in the remote past. We take this day to be the epoch of the long count and, following many scholars, take this to be the Julian date 8/9/3114 BC, which has a Julian day number of 584 285.

The Mayan calendar has interesting mathematical features involving interlocking cycles; we will explore these first.

The long count

The long count measures, the time elapsed from the start of the great cycle in a set of units:

1 pictun	= 20 baktuns	=	2 880 000 days
1 baktun	= 20 katuns	=	144 000 days
1 katun	= 20 tuns	=	7200 days
1 tun	= 18 uinals	=	360 days
1 uinal	= 20 kins	=	20 days

A day in the long count is thus specified by a series of numbers which define the number of days since the start of the great cycle. It is convenient to denote the numbers of the different units by six parameters: P denotes the number of pictuns, B of baktuns, K of katuns, T of tuns, U of uinals, and D of kins (or days).

The calendar round

The calendar round specifies a date by four numbers which I shall denote by A, B, C, D:

A lies in the range 1 to 13 and defines the position of the day in the trecena of 13 days.

B lies in the range 1 to 20 and defines the position of the day in the veintena of 20 days.

C lies in the range 0 to 19 (or 0 to 4 for the last uinal) and defines the position of the day in the current uinal of the current haab.

D lies in the range 0 to 18 and defines the current uinal of the current haab.

Together, *A* and *B* define the position in the 260-day 'tzolkin' cycle. Likewise, *C* and *D* define the position of a day in the current 365-day haab. The haab is divided into 18 uinals of 20 days each. It is convenient to refer to the remaining five days as forming a 19th uinal.

Not all 94 900 (13 × 20 × 365) combinations of the four numbers which specify a day in the calendar round are possible. The sequence of numbers repeats after 18 980 days (about 52 years); 18 980 is the lowest common multiple of 365, 20, and 13; and only one fifth of the possible combinations are possible.

The values of *A*, *B*, *C*, and *D* for the epoch of the long count are 4, 20, 8, 17.

The specification of the calendar round only fixes a day within the current 18 980-day cycle. A further quantity (which I call simply the cycle number and denote by *Y*) can be specified; this gives the number of complete 18 980-day cycles elapsed since the start of the great cycle. It must be emphasized that the Mayans did not use this device; instead, they employed the long count to define a date in the current calendar round and so fix it within a longer time span.

The tzolkin

The trecena, *A*, and the veintena, *B*, fall in the ranges:

$$1 \leq A \leq 13 \quad \text{(trecena)}$$
$$1 \leq B \leq 20 \quad \text{(veintena)} \tag{27.1}$$

All 260 (13 × 20) combinations of the values of *A* and *B* are possible. Together, *A* and *B* define a day, *M*, in the current tzolkin such that:

$$0 \leq M \leq 259 \tag{27.2}$$

Suppose that the number of trecenas which precede the current trecena is '*a*', and the number of veintenas which precede the current veintena is '*b*', so that:

$$0 \leq a \leq 12$$
$$0 \leq b \leq 19 \tag{27.3}$$

It is easy to see that:

$$M = 13a + A - 1 = 20b + B - 1 \tag{27.4}$$

It follows that:

$$13a - 20b = B - A = r = -r(13*3 - 20*2) \tag{27.5}$$

and we may show from equation 27.3 that:

$$-12 \leq r \leq 19 \tag{27.6}$$

Mathematicians may note that the term in brackets in equation 27.5 is obtained from the penultimate convergent of 13/20, as defined in the theory of continued fractions; it is equal to −1.

It follows from equation 27.5 that:

$$13(a + 3r) = 20(b + 2r) = Q \tag{27.7}$$

It is clear that Q is divisible by both 13 (since $Q = 13(a + 3r)$) and 20 (since $Q = 20(b + 2r)$), so that $(a + 3r)$ is divisible by 20 and $(b + 2r)$ by 13. It then follows by dividing equation 27.7 by 13 and 20 that:

$$(a + 3r)/20 = (b + 2r)/13 = t \tag{27.8}$$

where t is an integer. It then follows that:

$$a = 20t - 3r$$
$$\text{and} \quad b = 13t - 2r \tag{27.9}$$

We must choose t so as to ensure that a and b are positive; bearing in mind the range of values open to r given by equation 27.6, we express a and b by:

$$a = \text{MOD}(60 - 3r, 20)$$
$$\text{and} \quad b = \text{MOD}(39 - 2r, 13) \tag{27.10}$$

It then follows from equation 27.4 that:

$$M = 13\text{MOD}(60 - 3r, 20) + A - 1$$
$$\text{and} \quad M = 20\text{MOD}(39 - 2r, 13) + B - 1 \tag{27.11}$$

and also that:

$$A = 1 + \text{MOD}(M, 13)$$
$$\text{and} \quad B = 1 + \text{MOD}(M, 20) \tag{27.12}$$

Taking the first expression in equation 27.11 for M, we find that:

$$M = 13\text{MOD}(60 + 3(A - B), 20) + A - 1 \tag{27.13}$$

These results enable us to calculate A and B, given M, and conversely to calculate M, given A and B.

The haab

The analysis of the haab proceeds in a similar manner. The day of the uinal, C, and the uinal number, D, fall in the ranges:

$$0 \leq C \leq 19$$
$$0 \leq D \leq 18 \tag{27.14}$$

We may define a day of the haab N so that:

$$0 \leq N \leq 364 \tag{27.15}$$

and it follows that:

$$C = \text{MOD}(N, 20) \tag{27.16}$$

$$D = N/20 \tag{27.17}$$

$$N = 20D + C \tag{27.18}$$

These expressions enable us to calculate C and D, given N, and vice versa.

Cycles

A cycle of 18 980 days contains 73 ($= 18980/260$) complete tzolkins and 52 ($= 18980/365$) complete haabs. Suppose that R is the day of the cycle so that:

$$0 \leq R < 18\,980 \tag{27.19}$$

and suppose that m is the current tzolkin and n is the current haab such that:

$$0 \leq n \leq 51$$
$$0 \leq m \leq 72 \tag{27.20}$$

Then it is easily seen that:

$$R = 260m + M = 365n + N \tag{27.21}$$

so that:

$$365n - 260m = M - N = q \tag{27.22}$$

Since the left hand side of equation 27.22 is divisible by 5, so also is q. This requirement is equivalent to the observation that only one fifth of the possible combinations of A, B, C, and D are valid. We may thus write $5p$ for q and it then follows that:

$$73n - 52m = p = p(73 * 5 - 52 * 7) \tag{27.23}$$

and hence that:

$$73(n - 5p) = 52(m - 7p) \tag{27.24}$$

so that:

$$\frac{(n-5p)}{52} = \frac{m-7p}{73} = u \tag{27.25}$$

which gives:

$$n = 52u + 5p$$
$$m = 73u + 7p \tag{27.26}$$

We must chose u so as to ensure that n and m are in the range given by equation 27.20; to do this we again use the MOD function and write:

$$n = \text{MOD}(364 + 5p, 52)$$
$$m = \text{MOD}(511 + 7p, 73) \tag{27.27}$$

so that:

$$R = 365\text{MOD}(364 + 5p, 52) + N$$
$$\text{and} \quad R = 260\text{MOD}(511 + 7p, 73) + M \tag{27.28}$$

Taking the first of the expressions for R in equation 27.29, and substituting $M - N$ for 5p from equation 27.22, we find:

$$R = 365\text{MOD}(364 + M - N, 52) + N \tag{27.29}$$

⚃ To convert a day number to a calendar round date

We must first calculate, using day J, the number of days, L, elapsed since the start of the last great cycle. This is given by:

$$L = J - 584\,285 \tag{27.30}$$

Then Y, the number of cycles of 18 980 days that have been completed, is given by:

$$Y = L/18980 \tag{27.31}$$

The remainder represents the day number, R, in the current calendar round:

$$R = \text{MOD}(L, 18980) \tag{27.32}$$

We may now use equation 27.12 to obtain A, the trecena number. However, we must arrange that $A = 4$ when $R = 0$ (since this is the trecena number of the first day of the great cycle), and that A is in the range 1 to 13. This is easily done by:

$$A = 1 + \text{MOD}(R + 3, 13) \tag{27.33}$$

Likewise, we may calculate the veintena number B (which is 20 when R is 0) by:

$$B = 1 + \text{MOD}(R + 19, 20) \tag{27.34}$$

The days of the haab, N, are numbered 0 to 364 and are similarly given by:

$$N = \text{MOD}(R + 348, 365) \tag{27.35}$$

since when $R = 0$, $N = 348$ (or $C = 8$, $D = 17$). From N, we may easily calculate C and D using equations 27.16 and 27.17.

These formulae are collected together and presented as steps in an algorithm, for use with a pocket calculator or for programming into a computer. Therefore, to convert day number J into a calendar round date specified by A, B, C, D in cycle Y:

$\mathcal{A} l g o r i t h m \ K$

---

1 $L \Leftarrow J - 584\,285$	(27.30)
2 $Y \Leftarrow L / 18\,980$	(27.31)
3 $R \Leftarrow \text{MOD}(L, 18980)$	(27.32)
4 $A \Leftarrow 1 + \text{MOD}(R + 3, 13)$	(27.33)
5 $B \Leftarrow 1 + \text{MOD}(R + 19, 20)$	(27.34)
6 $N \Leftarrow \text{MOD}(R + 348, 365)$	(27.35)
7 $C \Leftarrow \text{MOD}(N, 20)$	(27.16)
8 $D \Leftarrow N / 20$	(27.17)

Several of these steps can be compressed, and the algorithm made more efficient; this is left to the user.

▣ To convert a calendar round date to a day number

We may easily calculate the number of days which precede the start of the current calendar round from Y, the cycle number. If this is not available, the conversion is not possible.

We may assume that J is the sum of three terms:

$$J = T_1 + T_2 + R \tag{27.36}$$

T_1 is the day number of the first day of the current great cycle; T_2 is the number of complete 18 980-day cycles which have elapsed in the great cycle; R is the number of days from the start of the current calendar round to the current date expressed as A, B, C, D. We shall assume that this set of numbers represents a valid date; not all sets are valid.

In order to calculate R, we use the expression for R given by equation 27.30. This also requires the values of M and N which may be calculated from equations 27.13 and 27.18. The expression for N tacitly assumes that the first day of the first year of the great cycle was described by $C = 0$, $D = 0$. In fact, it was 8, 17, corresponding to a value of N of 348. We must accordingly subtract 348 from N. For a similar reason we must subtract 159 from M to get:

$$N = \text{MOD}(N + 17, 365) \tag{27.38}$$

$$M = \text{MOD}(M + 101, 260) \tag{27.39}$$

Note that $17 = 365 - 348$ and $101 = 260 - 159$.

We may now calculate R from M and N using equation 27.28. Finally, we may set T_1 to 584 285 and T_2 to 18 980Y.

These formulae are collected together and presented as steps in an algorithm. Therefore, to convert a calendar round date specified by A, B, C, D in cycle Y to a day number, J.

Algorithm 2.

1	$M \Leftarrow 13\text{MOD}(60 + 3 * (A - B), 20) + A - 1$	(27.13)
2	$M \Leftarrow \text{MOD}(M + 101, 260)$	(27.38)
3	$N \Leftarrow 20 * D + C$	(27.18)
4	$N \Leftarrow \text{MOD}(N + 17, 365)$	(27.37)
5	$R \Leftarrow 365\text{MOD}(364 + M - N, 52) + N$	(27.29)
6	$J \Leftarrow 584\,285 + 18\,980Y + R$	(27.36)

Again, several of these steps can be compressed and the algorithm made more efficient; this is left to the user.

To convert a day number to a long count date

Since the day number, L, of the start of the last great cycle is 584 285, the number of days elapsed since the start of this is given by equation 27.31.

If we divide L by 20, the number of complete uinals within the period is given by the quotient, and the remainder gives the number of odd kim or days. If we continue and divide the first quotient by 18, we obtain analogously the number of complete tuns as the quotient and the number of odd uinals as the remainder. We may proceed in this way to find the number of complete katuns, baktuns, and pictuns.

⬚ To convert a long count date to a day number

If we multiply the number of pictuns by 20, we obtain the number of baktuns contained in them; we may add to this the number of baktuns and multiply the result by 20 to obtain the number of katuns. In this way we may calculate the number of days denoted by the long count specification. This is simply expressed by:

$$J = 584\,285 + ((((P * 20 + B) * 20 + K) * 20 + T) * 18$$
$$+ U) * 20 + D \tag{27.39}$$

This sort of procedure will be familiar to anybody used to converting pounds, shillings, and pence to pence!

⬚ To interconvert calendar round and long count dates

No special algorithms are required for these conversions. First, convert the calendar round or long count date to a day number; then, convert the day number to a long count or calendar round date. Use the algorithms already given.

PART IV

EASTER

A Short History of Easter

☒ Easter in the Christian religion

Easter is the time when Christians celebrate their central mystery of the Crucifixion and Resurrection of Jesus Christ. The word 'Easter' (German, 'Oster')
itself derives from the name of the old Northern god of the spring, Eostre, but
a different word, deriving from the Hebrew for Passover, 'Pesah', is used in
almost all other European languages and in the English adjective, 'paschal'.
There is no etymological connection between paschal and Christ's Passion on
the Cross.

The Gospels are unanimous that Christ was crucified when Pontius Pilate
was the Roman procurator. This is corroborated by two historians, writing a
little while after the Crucifixion: both Tacitus and Josephus put the event in
the range AD 27 to AD 37. The Gospels are also agreed that Jesus was crucified
on the day after the Last Supper; this has been taken to represent the Passover
feast which, according to Jewish custom, was eaten on the afternoon of the
14th day of Nisan. There is a discrepancy between the testimony of the Gospel
of St John (which suggests that the Crucifixion took place before the Passover
feast) and the other three Gospels, which has generated much controversy.
Finally, the Gospels are agreed that the Crucifixion took place on the day of
preparation, the Friday, which precedes the Jewish Sabbath honoured on the
Saturday.

We therefore need to consult astronomical tables and select a Thursday
within the years specified which was also the 14th day of a moon. This yields
possible, maybe probable, dates for the Crucifixion of Friday, 7 April AD 30 or
3 April AD 33. The former date is preferred on astronomical grounds and the
latter on political grounds. If Jesus was born in 4 BC, and died in AD 30, he
would have been 33 or 34 years old at the time of his Crucifixion.

☒ The Quartodeciman heresy

There is no mention of the celebration of Easter in the books of the New
Testament and St Paul himself had little time for dates and anniversaries,

but a practice soon arose of celebrating the Eucharist on Sundays, the day of the Resurrection.

Some early converts from Judaism to Christianity continued to celebrate the Passover, whereas other Christians may have taken over a vernal festival from local pagan religions. Nevertheless, by the second century, the observance of an annual festival was already well established; there is mention in a letter written in about AD 200 of an Easter celebration with the burning of wax candles.

Christianity is a historical religion—that is to say, its adherents consider it to be based upon real historical events. Ultimately, therefore, dates were important to it—the Gospels abound in historical references designed to place the events it describes in a historical context. Consequently, by the middle of the second century, the problem of exactly when Easter should be celebrated had arisen.

The early Christian converts saw the Mysteries of the Resurrection as the fulfilment of the Jewish Passover which begins on the first full moon following the vernal equinox, but discrepancies arose between the practices of Christians converted from Judaism and others in Asia Minor, on the one hand, and the Roman and Alexandrian Christians on the other. Among other doctrinal differences, the former stressed the day of the month, the 14th of Nisan, and they came to be known as the Quartodecimans (that is, the 14th people), while the latter (the Quintadecimans) stressed the day of the week and celebrated the Resurrection on the Sunday which follows Nisan 14th.

In about AD 155, the controversy came to a head and Polycarp, Bishop of Smyrna and a representative of the Quartodecimans, travelled to Rome to confer with the current Pope, Anicetus (AD 155–166). Anicetus maintained that he had to abide by the customs of his predecessors and he and Polycarp parted amicably, having agreed to differ. Polycarp was held in great respect by those who met him and although he could never have met Jesus Christ himself, it is likely that he had known several people, including St John, who had. He was martyred in a bout of anti-Christian sentiment shortly after his visit to Rome, and later canonized; he was over 86 when he was burnt to death.

Nevertheless, by about AD 191, not only had the Quartodecimans quarrelled among themselves but a Quartodeciman heretic called Blastus was aggravating the Church authorities in Rome; the whole controversy was beginning to make the early Christian Church a laughing stock among the pagans. One reason for argument was that if Easter was tied to the date of the Jewish Passover, the Christians would always need to consult the Jewish authorities to ascertain the date, for only they knew how to tell when the Passover should be.

In AD 197, Victor, Pope from 189 to 198, had acrimonious discussions with the metropolitan of Asia Minor and the representative of the Eastern Church, Polycrates of Ephesus. Councils were held all over Christendom—in Palestine, Rome, Gaul, and Corinth. Most agreed to observe the Resurrection on the Sunday; the Church at Alexandria followed suit, and only Asia Minor resisted. Victor wrote letters to the bishops in Asia Minor, threatening to excommunicate those who adhered to Quartodeciman practices. Victor received a defiant answer from Polycrates: 'seven of my kinsmen were bishops and I am the eighth and my kinsmen always kept the day when God's people put away the leaven'. Moreover, Victor received no support for his threats of excommunication from the other, less draconian, Western bishops, notably Irenaeus, who headed the Council in Gaul (and had been greatly impressed by the saintly Polycarp in his youth). The threats were not carried out. The bishops blamed Victor for turning a matter of custom into a matter of faith. Some Churches continued in their own Quartodecimanian ways.

Later, in AD 314, the Council of Arles confirmed that Easter should be celebrated on the same day throughout the world, but without detailing how this might be accomplished.

ⅩⅠ The Council of Nicaea

In AD 325, the Church was being torn apart by many heresies—not only that of the Quartodecimans. The Emperor Constantine, though not then a baptized Christian himself, was sympathetic to the Church. One of the methods by which the Roman emperors kept their empire together was to declare themselves gods and insist on being worshipped. The Christians would have nothing to do with this and their increasing popularity began to threaten the integrity of the empire. Constantine, realizing that he could not beat them, decided to join them. He thus had an interest in resolving the various disputes and in defining Christian belief. He therefore called the first Oecumenical Council, the Council of Nicaea. Three hundred and eighteen representatives from all over Christendom met in the Imperial Palace at Nicaea (Iznik) in Bithynia (a province in north-western Turkey); they sat from 20 May to 25 July in AD 325. Constantine himself addressed the meeting, which was at times acrimonious—although it ended with a great banquet.

As well as giving us the Nicene Creed, the Council reaffirmed the position of the Western Church over the date of Easter. Records of the discussion itself are not extant, but a letter from the Council to the Church at Alexandria, and a circular letter from Constantine himself, confirmed that the dispute was over, that everyone had agreed to celebrate Easter on the same day, and that the Egyptian Church would be responsible for telling them when to do it.

This, alas, did not end the controversy; despite the excommunication of those still holding to the Quartodeciman practice in AD 341 (by the Council of Antioch), and despite further condemnation in AD 364 (at the Council of Laodicaea) and AD 381 (at the Council of Constantinople), the heresy lingered on in the more remote parts of Christendom, only finally ending several centuries later.

Later in this fourth, momentous century, Constantine was converted to Christianity and transferred his capital from Rome to Byzantium, which was renamed Constantinople. Later, the Roman Empire itself split, and in 410 Alaric sacked Rome. St Augustine of Hippo wrote *The city of God* to rebut the charge that Christianity was the cause of this shocking catastrophe; the Dark Ages if not the end of the world were nigh.

▨ Early Easter canons

If Christianity was to be independent of the Jewish religion it was necessary that it should develop its own method for working out the date for Easter. In the early years, before the Council of Nicaea, some depended on Alexandria to tell them when Easter should be celebrated, and some on Rome, but few early accounts of the methods used survive. The bishops of Alexandria, that home of Greek astronomy, started the custom of writing paschal letters in which they announced the date of the forthcoming Easter. The modern practice of bishops sending Lenten letters to their clergy originated from this practice.

The date of the Jewish Passover was related to the date of the vernal equinox and to the age of the moon, initially established by direct observation. One problem with this arrangement was that the date of Easter could vary with longitude. The exact time of any astronomical event, such as the moment that the sun enters the first point of Aries (that is, the vernal equinox), depends on the local time of the observer. Another problem arose from the institution of leap days in the Julian calendar; these caused the date assigned to the vernal equinox to oscillate around the actual event.

To circumvent these difficulties, astronomical observation had to be abandoned. In its place various cycles of greater or less accuracy were used to predict the astronomical events. The idea was to use, instead of the real moon, a notional or theoretical moon with a well-defined period which matched, at least approximately, the true synodic period. It would then, if the model were accurate, be possible to calculate the dates of new and full moons. Likewise, a fictional vernal equinox on some fixed date had to be assumed. These would enable a date to be calculated for Easter which would be valid worldwide.

According to a paschal letter written by Dionysius of Alexandria in about AD 260, the bishops there were maintaining that Easter should not be celebrated till after the vernal equinox, which they assumed was on 21 March, and were using the octaeteris (see Chapter 6) to calculate the date of the first full moon after the equinox. The octaeteris guarantees that the dates of this would repeat every eight years, and it thus provides a very simple method suitable for use by any country priest. It is not, however, an accurate representation of the real moon. The bishops might have used the more accurate Metonic cycle, which had been known for a thousand years—but for some reason did not.

The Romans used a method designed by Hippolytus (who died in about AD 236), which was essentially similar to the Octaeteris, but employed a cycle of 16 years; they believed that the equinox fell on 25 March. The inscription on a statue of Hippolytus, unearthed in Rome in 1551, provided full details of this canon—the dates of the moons repeated every 16 years, but on different days of the week; a combination of day of the week and date repeated every 112 years. However, the scheme was no more accurate with respect to the real moon than the octaeteris.

Several more attempts were made to improve the canon. These included the use of an 84-year cycle by an otherwise unknown Roman, Augustalis. In this method, the dates of Easter Sunday repeated every 84 (3 × 4 × 7) years, after which the new moon would appear a mere 1.3 days late. Augustalis' method also first introduced the idea of the epact—the age of the moon at the start of the calendar year. This became of crucial importance after 1582, as will be seen.

The method implicitly accepted by the Council of Nicaea was based on the practices of Rome and Alexandria and incorporated the following rule: Easter Sunday, the anniversary of the Resurrection, was to be celebrated on the first Sunday which came after the 14th day of the paschal moon (the days being counted from the appearance of the new moon). The paschal moon was the first whose 14th day fell on or after the vernal equinox. This rule was designed to ensure, among other things, that Easter day was never on or before the Jewish Passover; the Roman Christians wished to dissociate themselves from Judaism. The bishops of Rome and Alexandria were to inform their colleagues throughout Christendom of the correct date each year in advance.

At the time, Roman and Alexandrian practices differed, as already stated, in that Rome believed the vernal equinox to occur on 25 March, whilst the Alexandrians held it to be 21 March. The matter could have been resolved by observation (when in some years neither would be found to be correct) but apparently was not. While the Romans stuck to their 16-year Hippolytic cycle, the Alexandrians had adopted the Metonic cycle which had been introduced by Anatolius; the latter became Bishop of Laodicea in AD 268 and died there in 282.

From AD 328, St Athanasius, Bishop of Alexandria (who had won out over his rival Alexandrian, Arianus at Nicaea), began to publish paschal letters based on the Metonic cycle. At first, the collaboration with Rome went well, but as time went on, there were occasions when the Romans and Alexandrians celebrated Easter on different days, as in AD 387. Part of the reason for this was that the Romans did not know how to operate the Metonic cycle.

The Alexandrians prepared tables of the dates of Easter, covering an extended series of years. Eventually, a set covering the years 380 to 479 were sent by Bishop Theophilus (385–412) to the emperor, Theodosius I (379–95); this was the same Theophilus who had finally destroyed the Library at Alexandria. Likewise, Theophilus' successor, Cyril, sent a set covering 437 to 531 to Theodosius II (401–50). These tables enabled the Roman computists to figure out how the Alexandrians did their calculations.

An important step in establishing a definitive and universally acceptable method was taken by Victorius, Bishop of Aquitaine, at the request of Hilary, Deacon of Rome (who later became Pope). Hilary had asked Victorius to examine the reasons why the Roman and Alexandrian canons differed occasionally. Victorius discovered that the dates of Easter Sunday calculated according to the Alexandrian canon repeat every 532 ($7 \times 19 \times 4$) years—the paschal cycle. He went on to propose a method based on this discovery and a table of dates for Easter spanning 532 years from AD 28. Hilary published Victorius' Easter method or canon in AD 465, and the 532-year cycle is sometimes called the 'Victorian cycle'.

Victorius' canon did not receive universal acclaim because it resulted in the recommended date of Easter sometimes lying outside the expected traditional range and because it did not agree with the method used in Alexandria.

⌘ The Dionysian canon

The tables sent by Cyril were due to come to an end in 531 and the matter was still not resolved. Pope John I (523–6) asked two of his secretaries, Primicerius Bonifatius and Secundicerius Bonus, to enlist the Abbot of Scythia, Dionysius Exiguus ('Dennis the Little'; so called on account of his self-demeaning manner) to look into the matter.

Dennis did several things. First, and momentously, he introduced a new era—the Christian era; secondly, he set out clear methods for calculating the date of Easter for any year, which were again based on the Metonic cycle as used in Alexandria; this was the Dionysian canon, to be used for a thousand years.

Dennis knew that there was a new moon on 23 March AD 323 and he also believed (wrongly) that there was an equinox on the same day. This coincidence seemed a good reason to assign this year to the first of a Metonic cycle. He then worked out from this assumption that the first new moon of AD 325 fell on 1 January which by a happy coincidence was the year of the Council of Nicaea. Finally he calculated that 1 BC, the year he had decided was that of the Nativity, was also the first year of a cycle.

Dennis invented his new era so that he need not associate Christianity with the era of Diocletian (discussed in Chapter 16), which had been used in the Alexandrian system for calculating Easter. His strategy for calculating the date of Easter was widely approved and adopted. Isidore of Seville continued the tables for another 95 years, and the venerable Bede later completed them to the end of the second Victorian period (AD 1064).

Bede (673–735), a Benedictine monk from the monasteries of Wearmouth and Jarrow in County Durham, was one of the few Englishmen to make a significant contribution to the development of the calendar. He originally wrote a small pamphlet, 'Liber de temporibus', for his students. This was well received, but, it is said, little understood. Bede expanded it into one of his major works *de temporum rationale*, which was finished in AD 725. This comprehensive treatise set the standard for generations of computists to come; in it, Bede used the Christian era to date events. In another of his works, on the history of Christianity in England, he frequently berates his countrymen for not observing Easter on the correct date. His apparently pedantic attitude stemmed from his belief that the terrible pillaging and massacres instigated by bloodthirsty tribal kings, which had become endemic in England after the last Roman legions had left, could be curbed only by uniting all men behind a common set of Christian beliefs and practices. He may have been right, though violence did not end once orthodox Christianity had become established.

Despite approval of the Dionysian canon from both the Eastern and Western Churches, and the availability of tables for calculating the date of Easter, the furthest outposts of Christendom were slow to adopt it. Spain only did so in AD 587, and Gaul waited till the close of the eighth century.

The Celts, at the fringes of the Christian world in Britain, found communication with Rome difficult—but the matter came to a head in the seventh century. King Oswy of Northumbria was following the Celtic tradition but his wife, Eanfleda, adhered to Rome; this sometimes resulted in them observing Easter on different days. Eventually, this difference caused strife and, in 664, the Synod of Whitby was called to settle the matter. The king gave way and the English Church was, at last, united—although the monks at Iona held out till AD 715.

⌛ The Gregorian reform

Unfortunately, the basis of this otherwise admirable scheme was faulty. The length of the Julian year did not properly match that of the true tropical year, and the average length of a notional lunation did not match the mean synodic period of the moon. As the centuries passed, the date of the actual vernal equinox fell earlier and earlier before the assumed calendar date of 21 March. Likewise, astronomical new moons were observed earlier and earlier before the calendar dates of the notional new moons. It began to be obvious that the intentions of the Council of Nicaea were being thwarted.

From the thirteenth to the sixteenth century, there followed a long and dismal series of attempts to remedy the matter (which are described more fully in Chapter 15). Finally, Pope Gregory published his bull and a new calendar was launched in 1582, together with a new canon for the calculation of the date of Easter.

The Protestant countries objected to the new canon. Among other reasons, they believed that to fulfil the rule handed down from the Council of Nicaea, the date of Easter should be worked out with reference to the real moon (a naive and retrogressive point of view). The evangelical states went so far as to base their calculations on observations made at the longitude of Uraneborg, where Tycho Brahe had his observatory. All went well for a time but, in 1724, their date and the new Gregorian date conflicted and riots between Protestants and Catholics ensued. As the difficulties attendant on having different calendars in the different states of Europe become intolerable, the Protestant countries saw sense and followed, one by one, the Catholics in adopting the new calendar. England reformed its calendar in 1752 and Russia in 1918.

The Eastern Church never adopted the Gregorian reform. Instead, at a congress held in Constantinople in 1923, it resolved on a reform of its own in which centurial years are leap years only if they leave a remainder of two or six when divided by nine, and in which the date of Easter Sunday is calculated with reference to the actual moon observed in the meridian of Jerusalem.

Some see the movement of the date of Easter as a great inconvenience. Although the Vatican has stated that it has no objection to a fixed date for Easter, and although the Easter Act of 1928 allows an order in Council to fix the date of Easter in Great Britain as the first Sunday after the second Saturday in April, nothing has so far come of reform.

It cannot be emphasized too strongly that both the Dionysian and Gregorian canons calculate the date of Easter with reference to a notional, ecclesiastical moon, and a notional vernal equinox. The phases of the ecclesiastical moon may come before or after the corresponding phases in the real moon by as much as three days because the period of an actual lunation varies. Even in recent years, this can cause consternation among the ignorant as, for instance, in 1845, when Easter Sunday was celebrated on the same day that a full moon could be seen in the sky. Likewise, the vernal equinox does not always occur on the date assigned to it.

CHAPTER 29

The Date of Easter Sunday

⌛ Introduction

I am reminded of the author who, wishing to know how to calculate the date of Easter, telephoned the Royal Observatory at Greenwich. He was told that those august astronomers consulted the tables at the back of the Prayer Book. This not altogether satisfactory, albeit apocryphal, response prompted this chapter, in which I discuss how the date of Easter Sunday is calculated first according to the Dionysian canon in the Julian Calendar and then according to the Gregorian canon. This involves a discussion of golden numbers and epacts and how these and dominical numbers are used in the calculation. (Dominical letters and numbers have already been described at length in Chapter 24.)

The traditional method involves the use of tables, worked out in advance, such as may be found in the Book of Common Prayer of the Anglican Church. There, they are adequately explained and are not discussed further here. Tables published from 1662 to 1752 are for use with the Julian calendar; those published after 1752 are for use with the Gregorian calendar.

According to the Dionysian canon, Easter falls on the first Sunday that follows the paschal full moon. The paschal full moon is defined as the first full moon that occurs on or after the vernal equinox which is taken, notionally, to fall on 21 March.

The moon referred to is not, however, the real moon up in the heavens. It is, instead, a notional or theoretical moon which, it is supposed, keeps more or less in step with the real moon. This notional moon is governed by relatively simple rules and the prediction of the date of its new moons does not require either observation of the heavens or elaborate astronomical calculations. The rules that govern its behaviour are based on the Metonic cycle, and first we must discuss how this is used.

🔟 The Metonic cycle

The form of the Metonic cycle used to describe the notional moons for the calculation of the date of Easter differs a little from that used in the Jewish calendar and from that originally prescribed by Meton.

The 19-year Metonic cycle contains 12 common years each consisting of 12 months or lunations and seven embolismic years each containing 13 lunations. The 19 years thus include 235 lunations ((12 × 12) + (7 × 13)) in all. The common years each contain six lunations of 29 days each (hollow lunations) and six of 30 days (full lunations). Six of the embolismic years contain, besides a similar complement of 29 and 30 day lunations, an extra, intercalated, full lunation of 30 days; in the seventh embolismic year, the extra lunation is hollow with 29 days. (This exception was known as the 'saltus lunae', the jump of the moon.)

Thus, the 235 lunations contained 6935 days ((19 × 6 × 29) + (19 × 6 × 30) + (6 × 30) + 29) in all. However, 19 Julian years contain 6939.75 days (19 × $365\frac{1}{4}$). The extra 4.75 days were accommodated by inserting an extra leap day every four years into whichever lunation contained 24 February. This lunation would then contain 31 or 30 days in leap years (according to whether it ordinarily had 30 or 29 days). These extra leap days brought the cycle into complete harmony with the Julian calendar.

The 12 common lunar years thus contain 354 days each, and the seven embolismic years, 384 days each—except the last, which contains 383 days. Obviously there is no constant alignment between the start of these lunar years and the Julian years of 365 days (ignoring any leap day). The first lunation of a Julian year is taken to be the first one that *ends* in it. Thus, lunations that straddle two Julian years are, by convention, taken to belong to the year in which they end so that some lunations start in December of one year and end in January of the next.

It may be seen that some 19-year cycles contain four leap years and some, five leap years; the cycle of lunations recurs on the same days of the year every 76 years (4 × 19)—the Calippic cycle.

The golden number

Of prime importance for the calculation of the date of Easter is the position of the relevant year in the Metonic cycle of 19 years. This position defines the golden number of the year. The calculation of this golden number requires the specification of the first year of some cycle. This was settled by Dionysius Exiguus, as explained in Chapter 16, so that 1 BC was the first year of a cycle. It follows that the golden number, G, of any year AD Y is given by the simple equation:

$$G = 1 + \text{MOD}(Y, 19) \tag{29.1}$$

For instance, the golden number of the year 1066 is $1 + \text{MOD}(1066, 19) = 3$. The definition of the golden number and this formula for its calculation are equally valid for the Gregorian calendar.

The lunar almanac

The next problem is to see how the 120 full lunations of 30 days each and the 115 hollow lunations of 29 days were distributed among the 19 years of the cycle. Note that each year was assumed to consist of exactly 365 days. The leap days were inserted after the lunations had been distributed, as already explained.

There was some variation amongst different computists in the distribution of the lunations; I shall describe the convention described by Clavius. The variations used were not such as to affect the date calculated for Easter.

The distribution given by Clavius is shown in TABLE 29.1 which is known as a lunar almanac. It consists of a list of all 365 days of a year. Against the dates are written the golden numbers of the years in which a new moon is deemed to have occurred on that date. For instance, the XIX against 5 January indicates that a new moon is scheduled for that day in the 19th year. Thus, the almanac shows the dates of all 235 notional new moons in the 19-year cycle. Because there are 365 dates in the almanac and only 235 new moons, some dates are left blank.

The table was constructed by first entering the new moons in the year of golden number III, with the first on 1 January. This, it may be calculated, ensures that a new moon fell on 23 March AD 323 and on 1 January in the year of the Council of Nicaea. The reasons for this apparently arbitrary choice for the start of the 19 year cycles has been explained above.

Full and hollow lunations are then entered, generally in alternation. For instance, the second III in the table is on 31 January, 30 days after the first. When the counting passes into the following year, the next golden number is introduced, with golden number I following XIX. So, for instance, IV occurs on 20 January, 30 days after the last III on 21 December.

It may be seen from TABLE 29.1 that the last lunation of II, which starts on 2 December, brings the cycle to an end on 31 December, so that the first lunation of III starts on 1 January again. It should also be noted that the golden numbers repeat in a fixed order throughout the year. Golden number G is followed after one or two days by golden number $1 + \text{MOD}(G + 7, 19)$.

If the reader would care to count the number of days between successive new moons in each of the years, he would find the lengths of lunations shown

TABLE 29.1 *The Julian lunar almanac*

Day of month	Jan C G	Feb C G	Mar C G	Apr C G	May C G	Jun C G
1	A III	D	D III	G	B XI	E
2	B	E XI	E	A XI*	C	F XIX
3	C XI	F XIX	F XI	B	D XIX	G VIII
4	D	G VIII	G	C XIX*	E VIII	A XVI
5	E XIX	A	A XIX	D VIII*	F	B V
6	F VIII	B XVI	B VIII	E XVI	G XVI	C
7	G	C V	C	F V	A V	D XIII
8	A XVI	D	D XVI*	G	B	E II
9	B V	E XIII	E V*	A XIII	C XIII	F
10	C	F II	F	B II	D II	G X
11	D XIII	G	G XIII*	C	E	A
12	E II	A X	A II*	D X	F X	B XVIII
13	F	B	B	E	G	C VII
14	G X	C XVIII	C X*	F XVIII	A XVIII	D
15	A	D VII	D	G VII	B VII	E XV
16	B XVIII	E	E XVIII*	A	C	F IV
17	C VII	F XV	F VII*	B XV	D XV	G
18	D	G IV	G	C IV	E IV	A XII
19	E XV	A	A XV*	D	F	B
20	F IV	B XII	B IV*	E XII	G XII	C
21	G	C I	C	F I	A I	D IX
22	A XII	D	D XII*	G	B	E
23	B I	E IX	E I*	A IX	C IX	F XVII
24	C	F	F	B	D	G VI
25	D IX	G XVII	G IX*	C XVII	E XVII	A
26	E	A VI	A	D VI	F VI	B XIV
27	F XVII	B	B XVII*	E	G	C III
28	G VI	C XIV	C VI*	F XIV	A XIV	D
29	A		D	G III	B III	E XI
30	B XIV		E XIV*	A	C	F
31	C III		F III*		D XI	

TABLE 29.1 *Continued*

Day of month	Jul C G	Aug C G	Sep C G	Oct C G	Nov C G	Dec C G
1	G XIX	C VIII	F XVI	A XVI	D	F XIII
2	A VIII	D XVI	G V	B V	E XIII	G II
3	B	E V	A	C XIII	F II	A
4	C XVI	F	B XIII	D II	G	B X
5	D V	G XIII	C II	E	A X	C
6	E	A II	D	F X	B	D XVIII
7	F XIII	B	E X	G	C XVIII	E VII
8	G II	C X	F	A XVIII	D VII	F
9	A	D	G XVIII	B VII	E	G XV
10	B X	E XVIII	A VII	C	F XV	A IV
11	C	F VII	B	D XV	G IV	B
12	D XVIII	G	C XV	E IV	A	C XII
13	E VII	A XV	D IV	F	B XII	D I
14	F	B IV	E	G XII	C I	E
15	G XV	C	F XII	A I	D	F IX
16	A IV	D XII	G I	B	E IX	G
17	B	E I	A	C IX	F	A XVII
18	C XII	F	B IX	D	G XVII	B VI
19	D I	G IX	C	E XVII	A VI	C
20	E	A	D XVII	F VI	B	D XIV
21	F IX	B XVII	E VI	G	C XIV	E III
22	G	C VI	F	A XIV	D III	F
23	A XVII	D	G XIV	B III	E	G XI
24	B VI	E XIV	A III	C	F XI	A XIX
25	C	F III	B	D XI	G XIX	B
26	D XIV	G	C XI	E XIX	A	C VIII
27	E III	A XI	D XIX	F	B VIII	D
28	F	B XIX	E	G VIII	C	E XVI
29	G XI	C	F VIII	A	D XVI	F V
30	A XIX	D VIII	G	B XVI	E V	G
31	B	E		C V		A XIII

The occurrence of a golden number, G, on a date indicates that a notional new moon occurs on that date in a corresponding year. The calendar letter, C, of each date is also shown. The paschal new moons are marked *.

TABLE 29.2 *Distribution of full and hollow lunations among the 19-year cycle.*

Golden number year/cycle	Lunation													Total
	1	2	3	4	5	6	7	8	9	10	11	12	13	
III	30	29	30	29	30	29	30	29	30	29	30	29		354
IV	30	29	30	29	30	29	30	29	30	29	30	29		354
V	30	29	30	29	30	29	30	29	30	30	29	30	29	384
VI	30	29	30	29	30	29	30	29	30	29	30	29		354
VII	30	29	30	29	30	29	30	29	30	29	30	29		354
VIII	30	29	30	30	29	30	29	30	29	30	29	30	29	384
IX	30	29	30	29	30	29	30	29	30	29	30	29		354
X	30	29	30	29	30	29	30	29	30	29	30	29		354
XI	30	30	29	30	29	30	29	30	29	30	29	30	29	384
XII	30	29	30	29	30	29	30	29	30	29	30	29		354
XIII	30	29	30	29	30	29	30	29	30	29	30	29	30	384
XIV	30	29	30	29	30	29	30	29	30	29	30	29		354
XV	30	29	30	29	30	29	30	29	30	29	30	29		354
XVI	30	29	30	29	30	29	30	29	30	30	29	30	29	384
XVII	30	29	30	29	30	29	30	29	30	29	30	29		354
XVIII	30	29	30	29	30	29	30	29	30	29	30	29		354
XIX	30	29	30	30	29	30	29	29	29	30	29	30	29	383
I	30	29	30	29	30	29	30	29	30	29	30	29		354
II	30	29	30	29	30	29	30	29	30	29	30	29	30	384

Note that the first lunation (of 30 days) begins (except for year III) in the previous calendar year.

The lunations assigned to a golden number define a lunar year; the last column gives the number of days in each such lunar year (which does not include any leap day).

The embolismic years have 13 lunations and the others, 12.

in TABLE 29.2. Some of the lunations start in December of one year and end in January of the next; in TABLE 29.2, these are assigned to the year in which they end. The table is seen to contain mainly full lunations alternating with hollow ones, but there are exceptions. Some years contain only 12 complete lunations, whereas others contain 13; the latter are the embolismic years.

Dionysius Exiguus, or the early church computists who devised the system, arranged the 120 full and the 115 hollow lunations in the order shown in TABLE 29.2 with several criteria in mind. Firstly, every year had to have exactly 365 days. Then, they required that:

1. The first lunation of each year should have 30 days and begin in the previous year (*except for year I*).

2. There should never be more than one golden number on the same day in the table.

3. All the paschal lunations (whose new moon occurs from 18 March to 5 April) should contain just 29 days.

4. Full and hollow lunations should generally alternate, particularly at the beginning and end of each year.

It is unnecessary to discuss in detail the extent to which these criteria are satisfied in TABLE 29.2, and I will only remark that the first three are met without exception, and that the full and hollow months alternate regularly in all the non-embolismic years. It may also be noted that the common lunar years have exactly 354 days each and the embolismic years have 384 days (except year XIX which has 383 days).

TABLE 29.1 can be used to find the Julian date of the notional new moons of any year, if we know its golden number. The 13th day after the new moon is the notional date of the corresponding full moon. We could thus easily, but tediously, construct a corresponding almanac of full moons rather than new moons, and this would give us the dates of the full moons in any year.

⌛ The date of Easter

As we have mentioned, the date of Easter was originally worked out with tables; instead, we will calculate it using the rule, according to the Dionysian canon, that Easter Sunday is the first Sunday that follows the paschal full moon. The paschal full moon is defined to be the first one that occurs on or after the vernal equinox which is taken, notionally, to fall on 21 March.

A full moon occurs 13 days after its corresponding new moon, counting the day of the new moon as the first day. Thus, the paschal new moon must fall no earlier than 13 days before 21 March, that is, on or after 8 March.

It may be seen from TABLE 29.1 that there is in fact a new moon on 8 March in year XVI, so that Easter Sunday must fall on a Sunday after 21 March; its earliest possible date is 22 March. The paschal new moons for the other 18 years of the cycle fall on dates after 8 March, as shown in TABLE 29.1; the latest falls on 5 April in year VIII, with the corresponding full moon on 18 April. If this happens to be a Sunday, Easter Sunday must be seven days later than this, on 25 April—its latest possible date.

Thus, Easter Sunday could fall on any day from 22 March to 25 April, a period of 35 days. Likewise, the paschal full moon can fall on any day from 21

March to 18 April, and the paschal new moon on any day from 18 March to 5 April.

To explain how Easter may be calculated, we first construct the paschal table shown in TABLE 29.3. In this, column G shows the golden number of each year of a cycle, and the date column shows the dates of the corresponding paschal full moons, arranged in temporal order. These dates must lie in the range 21 March to 18 April, as just explained. (They are easily found from TABLE 29.1 by noting the dates which are 13 days after the corresponding paschal new moon.) Column R shows the 'Day of March' for every date; this is simply the number of the day counting from 1 March.

To find the date of Easter Sunday for any year we must first find the golden number, as already explained, and the dominical letter, as explained in Chapter 24. We next locate the golden number in TABLE 29.3, and then scan down the date column from this line till we reach a date whose calendar letter (as given in column C) is the same as the dominical letter of the year in question. This gives the date of Easter Sunday.

As an example, consider the year 1500. Its golden number is 1 + MOD(1500,19) = 19 and its dominical letter is D. The paschal full moon in year XIX fell on 17 April; the calendar letter of this date is B; the first day after 17 April whose calendar letter is D is 19 April. This is Easter Sunday for 1500.

It is possible, however, to derive a formula for this date, so that we can dispense with the table. This is most easily expressed in terms of the epact.

Epacts

The epact of a year is defined as the age of the notional moon on the first day of the year, 1 January; if there is a new moon on this day, as in year III, the epact is zero. The epact can be used to calculate the date of Easter in the Julian calendar and was cleverly adapted by Lilio and Clavius to perform a similar role in the Gregorian calendar.

As noted, the epact of year III is 0, since a new lunation begins on 1 January; the epact increases by 11 days for each succeeding year, but when it comes to exceed 30, 30 is subtracted. There is an exception when we pass from year XIX back to year I, for then the epact increases by 12 from 26 to 38, or 8 after subtracting 30. This progression in the value of the epacts can be verified from TABLE 29.3, if we remember that the notional moon that is relevant is the first of each year which ends in the year in question. There is a unique epact for every golden number.

It is clear from this definition of the epact, E, that it may be calculated from the golden number, G, by the formula:

$$E = \text{MOD}(11 * G - 3, 30) \tag{29.2}$$

TABLE 29.3 *The Dionysian paschal table*

R	Date	C	G	E	Sum
21	21 March	C	XVI	23	44
22	22	D	V	22	44
23	23	E		(21)	(44)
24	24	F	XIII	20	44
25	25	G	II	19	44
26	26	A		(18)	(44)
27	27	B	X	17	44
28	28	C		(16)	(44)
29	29	D	XVIII	15	44
30	30	E	VII	14	44
31	31	F		(13)	(44)
32	1 April	G	XV	12	44
33	2	A	IV	11	44
34	3	B		(10)	(44)
35	4	C	XII	9	44
36	5	D	I	8	44
37	6	E		(7)	(44)
38	7	F	IX	6	44
39	8	G		(5)	(44)
40	9	A	XVII	4	44
41	10	B	VI	3	44
42	11	C		(2)	(44)
43	12	D	XIV	1	44
44	13	E	III	0	44
45	14	F		(29)	(74)
46	15	G	XI	28	74
47	16	A		(27)	(74)
48	17	B	XIX	26	74
49	18	C	VIII	25	74
50	19	D			
51	20	E			
52	21	F			
53	22	G			
54	23	A			
55	24	B			
56	25	C			

Key

R 'Day of March' of paschal full moon.

Date Calendar date of paschal full moon.

C Calendar letter of date.

G Golden number whose paschal full moon falls on date.

E Epact corresponding to golden number.

Sum Sum of epact and 'Day of March'.

Epacts within parentheses do not occur but are entered to demonstrate the continuity of their progression.

The epacts for all 19 golden numbers are shown in column E of TABLE 29.3. Since there are only 19 golden numbers, there are only 19 possible values for the epact, and these are seen to fall in the range 0 to 29.

Column S gives the sum of the epact and the 'Day of March', and we see that this is either 44 or 74; it is 74 only for the years VIII, XI, and XIX. It is clear that the date of the paschal new or full moon may be deduced easily from the epact, which in turn may be calculated from the golden number.

The fact that the epact progresses by 12 instead of 11, in passing from year XIX to year I, may explain why the computists moved the hollow embolismic month from year II to year XIX. If they had not done this, the sum for year XIX would have been 73 instead of 74.

We can now see how the tables may be dispensed with entirely and the date of Easter obtained by pure calculation. From the year, Y, first calculate the golden number, G, the dominical number, N (see Chapter 24), and the epact, E, using the formulae already given. We now calculate the 'Day of March', R, of the day following the paschal full moon:

$$R = 44 + 1 - E \quad \text{if} \quad E < 24$$
$$\text{or} \quad R = 74 + 1 - E \quad \text{if} \quad E \geq 24$$

It is possible to combine these two relations and show, as may be readily verified, that R is given, quite generally, by:

$$R = 22 + \text{MOD}(30 + 23 - E, 30)$$
$$\text{or} \quad R = 22 + \text{MOD}(53 - E, 30) \tag{29.3}$$

The calendar number (see Chapter 24) of this day is given by:

$$C = 1 + \text{MOD}(R + 2, 7) \tag{29.4}$$

Easter Sunday may be this day if R is a Sunday ($N = C$), otherwise it is the next Sunday which follows. We suppose that this Sunday occurs d days later than R and hence on a day whose 'Day of March' is $R + d$.

$$\text{If} \quad N = C \quad \text{we have} \quad d = 0$$
$$N > C \qquad\qquad d = N - C$$
$$N < C \qquad\qquad d = 7 + N - C$$

or quite generally:

$$d = \text{MOD}(7 + N - C, 7) \tag{29.5}$$

Thus the 'Day of March' of Easter Sunday, S, is:

$$S = R + \text{MOD}(7 + N - C, 7) \tag{29.6}$$

For example, if $Y = 1500$, we find that $G = 19$, $N = 4$, and $E = 26$. These give $R = 49$, $C = 3$, $d = 1$, and hence, $S = 50$, which indicates that Easter Sunday falls on 19 April.

An algorithm for Easter Sunday by the Dionysian canon

The component parts of the complete algorithm for finding the 'Day of March' of Easter Sunday, S, according to the Dionysian canon for any Julian year, Y, are assembled together and simplified to give the following algorithm. (Note that the formulae for the calculation of the dominical number of the year are discussed in Chapter 24.)

Algorithm M

1 $P \Leftarrow Y + Y/4 + 4$	(26.12)
2 $N \Leftarrow 7 - \text{MOD}(P, 7)$	(26.14)
3 $G \Leftarrow 1 + \text{MOD}(Y, 19)$	(29.1)
4 $E \Leftarrow \text{MOD}(11 * G - 3, 30)$	(29.2)
5 $R \Leftarrow 22 + \text{MOD}(53 - E, 30)$	(29.3)
6 $C \Leftarrow 1 + \text{MOD}(R + 2, 7)$	(29.4)
7 $S \Leftarrow R + \text{MOD}(7 + N - C, 7)$	(29.6)

This may be simplified to give a shorter and more efficient version:

Algorithm N

1 $A \Leftarrow \text{MOD}(Y, 19)$

2 $B \Leftarrow 22 + \text{MOD}(225 - 11 * A, 30)$

3 $S \Leftarrow B + \text{MOD}(56 + 6 * Y - Y/4 - B, 7)$

This algorithm is valid for any year in the Christian era. It is a simple matter to convert S, the 'Day of March' of Easter Sunday, to a date in March or April, D/M/Y, by:

4 $M \Leftarrow 3 + S/32$

5 $D \Leftarrow 1 + \text{MOD}(S - 1, 31)$

⏳ The Gregorian canon

So far I have only considered the Easter in the Julian calendar, according to the Dionysian canon. In the Gregorian calendar, the situation is more complicated on account of the adjustments introduced by Pope Gregory. To understand these, first let us note that while the average length of a Julian year is 365.25 days, the true length of the astronomical mean tropical year is about 365.2422 days; the Julian year is about 11 minutes too long. This means that the true vernal equinox fell on earlier and earlier calendar dates as the centuries rolled by; by 1582 it was observed to occur on 11 March.

Similarly, in a Metonic cycle there are 235 lunations which contain between them 6939.75 ($= 19 \times 365\frac{1}{4}$) days. Thus, while the average period of a lunation was 29.53085 ($= \frac{6939.75}{235}$) days, the true value of the mean synodic revolution is about 29.53059 days; the notional lunation was on average 22 seconds too long. This meant that the astronomical new moons were observed to occur on calendar dates falling earlier and earlier in relation to the dates given in the lunar almanac (TABLE 29.3). By 1582, the real paschal full moon was observed to occur four days before that of the notional moon as predicted by the lunar almanac.

These discrepancies were not acceptable to the Church, who saw that Easter was being celebrated on dates more and more removed from those required to fulfil the original intentions of the Council of Nicaea. In time, it was deduced, Easter would be celebrated in the seasonal summer and eventually, the winter, and never according to the canon—if nothing were done to correct it.

The papal bull of Gregory XIII, published in 1582, directed two adjustments to be made which attempted to compensate for the discrepancy in the length both of the year and also of the lunations. These effected the lunar almanac.

Besides his desire to correct these discrepancies, Gregory was concerned to alter the calendar as little as possible and in particular, to ensure that Easter Sunday should not fall outside the range of dates (22 March to 25 April) prescribed by the Dionysian canon.

The solar adjustment

Two major changes were introduced to correct for the error in the length of the Julian year. First, the calendar was adjusted to bring 21 March back to the date of the astronomical vernal equinox. Secondly, the length of the year was shortened by dropping certain leap days.

Ten days were to be dropped from the calendar in one fell swoop in order to return the solar equinox from 11 March to 21 March—the date it occurred on in the year of the Council of Nicaea. To minimize the disruption to religious practices, it was decided to drop these 10 days in October, so that 15 October came immediately after 4 October in 1582. Various alternatives to this were mooted, but rejected. One involved suppressing all leap days for 40 years—an expedient similar to that used by Augustus when he put Caesar's calendar to rights.

The discrepancy in the length of the Julian year was corrected by dropping three leap years every 400 years. The new rule was to drop the leap year in centurial years divisible by 100, but not by 400 (for example, 1700, 1800, and 1900 were not to be leap years, but 2000 was; 1600 remained a leap year). In a period of 400 years there were, according to this rule, 146 097 days—so the average length of the Gregorian year is 365.2425 days. This is still about 52 seconds too long, but the vernal equinox would only lag behind 21 March by one day in about 1600 years.

Both of these changes had the side-effect of altering the dates of the notional new and full moons. The first change meant that the dates of all the notional new moons were advanced by 10 days; the second change meant that the average period of a notional lunation was decreased by three parts in 146 097 or by about 52 seconds.

The lunar adjustment

As mentioned, the adjustment to the date of the vernal equinox affected the dates of the new moons. After the 10 days had been dropped, the astronomical new moons would have occurred six days after the notional dates, instead of four days before. It would seem that Clavius needed to move the dates of the new moons forward by six days. However, he decided to move them on by seven days instead. His motive for including the extra day was to ensure that the intentions of the Council of Nicaea were fulfilled; these, it was believed, stipulated that Easter Sunday should never fall on or before the Jewish Passover. The extra day ensured this. The change was effected by moving all the entries in the lunar almanac (TABLE 29.1) to a position seven days later.

Next, it was necessary to compensate for the decrease in the number of leap years. Every time a leap year was dropped, the date of a given astronomical event would be increased by one day. In particular, the astronomical new moons would occur on a calendar day one day later than otherwise expected. It was therefore essential to increase the date of each notional new moon by one day. Again, this involved shifting all the golden numbers in the lunar Almanac down one, to a position one day later. This adjustment is known as the solar equation and I will denote it by SOL.

That was not all. The actual mean period of the notional lunations had been too long and it was necessary to reduce it by one day in about 313 years. This was done by shifting the golden numbers up in the lunar almanac (to an earlier date) by eight days every 2500 years. The first shift was to occur in 1800, and then eight times every 2500 years—one day every third centurial year for 2100 years, and then once more after an interval of another 400 years. This adjustment is known as the lunar equation and I will denote it by LUN.

It is emphasized that the shifts in the dates of the golden numbers or new moons are cumulative and although introduced in the centurial years, apply to every year that follows.

The solar and lunar equations shift the golden numbers in the table in opposite directions, sometimes cancelling each other out. In fact there are four possibilities for each centurial year:

1. neither the solar nor lunar equations apply—in this case the golden numbers are not moved;

2. both the solar and lunar equations apply—in this case they cancel each other out and the golden numbers are not moved;

3. the solar equation applies but not the lunar equation—in this case the golden numbers move down one day (that is, to a later date);

4. the lunar equation applies but not the solar equation—in this case the golden numbers move up one day (that is, to an earlier date).

Since the solar equation applies more frequently (18 times in 2400 years) than the lunar equation (8 times in 2500 years), the golden numbers drift progressively downwards. The sequence of shifts repeats every 100 centuries. In each such period, the golden numbers will have shifted down by $43 = (25 \times 3) - (4 \times 8)$ days.

⚅ The solar equation

The solar equation (using the term 'equation' in a sense now archaic), SOL, gives the number of days by which the dates of the notional paschal moons must be increased on account of the decrease in the average length of the calendar year. SOL depends only on the number of centurial years passed since AD 1600 at the start of the year in question. I will denote the year in question by Y.

Begin by defining H as the number of centurial years passed from the start of the era by the year Y, so that $H = Y/100$

The number of such centurial years passed since AD 1600 is clearly $H - 16$. Three leap days are dropped in every four centuries, and the number of such complete four-century periods is $(H - 16)/4$. It follows that the number of dropped leap days in all of these 400-year periods is $3 * ((H - 16)/4)$. There may also be up to three centurial years following these before year Y is reached, and a leap day is dropped in each. The number of these extra centurial years is given by MOD $(H - 16,4)$. We thus see that the total solar equation (for H > 15) is given by:

$$\begin{aligned}
\text{SOL} &= 3*((H - 16)/4) + \text{MOD}(H - 16, 4) \\
&= H - 16 - (H - 16)/4 \\
&= H - H/4 - 12 \\
&= Q - 12
\end{aligned} \tag{29.7}$$

This is valid for any H > 0 and where:

$$Q = H - H/4 \tag{29.8}$$

⬚ The lunar equation

The lunar equation, LUN, gives the number of days by which the dates of the notional paschal moons must be decreased on account of the lunar adjustment alone, for year Y. LUN, like SOL, depends only on the number of centurial years passed after AD 1600.

Several different, but equivalent, formulae for LUN have been given. I shall detail the one that is fastest to compute on most computers; it was first published by Delambre (1749–1822) in 1821.

The number of complete 25-century periods passed since 1800, by the start of century H, is $(H - 18)/25$; as each such period contains eight corrections, the total number of corrections in them is $8 * ((H - 18)/25)$. The number of centuries remaining, Z, is given by $Z = \text{MOD}(H - 18, 25)$. Z may be as large as 24 and contain up to seven corrections. If $Z \le 24$ there are $Z/3$ corrections in these centuries, but if $Z \ge 24$, there are seven corrections. We may express the number of these corrections as: $Z/3 - Z/24$. To get the total lunar correction we must add a further one, since there was a correction of one day in 1800.

We thus see that the total lunar correction is given by:

$$Z = \text{MOD}(H - 18, 25) \tag{29.9}$$

$$\text{LUN} = 1 + 8 * ((H - 18)/25) + Z/3 - Z/24 \tag{29.10}$$

It is possible to simplify this by purely algebraic manipulations and show that for any century with H > 13:

$$\text{LUN} = (H - 15 - (H - 17)/25)/3 \tag{29.11}$$

⬚ The Gregorian lunar almanac

We now see that a Gregorian lunar almanac could be constructed from the original Julian version shown in TABLE 29.3 for any year after 1582.

First, each golden number must be moved down in the almanac by seven days to accommodate the amendments required to restore the vernal equinox to 21 March and to correct for the four-day discrepancy in the dates of the new moons. The result of this was an initial almanac which was valid from 1582 to 1699.

In 1700, the solar (but not the lunar) equation was applied, and the golden numbers in the almanac moved down a further day. Then, as the centuries passed, the golden numbers were moved down or occasionally up; these movements took place in the centurial years according to the dictates of the solar and lunar equations.

After 43 centuries it is found that the golden numbers have been moved down by 30 days. At this point, the golden numbers should all revert to the position they were at in 1582 and the series of movements begin again—though this will only be relevant in the distant future. Thus, the number of days by which the golden numbers are moved down is given by $\text{MOD}(\text{SOL} - \text{LUN}, 30)$. This is sometimes called the index number of the year and it lies in the range 0 to 29.

In this way we can see how 30 different lunar almanacs were generated, and that one of these applied to each year after 1582. The computists allocated an index letter, corresponding to the index number, to each year.

Ⅺ Clavius' adjustment

We have already seen how the original Dionysian Paschal table was constructed from the Dionysian lunar almanac; in exactly the same way, 30 different Gregorian paschal tables could be constructed from the 30 Gregorian lunar almanacs. The paschal tables are labelled with the same index letter as the almanac from which they are derived.

From all these tables an unfortunate but inevitable feature becomes apparent. We have seen in the Dionysian lunar almanac table how the range of dates from 8 March to 5 April contained an entry for each of the 19 possible golden numbers. In some of the new tables, however, only 18 of the golden numbers fell within this range of 29 days; in some of the tables, one of them fell a day later on the 30th day, 6 April. Since Clavius and the Pope wished Easter to lie in the same period of 29 days as it had for the last thousand years, some adjustment had to be made.

The origin of this anomaly lies in the fact that although in the Dionysian lunar almanac all 19 golden numbers occur within a space of 29 days, the period between consecutive lunations is 30 days; some days in the almanac are blank. Examination of the lunar almanacs reveals that the missing golden numbers are invariably located on the 30th day, 6 April. Clavius' strategy was to move them back from 6 April to 5 April; but this gave rise to another problem—sometimes there was already another golden number on 5 April and it was an important rule that two paschal new moons belonging to two different years of the Metonic cycle should never occur on the same day. Accordingly, when this happened, Clavius also moved the golden number found on 5 April back to 4 April. This did not cause further problems since 4 April in these circumstances was invariably blank.

If you were to construct the full set of 30 Gregorian lunar almanacs, you could easily verify that the second shift of a golden number from 5 April to 4 April was only required when the golden number to be shifted happened to be XII or greater, and that when this was the case, 4 April was blank, ready to accommodate it. There is a simple relation between the two golden numbers found on 5 and 6 April: if G is that found on the 6th, that found on the 5th is $11 + G$.

We thus see that two sorts of adjustment were made:

1. A golden number was moved from 6 April back to 5 April.
   This was required for the 11 golden numbers in the range 9 to 19 when their new moons fell on 6 April.

2. Two golden numbers were moved from 5 and 6 April back to 4 and 5 April respectively. This was required when the eight golden numbers in the range 1 to 8 fell on 6 April and the eight golden numbers in the range 12 to 19 fell on 5 April.

🔢 Gregorian epacts

Perhaps because they did not want to burden themselves with 30 different almanacs, Lilio and Clavius abandoned the use of the lunar almanac as employed in the Dionysian Canon (and which I have exhaustively just described). Although epacts were not used in the old method of calculating Easter according to the Dionysian canon, they could have been (as I have previously demonstrated), and Lilio and Clavius used the notion to full advantage in the Gregorian canon.

As we have seen, in the Dionysian canon, the sum of the 'Day of March' of a paschal full moon and the epact for the corresponding year added up to 44 or 74 according to the Golden number. Now, suppose that all the golden numbers (which mark the date of a new moon) in the lunar almanac are moved down (to later dates) by x days. In accordance with this, the new moon of the first lunation will be moved down an identical number of days, and this means that the moon at the turn of the year will be x days younger; this in turn implies that the epact will be x days smaller. Clearly, the 'Day of March' of any golden number in the almanac will increase by *x*, and the epact will decrease by the same amount; therefore, their sum must remain constant at 44 or 74.

As with the Dionysian canon, the date of the paschal new or full moon can be easily determined for any year, given its epact.

Before we go on to derive explicit formulae, we must note two exceptions, already alluded to in the discussion of the Gregorian Lunar almanac. We saw that in certain years the rule just discussed would give a date of 6 April for the paschal new moon and that the effect of the adjustments necessary to avoid this must be considered.

If the paschal new moon falls on 6 April, the full moon falls on 19 April—the 50th 'Day of March'. In these circumstances the new moon must be pushed back, according to rule A, to 5 April, and the full moon, to 18 April. This has the effect of reducing the sum of the epact and the 'Day of March' by one. Lilio and Clavius conceived the notion of increasing the corresponding epact by one to allow for this. Since the epact in such years is easily shown to be 24, it must be increased to 25.

If, in addition, the golden number pushed from the 6th to the 5th is less than IX, the golden number on the 5th must be pushed back to the 4th. Lilio and Clavius decided that in these circumstance the epact should be increased from 25 to 26.

We thus have two rules (which I shall call Clavius' adjustments) whereby the calculated epact must be adjusted:

1. If the calculated epact is 24, increase it to 25.
2. If the golden number is greater or equal to XII, and the calculated epact is 25, increase it to 26.

In practice it is more convenient to suppose that the epact does not change and that, instead, the date of the corresponding moon is put back.

We have seen that the epact in the Dionysian canon is defined in terms of the golden number by the formula: $\text{MOD}(11 * G - 3, 30)$. This Dionysian epact must first be decreased by seven to realign the astronomical new moons with the notional ones, as already explained. The appropriate formula for the epact is now $\text{MOD}(11 * G - 3 - 7, 30)$ or $\text{MOD}(11 * G - 10, 30)$.

Next, this base epact must be decreased by the solar equation and increased by the lunar equation. The resulting decrement, SOL − LUN, must be reduced by a multiple of 30 if necessary; we thus obtain the formula:

$$\text{MOD}(11 * G - 10, 30) - \text{MOD}(\text{SOL} - \text{LUN}, 30)$$

However, if this exceeds 29, 30 must again be subtracted. We thus obtain an expression for the epact (to which Clavius' adjustment will be applied later):

$$E = \text{MOD}(57 + 11 * G - Q + (H - (H - 17)/25)/3, 30) \qquad (29.12)$$

The constant, 57, arises from the component parts of the expression and from the necessity of adding a sufficiently large multiple of 30 to ensure that the argument of the MOD function is never negative.

It is now necessary to calculate Clavius' adjustment; this can be done in several different ways; perhaps the most direct and easiest to understand is the formula:

$$V = E/24 - E/25 + (G/12) * (E/25 - E/26) \qquad (29.13)$$

It is easily verified that V takes the value of 1 if $E = 24$ or if $G \geq 12$ and $E = 25$, and in all other cases is zero. The correction may be applied by adding it to the epact, or, to produce the same effect, subtracting it from the 'Day of March', R, of the paschal full moon.

Another method relies on finding an expression, U, which depends on the epact and which takes its two greatest values, 29 and 28, when the epact is 24 and 25 respectively. To do this consider the expression from equation 29.3:

$$U = \text{MOD}(53 - E, 30) \qquad (29.14)$$

We may observe that with $E = 24$ we have $11(29 - U) = 0$, and when $E = 25$ we have $11(29 - U) = 11$. For all other values of E, U is greater than 19. We now consider the value of:

$$V' = (G - 1 + 11U)/319 \qquad (29.15)$$

We find that V' is equal to 1 (it can never exceed 1) only if:

$$G - 1 + 11U \geq 319$$

and this only happens if:

$$G - 1 \geq 319 - 11U = 11(29 - U).$$

This only occurs when $E = 24$ (with any value of G) or $E = 25$ and $G \geq 12$. These are precisely the conditions in which we need to apply Clavius' adjustment. We may thus subtract V' from the 'Day of March' of the paschal full moon instead of V.

Once the adjusted epact has been calculated, the 'Day of March' and the date of Easter Sunday may be calculated as for the Dionysian canon with the use of equations 29.3 and 29.6. This, of course, requires the calculation of the calendar number of R with equation 29.4.

This completes the calculation of the date of Easter Sunday for any year in the Gregorian calendar. TABLE 29.4 shows how the dates of the paschal new and full moons are assigned to the epacts.

TABLE 29.4 *The dates of the notional paschal new and full moons corresponding to various values of the epact*

Epact	Paschal new moon		Paschal full moon	
	Day of March	Calendar letter	Day of March	Calendar letter
23	March 8	D	March 21	C
22	9	E	22	D
21	10	F	23	E
20	11	G	24	F
19	12	A	25	G
18	13	B	26	A
17	14	C	27	B
16	15	D	28	C
15	16	E	29	D
14	17	F	30	E
13	18	G	31	F
12	19	A	32	G
11	20	B	33	A
10	21	C	34	B
9	22	D	35	C
8	23	E	36	D
7	24	F	37	E
6	25	G	38	F
5	26	A	39	G
4	27	B	40	A
3	28	C	41	B
2	29	D	42	C
1	30	E	43	D
0	31	F	44	E
29	32	G	45	F
28	33	A	46	G
27	34	B	47	A
25'26	35	C	48	B
24 25	36	D	49	C

25' is taken to represent an epact of 25 when the golden number is XII or greater.

Clavius and other commentators go on to show how the complete lunar almanac may be constructed for a year of any index. This is not strictly necessary if we only wish to calculate the date of Easter, and we do not tackle this problem here.

At this point the reader may be impressed by the ingenious and relatively simple method of dealing with the inaccuracies inherent in the Dionysian canon, (though I sympathize with some less stout souls who are appalled by its complexity) particularly when the calculation is done with the aid of pre-prepared tables as can be found in any Anglican prayer book. One's admiration is enhanced when one considers the comparatively primitive state of sixteenth-century mathematics.

FIG. 29.1 The Easter canon of Hyppolytus. This was discovered in 1551 and is now in the Vatican Library.

⚷ An algorithm for Easter Sunday by the Gregorian canon

The calculation of the 'Day of March' of Easter Sunday in the Gregorian canon closely follows the method used in the Dionysian canon. The differences to note are the new formulae for the dominical number (equations 24.13 and 24.14) and for the epact (equation 29.12) and also the need to make Clavius' adjustments (equations 29.13 or 29.15). The formulae for the golden number is the same (equation 29.1).

It is convenient to collect all the equations required together and cast them as steps in an algorithm:

$\mathcal{A}$ l g o r i t h m O

1	$P \Leftarrow Y + Y/4 - Y/100 + Y/400 - 1$	(24.13)
2	$N \Leftarrow 7 - \text{MOD}(P,7)$	(24.14)
3	$H \Leftarrow Y/100$	
4	$Q \Leftarrow H - H/4$	(29.8)
5	$G \Leftarrow 1 + \text{MOD}(Y,19)$	(29.1)
6	$E \Leftarrow \text{MOD}(57 + 11*G - Q + (H - (H-17)/25)/3, 30)$	(29.12)
7	$U \Leftarrow \text{MOD}(53 - E, 30)$	(29.14)
8	$V' \Leftarrow (G - 1 + 11*U)/319$	(29.15)
9	$R \Leftarrow 22 + U - V'$	
10	$C \Leftarrow 1 + \text{MOD}(R + 2, 7)$	(29.4)
11	$S \Leftarrow R + \text{MOD}(7 + N - C, 7)$	(29.6)

Here we have used the second of the two formulae for Clavius' adjustment (V'). We have also subtracted it from R, the 'Day of March' rather than add it to the epact itself. The algorithm is valid for any year in the Christian era.

The algorithm may be simplified to give a shorter and more efficient version:

$\mathcal{A}l\,g\,o\,r\,i\,t\,h\,m$ $\mathcal{P}$

1   $A \Leftarrow Y/100$

2   $B \Leftarrow A - A/4$

3   $C \Leftarrow \text{MOD}(Y, 19)$

4   $D \Leftarrow \text{MOD}(15 + 19 * C + B - (A - (A - 17)/25)/3, 30)$

5   $E \Leftarrow D - (C + 11 * D)/319$

6   $S \Leftarrow 22 + E + \text{MOD}(140004 - Y - Y/4 + B - E, 7)$

This is valid for all years in the Christian era up to AD 100 000. By this date, the intentions of the Council of Nicaea will have been thwarted yet again but if, for some strange reason, you require the date of Easter beyond this year, the constant 140 004 may be replaced by another of the form $4 + 7X$ where $7X \geq 5 * Y_m/4$ and Y_m is the maximum year required.

⌛ A brief history of the calculation

Before the development of an adequate understanding of algebra, the only way of working out the date of Easter was with the aid of tables. These had to be tediously prepared by the medieval computists—the experts in the field. The difficulty of this was compounded, before the eleventh century, by the lack of an adequate number system. In fact, some authorities opine that the problem of the date of Easter provided the main motivation for the development of arithmetic in the Middle Ages.

Eventually, methods of doing sums or carrying out the common arithmetic operations using the new number system brought by the Arabs from India became widespread, and algebraic formulae began to be used to express arithmetic recipes.

Even so, Clavius provided extensive tables in his great work and few algebraic formulae. Possibly the first algebraic algorithm is in an unpublished work by the English mathematician, Thomas Harriot (1560–1621), but it was not until the start of the nineteenth century that algebraic formulae for calculating the date of Easter according to the Gregorian canon became generally available. Several have now appeared in print—some correct, some not.

The first modern method was given by the great German mathematican, Carl Friedrich Gauss (1777–1855), which he published when he was a young man in 1800. Gauss' method was defective in that it ignored the final 400-year gap between applications of the lunar equation—perhaps because it would not be relevant till AD 4200. It is corrected in a marginal note in his own copy of the paper and in later work. It is said that Gauss was motivated to work on the problem by his mother who could not remember the date on which he was born, save that it was a Wednesday, eight days before Ascension Day.

A few years later the French astronomer, Delambre, published an algorithm and an extensive discussion of the subject. In various respects these formulae were incomplete in that Clavius' adjustment required special treatment rather than to be incorporated in the formulae. The work of Gauss and Delambre and errors in their earlier papers are discussed by Oudin (see Further Reading p. 417).

A useful discussion was provided by De Morgan in 1845 on the occasion of there being a full moon on Easter Sunday. However, the first complete and correct algorithm, by an anonymous American, was published in the English scientific journal, *Nature*, in 1876. This has been reprinted in several places, usually with little explanation. The most compendious discussion was published posthumously in 1877 by S. Butcher, who had been Anglican Bishop of Meath. In this work, Butcher proves that the various algorithms published by Gauss, by Delambre, and in *Nature*—as well as others—are all equivalent. W. S. B. Woolhouse wrote his well-known article in the eleventh edition of the *Encyclopaedia Britannica*, published in 1910; this algebraic treatment omits to provide formulae for dealing with Clavius' adjustment.

There have been several algorithms published in English more recently, though none of their authors provide a full discussion of the subject. Among them are the following: Uspensky and Heaslet discuss the problem in their mathematical text book *Elementary Number Theory* of 1939; T. H. O'Beirne published two different algorithms (based upon the *Nature* method) in his book of popular mathematics appropriately entitled *Puzzles and Paradoxes* of 1965; Knuth presented in 1967 an algorithm which he claims is based on Clavius' work and deals with Clavius' correction using a decision procedure (which is probably faster on most computers) rather than by algebraic formulae; Schocken presented a brief and incomplete discussion in 1976; and a useful pamphlet by Yallop, published by H. M. Nautical Almanac Office in 1985, gives Gauss' algorithm and a short bibliography.

The algorithms of *Nature*, O'Beirne, and Knuth, and the present method, give identical dates for Easter for the years from AD 1583 to AD 100 000. Moreover, all give dates for Easter Sunday for all years from 1583 to 2000 which are in complete agreement with those listed in *The Handbook of Dates*.

Meanwhile, tables continue to be printed in the Book of Common Prayer which will enable any reader to work out the date of Easter for a limited range of years, according to the century of publication—should he dislike algebra.

Parise has provided extensive tables of golden numbers, dominical letters, epacts, and dates of Easter—but has defined epacts in an unconventional way.

⌛ Postscript

> *"Fourteenth of March, I think it was" he said.*
> *"Fifteenth" said March Hare.*
> *"Sixteenth" said Dormouse.*
> *"Write that down" the King said to the Jury; and the Jury eagerly wrote down all the three dates on their slates, and then added them up, and reduced the answer to shillings and pence.*
>
> LEWIS CARROLL, *Alice in Wonderland*

The issue concerned who stole the tarts, but it sounds suspiciously like a debate over the date of Easter.

CHAPTER 30

A Book of Hours

⊞ The praises of God

St Benedict founded his order of monks in the sixth century. His 'rule for monks' required that the praises of God be sung seven times a day—an idea which probably originated from ancient Jewish practices. The times, the canonical hours, at which this should be done came to be known as matins and lauds, prime, tierce, sext, none, vespers and compline. Matins and lauds were sung without break in the middle of the night, with prime following at dawn; tierce, sext, and none followed at regular intervals during daylight (none eventually gave its name to noon); and vespers and compline were executed in the evening. The result was a near continuous hymn of praise. The actual texts to be sung at the canonical hours were taken from, or at least inspired by, the Book of Psalms and copied into the Divinum Officium (the book of the Divine Office).

As time went on many other items were added to the Divinum Officium and some of these became popular in the private devotions of the laity. By the tenth century it had become fashionable for people who could afford it, and who could read, to commission a book filled with simplified devotions addressed to the Virgin Mary (the Office of Our Lady). These books, or Horae, were frequently richly decorated with pictures that illustrated the biblical themes of the text. The Office of Our Lady was encouraged by the Dominicans (founded in about 1205), who saw it as a means of combatting heresy. By the fourteenth century it had become well established and the contents of the Horae settled into a tradition that lasted for 200 years.

But, as time progressed, the decorations in the books became more elaborate and naturalistic elements (plants, birds, and so forth) appeared. By the seventeenth century, when the continuous praise of God was less popular, the Horae were tending to resemble herbals rather than books of devotions.

The Horae, or Books of Hours, as they were called, usually began with a calendar or almanac to help the reader to follow the succession of Christian feasts and fasts in the course of the year.

⧖ *Les très riches heures*

One particularly fine Horae which has survived from the early fifteenth century was commissioned by Jean, Duc de Berry (1340–1416) from the three Limbourg brothers, Paul, Herman, and Jean. It is generally known as *Les très riches heures*. The calendar part of this contains 12 miniatures, one for each month, which illustrate scenes from medieval France appropriate to the season.

The Duc unfortunately died before the work was completed but in 1485 the Duc de Savoie commissioned Jean Columbe to complete it. The work was well known during the sixteenth century and inspired the work of several Flemish painters of this period. Later, with a decrease in the popularity of medieval paintings, it disappeared into obscurity till it was rescued by the founder of the Musée Condé at Chantilly in 1856.

Each of the 12 calendrical miniatures is surmounted by a semicircular array of signs and symbols which spell out the details of the calendar for the corresponding month. Only seven of these semicircles were completed (those of February, March, June, July, September, November, and December), though October is nearly complete.

The innermost, semicircular part of each design contains a painting, in bright heavenly blue, of Apollo, the sun-god, riding in his chariot, drawn by winged horses. Surrounding this there are three semicircular rings (which we may call A, B, and C), each divided into 30 or 31 (or 28 for February) equal parts, one for each day of the month, and a further ring, D, containing Latin text.

The innermost of these rings (A) contains the numbers of the days. Several of these numbers—4, 5, and 7 in particular—seem unfamiliar; they are in fact archaic versions of the Hindu–Arabic numbers which only reached Europe in the twelfth century.

The next ring (B) contains the first 19 letters of the alphabet, a to t (excluding j, which was only introduced into the alphabet in the fifteenth century). These letters represent the integers 1 to 19 in a notation similar to the old Milesian Greek system. Thus, a represents 1, b represents 2, and so forth, to t which represents 19. Some of the days have one of these 19 letters assigned to them and some do not. This ring represents a lunar almanac for the month, similar to that shown in TABLE 29.1. The letters (or the corresponding numbers) represent the possible golden numbers of a year, and they occur in the same order as the golden numbers in a lunar almanac shown in TABLE 29.1. Each golden number is assigned to a date on which a notional new moon is supposed to occur in the corresponding year. For instance, in

FIG. 30.1 September from *Les très riches heures*. Source: Ancient Art & Architecture Collection.

TABLE 30.1　*The number of days in the four seasons as indicated in* Les très riches heures *(TRH) and in modern ephemerides*

Season	Days	
	TRH	AD 1400
Vernal equinox to summer solstice	92¼	93.0
Summer solstice to autumnal equinox	94¼	93.5
Autumnal equinox to winter solstice	88	89.6
Winter solstice to vernal equinox	90	89.2
	365	365.3

The figures in the last column represent the number of days in each of the four seasons calculated for AD 1400.

February, the letter 'k' (the 11th letter of the alphabet) is assigned to the ninth day. This means that in a year whose golden number is 11, a notional new moon is supposed to occur on 9 February.

However, as well as a few obvious mistakes, more serious anomalies exist. For example, the number of days separating adjacent entries of the same golden letter vary from 28 to 31, whereas they are invariably 29 or 30 in TABLE 29.1. The sequence of golden numbers for March is identical to the sequence for July in TABLE 29.1. Neugebauer claims that the first new moon of a year with golden number 1 is scheduled for 19 January rather than 23 January as in TABLE 29.1. The question arises as to whether these anomalies represent genuine mistakes on the part of the Limbourg brothers or an alternative lunar almanac worked by some competent computist. The fact that only half of the calendrical entries were completed invites the speculation that errors were detected halfway through their production, and the problem as to how to resolve the matter was never settled.

The third ring (C), contains crescents symbolic of the new moons. These are drawn in positions corresponding to the day that is marked by a golden number in B. The fourth ring (D) contains an inscription in Latin which describes the information in B and C and gives the number of days in the month. For instance, the February inscription reads 'Primationes lune mensis February dies xxviii' or 'The new moons of February, 28 days'.

Moving outwards from the centre of the semicircles, we next encounter an area with a background of uniform heavenly blue which is heavily decorated with golden stars and pictorial representations of the signs of the zodiac.

These are drawn in the region of the semicircle corresponding to the dates in which the sun is to be found in the corresponding part of the ecliptic. The end of one sign and the beginning of the next is clearly marked by a radial line at the appropriate part of the month. For instance, in February, a line is drawn at a position corresponding to the middle of the 10th day of that month; the sun leaves the sign of Aquarius and enters that of Pisces on that day.

Further out still, there are two more inscribed rings (E and F). As the year progresses, the sun moves through each sign of the zodiac in turn, so that on each day of the year it is to be found within one or other of these 12 signs. Each sign is thus one twelfth part of a circle, or 30°, and the sun reaches the positions defined by the 30 angles on a defined day of the month. The angles (1 to 30) within each sign are indicated by divisions of ring F at positions corresponding to the day on which the sun reaches them. Each angular position is marked with the number of the angle in arabic numerals, like those in A. Ring E carries another Latin inscription. For instance, in February we find, written over the first 10 days of that month, 'Finis graduum Aquary' or 'The last degree of Aquarius', and over the remainder of the month: 'Initium Piscum gradus xix' or 'The beginning of Pisces, 19 degrees'.

It is possible to estimate the number of days allocated to each of the 12 signs of the zodiac (although the inscriptions are incomplete in some of the months, the divisions for each degree are drawn in for all 12 signs). The number of days allocated to the four seasons are given in TABLE 30.1.

It may be seen that the spacing roughly takes into account the variations in the speed of the sun around the ecliptic (mentioned in Chapter 2).

Neugebauer has drawn attention to the fact that numbers are represented, in these diagrams, in four different ways. First, the days of the month in A and the angles in F are written in the medieval form of arabic numerals; next, the golden numbers in B are given in the Greek manner of representing numbers by letters; then, the number of days in the month in D and the number of degrees of the sign present in E are given in roman numerals; and finally, the names of the months, September, October, November, and December spell out the number of the month in the old Roman calendar in Latin.

Although these calendrical reflections may be of interest, it is to be hoped that they will not prevent you from enjoying the delicacy and interest of the paintings themselves.

APPENDICES

Astronomical Constants

THE best modern estimates for the various units of time and astronomical periods, listed here, are taken from the 1992 Explanatory Supplement to the *Astronomical almanac*. The following definitions are also taken from that source. They are given to a precision exceeding that required for any realistic calendrical calculations.

It may be helpful to note that:

1	second	=	0.0000116 days
0.0086	seconds	=	0.0000001 days

The atomic second

The fundamental unit of time changed in the twentieth century after the development of the atomic clock. It was once based upon the mean solar day but the base unit of time in the International System of Units (SI) is now the second, defined as:

> *the duration of 9,192,631,770 periods of the radiation corresponding to the transition between the two hyperfine levels of the ground state of the Caesium-133 atom.*

There are, of course, still 60 of these seconds in a minute and 60 minutes in an hour. The day as a unit of time is now defined as 86 400 seconds or 24 hours; this is close to the mean solar day.

The day

MEAN SOLAR DAY

The average period of rotation of the earth about its axis with respect to the sun. The length of any particular day may vary by up to 50 seconds from this according to the season.

SIDEREAL DAY

The period of rotation of the earth about its axis with respect to the stars.

At the end of a sidereal day the earth has moved round its orbit by a degree or so and no longer presents the same face to the sun as it did at the start; it must rotate a little bit more to do so. This accounts for the fact that the mean solar day is about 4 minutes longer.

$$1 \text{ sidereal day} = 23.93447 \text{ hours} = 86164.1 \text{ seconds}$$
$$= 23 \text{ hours, } 56 \text{ minutes, } 4.1 \text{ seconds}$$

The sidereal day is almost constant. The friction in the tides gives rise to a very slow increase and other factors cause small, periodic, random, and secular changes. The secular change amounts to an increase of roughly 0.002 seconds per century. This corresponds to an increase of about 2 parts per million per century.

The month

MEAN SYNODIC PERIOD OF THE MOON (LUNATION)

The most important period of the moon is the mean synodic period (or lunation). This is the mean interval between conjunctions of the moon and the sun and corresponds to the mean period of the phases of the moon; it is the period of revolution of the moon about the earth relative to the sun.

It shows a small secular increase with time and is given in days by the formula:

$$29.5305888531 + 0.00000021621 T - 3.64 \times 10^{-10} T^2$$

where T is the date measured in centuries after AD 2000. This formula gives the values at different dates which are shown in TABLE I.I.

The tithi or 'lunar day' is one thirtieth of the mean synodic period:

$$1 \text{ tithi} = 0.98435 \text{ days} = 23 \text{ hours, } 37 \text{ minutes, } 28 \text{ seconds}$$

In addition, several other periods associated with the moon may be defined. Definitions and approximate mean values of some of these are given here.

SIDEREAL MONTH

The period of revolution of the moon in its orbit about the earth relative to the stars as seen from the earth: 27.32166 days.

TROPICAL MONTH

The time between successive passages of the moon through the first point of Aries: 27.32158 days.

DRACONIC MONTH

The period of revolution of the moon through the nodes of its orbit which mark the intersection of its plane of rotation with the ecliptic: 27.21222 days.

ANOMALISTIC MONTH

The interval between successive perigees of the moon: 27.55455 days.

TABLE 1.1 *The mean synodic period of the moon*

Epoch	Period in S.I. days
4000 BC	29.53057457 29 d 12 h 44 m 1.64 s
3000 BC	29.53057714 29 d 12 h 44 m 1.86 s
2000 BC	29.53057962 29 d 12 h 44 m 2.08 s
1000 BC	29.53058204 29 d 12 h 44 m 2.29 s
1 AD	29.53058439 29 d 12 h 44 m 2.49 s
1000 AD	29.53058665 29 d 12 h 44 m 2.69 s
2000 AD	29.53058885 29 d 12 h 44 m 2.88 s

Actual moon's period may differ by up to 7 hours according to the year. The current secular increase amounts to some 0.02 seconds or 0.8 parts per million per century.

PERIOD OF ROTATION OF THE NODES RELATIVE TO THE STARS

18.61 years.

THE SAROS

The period in which the sun, moon, and nodes of the moon's orbit return to almost the same relative positions. Eclipses tend to repeat after this time period: 223 lunations = 18.03 years.

The year

TROPICAL YEAR

The period between successive passages of the sun through the first point of Aries—the vernal equinox. This varies from year to year by about 20 minutes due to nutation and interaction with other planets. We can, however, define an average value, which currently shows a small secular decrease with time and is given in days by the formula:

$$365.2421896698 - 0.0000615359T - 7.29 \times 10^{-10}T^2 + 2.64 \times 10^{-10}T^3$$

where T is the time or epoch measured in centuries after AD 2000. This formula gives the values at different epochs shown in TABLE 1.2.

In addition, several other periods may be defined. Definitions and approximate mean values of some of these are given here.

THE SIDEREAL YEAR

The period of revolution of the sun in the ecliptic with respect to the stars: 365.25636 days.

TABLE I.2   *The length of the mean tropical year*

Epoch	Mean tropical year in S.I. days
4000 BC	365.2424992 365 d 5 h 49 m 11.93 s
3000 BC	365.2424625 365 d 5 h 49 m 8.76 s
2000 BC	365.2424177 365 d 5 h 49 m 4.89 s
1000 BC	365.2423664 365 d 5 h 49 m 0.46 s
1 AD	365.2423103 365 d 5 h 48 m 55.61 s
1000 AD	365.2422509 365 d 5 h 48 m 50.48 s
2000 AD	365.2421897 365 d 5 h 48 m 45.19 s

Any particular year may differ from the mean by up to 10 minutes on account of nutation. The secular decrease amounts to about 0.46 seconds per century or 1.4 parts per million per century.

ANOMALISTIC YEARS

The period between successive perihelions of the earth as it moves in its orbit about the sun: 365.25964 days.

THE LUNAR YEAR

The lunar year is the length of 12 mean lunations and is: 354.3671 days. This is 10.8751 days short of a tropical year.

THE MEAN NUMBER OF LUNATIONS IN A TROPICAL YEAR

This is simply the ratio of the length of the tropical year to the synodic period of the moon: 12.36827 lunations per year.

Since the lunation shows a small secular increase and the year a small secular decrease of comparable magnitude, this ratio is decreasing at the rate of about 0.6 parts per million per century.

THE NUMBER OF MEAN SOLAR DAYS IN A MEAN TROPICAL YEAR

This is the ratio of the length of a year to that of a day: 365.24219 days.

This ratio is subject to a small secular increase of 2 or 3 parts per million per century.

The Names of the Days of the Week

THE names of the days of the modern seven-day week in 69 languages are arranged here according to linguistic group and branch.

Note that the diacritical marks used in some languages are missing. The Russian, Greek, and Oriental names have been transliterated into Roman script. A few names are missing, not being given in the dictionaries consulted.

p	Mainly planetary names
t	Mainly ordinal names with Saturday as day 1
s	Mainly ordinal names with Sunday as day 1
m	Mainly ordinal names with Monday as day 1
d	Descriptive names
?	Uncertain
*	Mid-week

1 *Germanic languages* (Western branch)

Old English (p)	English (p)	Dutch (p)	German (p)	Old Frisian (p)	New Frisian (p)
Sunnendaeg	Sunday	Zondag	Sonntag	Sonnadei	Sneyn
Monandaeg	Monday	Maandag	Montag	Monadei	Moandie
Tiwesdaeg	Tuesday	Dinsdag	Dienstag	Tysdie	Tyesday
Wodensdaeg	Wednesday	Woensdag	Mittwoch*	Wernsdei	Wânsdei
Thuresdaeg	Thursday	Dondersdag	Donnerstag	Tunresdei	Tongersdei
Frigedaeg	Friday	Vrijdag	Freitag	Frigendei	Frêd
Saeternesdaeg	Saturday	Zaterdag	Samstag	Saterdei	Saterdei

2 *Germanic languages* (Scandinavian branch)

Swedish (p)	Norwegian (p)	Danish (p)	Modern Icelandic (s)	Old Icelandic (p)
Söndag	Søndag	Søndag	Sunnudagur	Sunnudagur

Måndag	Mandag	Mandag	Manudagur	Manudagur
Tisdag	Tirsdag	Tirsdag	Thridjudagur	Tyrsdagur
Onsdag	Onsdag	Onsdag	Midvikudagur*	Odinsdagur
Torsdag	Torsdag	Torsdag	Fimmtudagur	Thorsdagur
Fredag	Fredag	Fredag	Föstudagur	Frjadagur
Lördag	Lørdag	Lørdag	Laugardagur	Laugardagur

3 *Romance languages* (Derived from Latin)

Latin (p)	Italian (p)	Spanish (p)	French (p)
Dies Solis or Dies Dominica	Domenica	Domingo	Dimanche
Dies Lunae	Lunedi	Lunes	Lundi
Dies Martis	Martedi	Martes	Mardi
Dies Mercurii	Mercoledi	Miercoles	Mercredi
Dies Iovis	Giovedi	Jueves	Jeudi
Dies Veneris	Venerdi	Viernes	Vendredi
Dies Saturni or Sabato	Sabato	Sabado	Samedi

Romanian (p)	Catalan (p)	Portuguese (s)	Provencal (p)
Duminica	Diumenge	Domingo	Dimenge
Luni	Dilluns	Segunda-fiera	Dilus
Marti	Dimartis	Tercia-fiera	Dimars
Miercuri	Dimecres	Quarta-fiera	Dimercres
Joi	Dijous	Quinta-fiera	Dijous
Vineri	Divendres	Sexta-fiera	Divendres
Sambata	Dissabte	Sabado	Dissapte

4 *Celtic languages* (Brythonic branch)

Cornish (p)	Breton (p)	Welsh (p)
De Sul	Disul	Dydd Sul
De Lun	Dillun	Dydd Llun
De Merth	Dimeurz	Dydd Mawrth
De Mergher	Dimerc'her	Dydd Mercher
De Yow	Diziou	Dydd Iau
De Gwener	Digwenar	Dydd Gwener
De Sadorn	Disadorn	Dydd Sadwrn

5 *Celtic languages* (Goidelic branch)

Irish (d)	Gaelic (d)
Dé Domhnaigh	De Domhnaich
Dé Luain	De Luain
Dé Mairt	De Mairt
Dé Céadaoin	De Ciadain
Dé Ardaoin	Diar Daoin
Dé hAoine	Di hAoine
Dé Sathairn	Di Sathirne

6 *Slavic languages* (Eastern branch)

Russian (m)	Ukranian (m)
Nedelya	Nedilya
Ponedelnik	Ponedilok
Vtornik	Vivtorok
Sreda*	Sereda*
Chetverg	Chetver
Pyátnitsa	Pyatnitza
Subbota	Soboto

7 *Slavic languages* (Western branch)

Czech (m)	Polish (m)	Slovak (m)	Bulgarian (m)
Nedele	Niedziela	Nedela	Nedelya
Pondelí	Poniedzialek	Pondelok	Ponedyelnik
Utery	Wtorek	Utorek	Vtornek
Strèda*	Sroda*	Sreda*	Sryeda
Ctvrtek	Czwartek	Stvrtok	Chetvurtuk
Patek	Piatek	Piatok	Petuk
Sobota	Sobota	Sobota	Subota

8 *Slavic languages* (Southern branch)

Slovene (m)	Serbo-Croatian (m)
Nedelja	Nedjelja
Ponedilok	Ponedjeljak

Torek	Utorak
Sreda*	Sreda*
Cetrtek	Cetrtak
Petek	Petak
Sobota	Subota

9 Baltic languages

Latvian (m)	Lithuanian (m)
Svetdiena	Sekmadienis
Pirmdiena	Pirmadienis
Otrdiena	Antradienis
Tresdiena	Treciadienis
Ceturtdiena	Ketvirtadienis
Piektdiena	Penktadienis
Sestdiena	Sestadienis

10 Indo-Iranian languages (Iranian branch)

Persian (s)	Kurdish (s)
Yek-shambah	Yeksem
Du-shambah	Dusem
Sih-shambah	Sesem
Charah-shambah	Carsem
Panj-shambah	Pencsam
Juma	Ini
Shambah	Semi

11 Indo-Iranian language (Indic branch)

Sanskrit (p)	Panjabi (p)	Hindustani (p)	Bengali (p)
Ravivara	Etwar	Itwar	Rabibar
Somavara	Somwar	Somwar	Shambar
Mangalavara	Mangalwar	Mangalwar	Mangalbar
Saumyavara	Budhwar	Budhwar	Budhbaer
Brhaspativara	Visiwar	Brahspatwar	Brihaspatibar
Sukravara	Shukarwar	Shukrawar	Shukrabar
Sanivara	Shanitur	Shaniwar	Shanibar

Urdu (p)	Singhalese (p)	Gurkhali (p)	Romany (s)
Erwar	Erida	Aitabar	Koóroki
Somwar	Saduda	Sombar	Yek dpK
Mungul	Anggahanuvada	Mangalbar	Doóï dpK
Boodh	Badada	Budhabar	Trin dpK
Jumerat	Brahasparingda	Bihibar	Stor dpK
Juma	Sikurada	Sukrabar	Píansh dpK
			(dpK = dívvus pálla Koóroki)
Sunneecher	Sinsrurada	Sanicharbar	?

12 Other European languages (Various groups)

Indo-European languages			Unclassified[§]
Greek (s)	Albanian (p)	Armenian (m)	Euskera (?)
(Hellenic)	—	—	—
Kiriaki	Diéli	Giragi	Igande
Deutera	Hânë	Y-ergushepti	Astelehen
Triti	Martë	Y-erekshapti	Astearte
Tetarti	Mercurë	Chorek-shapti	Asteazken
Pempti	Enjete	Tetárte	Ostegun
Paraskeni	Prêmtë	Pémpte	Ostiral
Sabbaton	Shtune	Shabat	Larunbat

[§] Euskera is the language of the Basque people. The names given here are those of Unified Basque, the result of an attempt to unify the most common dialects. The names of Monday, Tuesday, and Wednesday mean the 'ist day', the 'middle day', and the 'end' of the week. Ostgun (Thursday) means 'the sky's day'. The meanings of the others is uncertain. In the Vizcaino dialect, Tuesday is the planetary Martizen, Saturday is Zapatu (Spanish 'Sabata'), and Sunday is Domeka (Spanish 'Domingo').

13 Finno-Ugric

Finnish (p)	Lapp (p)	Estonian (m)	Hungarian (m)
(Finnic branch)			(Ugric branch)
Sunnuntai	Ailek	Pühhapäev	Vasárnap
Maanantai	Manodag	Esmaspäev	Hétfö
Tiistai	Tisdag	Teispäev	Kedd
Keskiviikko*	Kaska-wakko*	Kolmapäev	Szerda*
Torstai	Tuoresdag	Neljapäev	Csötörtök

Perjantai	Perjedag	Rede	Péntek
Lauantai	Lawodag	Taupäev	Szombat

14 *Turkic languages* (South-western, Oghuz branch)

Turkish (s)
Pazar Günü
Pazartesi
Sali Günü
Çarsamba
Persembe
Cuma
Cumartesi

15 *Sino-Tibetan languages*

Tibeto-Burman group		Sinitic group	Thai group
Burmese (p)	Tibetan (p)	Chinese (s)	Thai (p)
Tanangganve	Zaldawi	Xing gi lian	van athit
Tanangla	Zamigmar	Xing gi yi	van chan
Angar	Zallakpa	Xing gi er	van angkhan
Buddhahu	Zaphurbos	Xing gi san	van phut
Kyasapade	Zapasang	Xing gi shi	van prahat
Sokkya	Zaspemba	Xing gi wu	van suk
Chana	Zangeema	Xing gi liu	van sao

16 *Mon-Khmer language*

Khmer (p)
Thnai àtit
Thnai càn
Thnai ankār
Thnai put
Thnai prahas
Thnai sòk
Thnai sau

17 *Semitic languages* (Various branches)

Arabic (s) (North Arabic)	Hebrew (s) (Canaanite)	Somali (s) (Cushite)	Hausa (s) (Chadic)	Amharic (s) (Ethiopic)	Arabic (d) (Badi calendar)
Yom ilHadd	Yom rishon	Isnün	Lahadi	Ehud	Jamal
Yom litneen	Yom sheni	Axadda	Litinis	Saino	Kamal
Yom ittalat	Yom shelishi	Salaasu	Talata	Makhsaino	Fidal
Yom larba	Yom revii	Arbaco*	Laraba	Ros	Idal
Yom ilkhamiis	Yom hamishi	Khamüs	Alhamis	Amus	Istijal
Yom ilguma	Yom shishi	Jumu	Jumma	Arb	Istiqlal
Yom issabt	Sabbath	Sabti	Asaba	Kidami	Jalal

18 *Niger-Congo languages* (various groups)

Swahili	(Benue-Congo, Bantu branch)	(W. Sudanic, Kwa branch)
Jumapili	Ssande	Ojo aiku
Jumatatu	Olwebbalaza	Ojo aje
Jumanne	Olwokubiri	Ojo isegun
Jumatano	Olwokusatu	Ojo riru
Alhamisi	Olwokuna	Ojo asesedaiya
Ijumaa	Olwokutaano	Ojo eti
Jumamosi	Olwomukaaga	Ojo abamata

19 *Other languages*

Japanese (?) (Independent)	Javanese (Austronesian)	Esperanto (p) (Synthetic)
Nichi-yobi	Redite	Dimanco
Getsu-yobi	Soma	Lundo
Ka-yobi	Anggara	Mardo
Sui-yobi	Budda	Merkredo
Moku-yobi	Wrespati	Juado
Kin-yobi	Sukra	Vendredo
Do-yobi	Sanischara	Sabato

The Names of the Days in the French Republican Calendar

Note that the Quintidis are all named after an animal of some sort and the Decadis are named after farm implements. The rest are plants, vegetables, or minerals.

	VENDÉMIAIRE	BRUMAIRE	FRIMAIRE	NIVÔSE	PLUVIÔSE	VENTÔSE	GERMINAL	FLORÉAL	PRAIRIAL	MESSIDOR	THERMIDOR	FRUCTIDOR
1 Primedi	Raisin	Pomme	Raiponce	Tourbe	Lauréole	Tussilage	Primevère	Rose	Luzerne	Seigle	Epeautre	Prune
2 Duodi	Safran	Céleri	Turneps	Houille	Mousse	Cornouiller	Platane	Chêne	Hémérocale	Avoine	Bouillon blanc	Millet
3 Tridi	Châtaignes	Poire	Chicorée	Bitume	Fragon	Violer	Asperges	Fougère	Trèfle	Oignon	Melon	Lycoperde
4 Quartidi	Colchiques	Bettrave	Nèfle	Soufre	Perce-neige	Troëne	Tulipes	Aubépine	Angélique	Véronique	Irraie	Escourgen
5 Quintidi	Cheval	Oie	Cochon	Chien	Taureau	Bouc	Poule	Rossignol	Canard	Mulet	Bélier	Saumon
6 Sextidi	Balsamine	Héliotrope	Mâche	Lave	Laurier-thym	Asaret	Blette	Ancolie	Mélisse	Romarin	Prèle	Tubéreuse
7 Septidi	Carottes	Figue	Chou-fleur	Terre végétale	Amadouvier	Alaterne	Bouleau	Muguet	Fromental	Concombre	Armoise	Sucrion
8 Octidi	Amarante	Scorsonère	Miel	Fumier	Mézéréon	Violette	Jonquille	Champignon	Martagon	Echalottes	Carthame	Apocyn
9 Nonidi	Panais	Alisier	Genièvre	Salpêtre	Peuplier	Marceau	Aune	Hyacinthe	Serpolet	Absynthe	Mûres	Réglisse
10 Decadi	CUVE	CHARRUE	PIOCHE	FLÉA	COIGNÉE	BECHE	COUVOIR	RATEAU	FAUX	FAUCILLE	ARROSOIR	ECHELLE

11 **Primedi**	Pomme-de-Terre	Salsifis	Cire	Granit	Ellebore	Narcisse	Pervenche	Rhubarbe	Fraise	Coriandre	Panis	Pastrèque
12 **Duodi**	Immortelle	Macre	Raifort	Argile	Brocoli	Orme	Charme	Sainfoin	Bétoine	Artichaut	Salicot	Fenouil
13 **Tridi**	Potiron	Topinambour	Cèdre	Ardoise	Laurier	Fumeterre	Morille	Bâton-d'or	Poix	Giroflée	Apricot	Epine-vinette
14 **Quartidi**	Réséda	Endive	Sapin	Grès	Avelinier	Vélard	Hêtre	Chamérisier	Acacia	Lavande	Basilic	Noix
15 **Quintidi**	Ane	Dindon	Chevreuil	Lapin	Vache	Chèvre	Abeille	Ver-à-soie	Caille	Chamois	Brebis	Truite
16 **Sextidi**	Belle-de-nuit	Chervis	Ajonc	Silex	Buis	Epinards	Laitue	Consoude	Oeillet	Tabac	Guimauve	Citron
17 **Septidi**	Citrouille	Cresson	Cyprès	Marne	Lichen	Doronic	Mélèze	Pimprenelle	Sureau	Groseille	Lin	Cardière
18 **Octidi**	Sarrasin	Dentelaire	Lierre	Pierre à chaux	If	Mouron	Ciguë	Corbeille-d'or	Pavot	Gesse	Amande	Nerprun204
19 **Nonidi**	Tournesol	Grenade	Sabine	Marbre	Pulmonaire	Cerfeuil	Radis	Arroche	Tilleul	Cerise	Gentiane	Tagette
20 **Decadi**	PRESSOIR	HERSE	HOYAU	VAN	SERPETTE	CORDEAU	RUCHE	SARCLOIR	ROURCHE	PARC	ECLUSE	HOTTE
21 **Primedi**	Chanvre	Bacchante	Brable-sucre	Pierre à plâtre	Thlaspi	Mandragore	Génier	Statice	Barbeau	Menthe	Carline	Eglantier
22 **Duodi**	Pêche	Azerole	Bruyère	Sel	Thimelé	Persil	Romaine	Fritillaire	Camomille	Cumin	Caprier	Noisette
23 **Tridi**	Navet	Garance	Roseau	Fer	Chiendent	Cochléaria	Marronnier	Bourrache	Chèvrefeuille	Haricots	Lentille	Houblon
24 **Quartidi**	Amaryllis	Orange	Oseille	Cuivre	Traînasse	Pâquerette	Roquette	Valériane	Caille-lait	Orcanète	Aunée	Sorgho
25 **Quintidi**	Boeuf	Faisan	Grillon	Chat	Lièvre	Thon	Pigeon	Carpe	Tanche	Pintade	Loutre	Écrivisse
26 **Sextidi**	Aubergine	Pistache	Pignon	Etain	Guède	Pissenlit	Arémone	Fusain	Jasmin	Sauge	Myrte	Bigarade
27 **Septidi**	Piment	Maçjonc	Liège	Plomb	Noisetier	Silve	Lias	Civette	Verveine	Ail	Colza	Verge d'or
28 **Octidi**	Tomate	Coing	Truffe	Zinc	Cyclamen	Capillaire	Peasée	Buglose	Thym	Vesce	Lupin	Maïs
29 **Nonidi**	Orge	Cormier	Olive	Mercure	Chélidoine	Frêne	Mertille	Sénevé	Pivoine	Blé	Coton	Marron
30 **Decadi**	TONNEAU	ROULEAU	PELLE	CRIBLE	TRAINEAU	PLANTOIR	GREFFOIR	LOULETTE	CHARIOT	CHALÉMIE	MOULIN	PANIER

Glossary

Abundant month See *Full month*.

Algorithm Here we use the term to denote a numerical recipe for calculating some number of interest.

Almanac A table showing the notable events or other information appertaining to every day of the year.

Aphelion See *Apogee*.

Apogee Point in the orbit of a planet (or moon) when it is furthest from the sun (or planet) or at its aphelion.

Apparent solar day The period between two successive transits of the sun. Judged by a clock, it varies by about 50 seconds in the course of a year. See also *Mean solar day* and *Equation of Time*.

Aries A constellation of stars. See also *First point of Aries*.

Arithmetic calendar A calendar that is based on arithmetic rules rather than astronomical events.

Astronomical calendar A calendar which requires for its operation, predictions of astronomical events such as lunations or equinoxes (see also *Arithmetic calendar*).

Astronomical new moon This term generally means the conjunction—when the moon is invisible. See also *New moon* and *Conjunction*.

Atomic clock A clock in which the 'tick' is the vibration of a special sort of light emitted by various atoms. The second (q.v.) is now defined in terms of an atomic clock.

Bissextile year A leap year (q.v.) in the Roman calendar. So called because the intercalated day was placed before VI a.d. Kal. Mar. and called bis VI a.d. Kal. Feb. Nowadays used as a synonym for a leap year.

Bronze Age A period which saw the use of bronze: in Western Europe about 2200–800 BC.

Calendar 1. A systematic way of labelling the days of the year by arranging them in shorter units such as months or weeks. **2.** A table similar to an almanac (q.v.). **3.** A table showing the day of the week of every day of every month of the year. **4.** A picture of some popular theme adorning a table of the days of the week. **5.** A table of events sorted according to the year in which they took place or are scheduled to take place.

Calendar letter The letter (A to G) which is assigned to a day of the year. See also *Calendar number* and *Regular*.

Calendar number A number corresponding to the calendar letter (q.v.) of a day.

Calippic cycle A period of 76 years consisting of four modified Metonic cycles.

Celestial sphere The sphere which appears to envelop the earth and to which the stars are fixed.

Centennial The hundredth anniversary of some event.

Centurial A centurial year in some era is a year which is divisible by 100. It is sometimes called a secular year (q.v.).

Chalak/chalakim A unit of time equal to 10/3 seconds which is employed in the calculations of the Jewish calendar.

Cist A prehistoric hollowed-out stone or tree. Often used as a coffin.

Clock 1. A device which chimes the hours and maybe the quarters. **2.** A device for telling the time which is not carried on the person.

Common year A year that is not a leap year (q.v.).

Computer 1. A medieval person assigned the task of calculating the dates of Church festivals. Sometimes called a computist. **2.** An electronic device for performing calculations, such as the calculation of the dates of Church festivals.

Computist Another word for a medieval computer (q.v.).

Concurrent A number related to the dominical number or letter (q.v.) of a year and used in calculations of the day of the week of any date. If N is the dominical number, the concurrent is $1 + \text{MOD}(13 - N, 7)$.

Conjunction Two bodies are in conjunction if they share the same celestial longitude and are on the same side of the earth.

Constellation A group of stars on the celestial sphere. The constellations are given fanciful names to aid discussion of the part of the celestial sphere at which they are located. It would be naive to expect the constellations to resemble in any way the object after which they are named: the Great Bear looks nothing like a bear.

Co-ordinate A measurement which defines the distance of a point in one direction from another point (the origin). See also *Co-ordinate system*.

Co-ordinate system A system whereby the position of a point, relative to a set of directions, can be specified by a set of numbers, its co-ordinates (q.v.). Several types of co-ordinate system are used: Cartesian, polar, spherical, and others. To define the system, directions of measurement must be defined, and also a point from which the co-ordinates are measured, the origin, must be chosen.

One example is the grid reference of a place on a map of England (which specifies its Cartesian co-ordinates); another is the latitude and longitude of a point on the earth's surface (which specify its spherical co-ordinates). (In this second example, latitude is measured from the equator and longitude from the meridian of Greenwich, England.)

Co-ordinated universal time (UTC) A time-scale in which the second is equal to a second of international atomic time but in which leap seconds (q.v.) are introduced from time to time to keep it in step with universal time (q.v.). These are required because of irregular variations in the rotation of the earth. It is often called Greenwich Mean Time, and is the time given by broadcast signals.

Cromlech A circle of prehistoric menhirs (q.v.) or large stones. A cromlech at one time was synonymous with a dolmen (q.v.), but this usage is now archaic.

Culmination An object is said to culminate when it passes through your meridian in making an upper transit. The times between successive culminations of a star define the sidereal day (q.v.).

Cycle A sequence of days, months, or years which is repeated endlessly.

Deficient month See *Hollow month*.

Day The period which includes day-time and night-time; the period of light separating two nights. See also *Mean solar day* and *Apparent solar day*.

Day of March The number of a day of the year starting the count on 1 March.

Decade A period of ten days. Decades replaced the week in the French Revolutionary calendar.

Decan As referred to by the Egyptians, one of a series of stars which made their heliacal rising (q.v.) at different times of the year and whose rising could be used to estimate the time at night.

Declination The angular measure north or south of the celestial equator of a star.

Diurnal Happening daily. A diurnal rhythm has a period of one tropical day.

Dolmen A prehistoric structure consisting of flat stones laid horizontally on vertical supports; they were often used as graves or sepulchres. They were once also called cromlechs (q.v.).

Dominical letter The letter of the first Sunday of a year when the letters A to G are assigned to the first seven days of the year. The dominical letter is sometimes called the 'Sunday letter'.

Dominical number The number of the letter (A = 1, and so on) of the dominical letter.

Dionysian canon A method of calculating the date of Easter devised by Dionysius Exiguus.

Ecclesiastical date A method of specifying the day of a year by referring it to a preceding religious festival.

Eclipse An eclipse of the moon occurs when the earth is interposed between the moon and the sun. The earth's shadow falls on the moon. The sun and moon are in opposition at an eclipse of the moon which can only occur when the moon is full and near to crossing the ecliptic.

An eclipse of the sun occurs when the moon is between the sun and the earth so that the sun is wholly or partially obscured by the moon. The sun and moon are in conjunction at a solar eclipse and the moon is near to crossing the ecliptic.

Ecliptic The plane of the sun's orbit.

Embolismic year A year containing an intercalated month.

Empirical An empirical calendar relies on observations of the moon or sun to keep it in synchrony with them.

Epact The epact of a year is the age of the notional moon in days on the first day of the year, 1 January.

Epagomenal day One of the five (or six) days at the end of the Egyptian wandering year which are needed to 'pad out' the number of days to 365 (or 366). Also used with reference to other calendars based on the Egyptian calendar. See also *Sanculotide*.

Epoch The day which is the first day of the first month of the first year of a calendar.

Eponym A (person's) name used to name something else. For example, the name of the next king is often used to start and name a new era. An example of this is the regnal (q.v.) dating system of England.

Era A system of numbering years which is based on some notable event which occurs in the first year.

Equation of time The time difference which must be added to mean solar time to give apparent solar time. It varies between about −14 minutes and +16 minutes as the year progresses.

Equinox The times of year at which the night and day are of equal duration. The vernal equinox occurs in the springtime (northern hemisphere) at the moment the sun crosses the equator of the celestial sphere from south to north. The autumnal equinox occurs in the autumn.

Extracalation The omission of a day or a month from a calendar so as to bring the start of the next month or year forward. See also *Intercalation*.

First point of Aries The point on the ecliptic where it crosses the celestial equator at the vernal equinox. It moves slowly westwards relative to the stars on account of precession (q.v.) and more rapidly on account of nutation (q.v.).

Fixed star A luminary which is not a planet; it appears to be fixed on the celestial sphere and does not move relative to them.

More recently it has been shown that some nearby stars move relative to more distant ones over the years. This movement is called their proper motion (q.v.).

Full month A month (q.v.) with 30 or more days. See also *Hollow month*.

Full moon The instant in the cycle of phases of the moon when the sun and the moon are in opposition on opposite sides of the earth. The visible area of the moon is then greatest.

Gnomon A pillar or rod, the direction of whose shadow tells the time of day, as in a sundial.

Golden number The position of a year in the 19-year Metonic cycle.

Great circle A circle which passes through the two ends of a diameter of a sphere. The equator, the ecliptic, and any meridian are great circles on the celestial sphere.

Greenwich Mean Time Until 1928, Greenwich Mean Time was a time-scale based on the rotation of the earth and regulated by observations made at the prime meridian at longitude 0° at Greenwich, near London, England. The term is now sometimes used to refer to universal co-ordinated time (q.v.).

Gregorian Canon A method of calculating the date of Easter promulgated by Pope Gregory XIII.

Haab A 365-day period used in the Mayan calendar.

Heliacal The heliacal rising of a star is its first appearance in the Eastern sky just before sunrise. As the days progress, the star rises earlier and earlier before sunrise till eventually it is setting at dawn. It is then invisible for a period till it makes its next heliacal rising.

Hollow month A month (q.v.) with but 29 days. See also *Full month*.

Horizon The circle at which the plane on which you seem to be standing appears to meet the celestial sphere. This circle is perpendicular to the Zenith (q.v.).

Hyppolytic cycle A 16-year cycle used by the Roman Church in the computation of the date of Easter.

Incommensurate The year and the lunation are both incommensurate with the day in that neither contains a whole number of days. Likewise, the year is incommensurate with the lunation.

Intercalation The insertion of a day or a month in a calendar so as to delay the start of the next month or year. The intercalated day or month sometimes breaks the normal sequence of names or numbers which must then be renumbered. See also *Extracalation*.

International atomic time (TAI) An international time-scale defined in terms of an atomic clock (q.v.).

Jour complémentaire The name of the epagomenal days of the French Revolutionary calendar.

Julian day number The number of a day counting from 1 January 4713 BC.

Leap day A day intercalated in a leap year (q.v.).

Leap seconds One or more leap seconds are introduced into or deleted from co-ordinated universal time (q.v.) to keep it in step with international atomic time (q.v.).

Leap year Generally used with reference to the Christian or Gregorian calendar. In a leap year an extra day, 29 February, is intercalated. So called from the description in the 1604 edition of the Anglican prayer book: 'On every fourth year, the Sunday letter leapeth'. See also *Common year*.

Lunar almanac A table giving the dates of the lunations for every year of the Metonic cycle.

Lunar equation The lunar equation is the number of days by which the dates of the notional paschal moons must be decreased on account of the lunar adjustment alone.

Lunation The period from one phase of the moon to its next occurrence; the synodic period. This can vary by up to seven hours from one lunation to another.

Lunisolar A lunisolar calendar attempts to keep its months in synchrony with the moon and its years with the sun.

Mean solar day The period between two successive transits of the fictitious mean sun which moves along the celestial equator at a constant rate. See also *Equation of Time* and *Apparent solar day*.

Megalith A large stone placed by man.

Menhir An upright megalith (q.v.) placed in position by man rather than nature. Generally of neolithic or Bronze Age origin.

Meridian The great circle perpendicular to the equator and passing through the celestial poles and an observer. Observers on the same longitude share the same meridian. A star or planet makes a transit as it passes through the meridian.

The prime meridian passes through Greenwich, near London, England at longitude 0°. This is used as a reference meridian in the definition of Greenwich Mean Time and other time-scales.

Mesolithic The period, starting about 12 000 BC and ending with the neolithic period. Stone tools improved during this period.

Metonic cycle A cycle of 19 years in which seven lunar months are intercalated to partly reconcile the lunar and solar years.

Millenarianism A belief that the end of the world is nigh.

Millennium A 1000-year period. In Christian thought, such a period is expected after the Second Coming and will be a time of peace and joy.

Molod The moment of conjunction of the theoretical moon employed in the Jewish calendar.

Month A period of about 30 days which is a component of most calendars. Originally, it was based on a rough estimation of the synodic period of the moon. The word is believed to derive via the Latin 'mensis' from an Indo-European root meaning 'to measure'.

Nadir The point on the celestial sphere immediately below you. See also *Zenith*.

Nakshatra A term from Hindu astronomy which denotes a point on the ecliptic that marks the passage of the moon.

Neolithic The New Stone Age: about 4500–2200 BC in Western Europe; earlier in parts of Asia. It saw the introduction of agriculture and pottery.

New moon The crescent moon as it first appears after a conjunction. See also *Astronomical new moon* and *Conjunction*.

Nodes The nodes of the moon's orbit are the two points where it crosses the ecliptic.

Noon The moment that the sun culminates or passes through the meridian. The times between successive noons vary according to the equation of time.

Nundinae Market day in the Roman calendar. It occurred every eighth day.

Nutation An effect whereby the direction of the pole of the earth 'wobbles' slightly with a period of about 18.6 years. It is caused mainly by interactions with the moon.

Opposition Two bodies are in opposition if they are at opposite sides of the earth and their celestial longitudes differ by 180°.

Orrery A working model of the solar system complete with some of the planets and their moons. The first modern orrery was built in 1712, for John Boyle, fourth Earl of Orrery.

Palaeolithic The period of man's existence to about 12 000 BC, which was followed by the mesolithic period (q.v.). Unpolished stone tools were in use in the later palaeolithic period.

Paschal cycle A cycle of 532 years after which the sequence of dates of Easter repeats. Sometimes called a Victorian cycle (q.v.).

Paschal full moon The Paschal full moon is the first full moon that occurs on or after the vernal equinox which is taken notionally to fall on 21 March.

Paschal new moon The new moon which precedes the paschal full moon (q.v.).

Perigee Point in the orbit of a planet (or moon) when it is nearest to the sun (or planet) or at its perihelion (q.v.).

Perihelion See *Perigee*.

Period An interval of time.

Periodic An event or sequence of events which repeats indefinitely is said to be periodic.

Planet A luminary which moves relative to the celestial sphere. In ancient times the sun, moon (as well as Mercury, Venus, Mars, Jupiter, and Saturn) were called planets. Today, it is not usual to call the sun or moon, planets. Uranus, Neptune, Pluto, and the host of asteroids were unknown to the ancients.

Pole The axis about which the earth rotates relative to the stars or celestial sphere, or vice versa, according to your point of view.

Precession An effect whereby the points of intersection of the ecliptic and the celestial equator move slowly round the ecliptic.

Proleptic Most calendars have a starting date. A calendar date before the start of the calendar is a proleptic date.

Proper motion The slow movement, relative to the fixed stars (q.v.), of another nearby 'fixed' star.

Quarter-remainder year A year of $365\frac{1}{4}$ days, mainly used in the context of Chinese calendars.

Regnal dating A method of specifying the year of a date by reference to the year of the reign of the then current king (see *Regnal year*, *Eponym*).

Regnal year The number which indicates how many years a king has reigned.

Regular The calendar number of the first day of a month. It was used in calculations of the day of the week of any date.

Regular calendar A calendar with a rule for intercalating leap days at constant intervals.

Resonance Two cycles (of days or years) of different periods resonate—the starts of the two cycles coincide at intervals which repeat regularly with a period (in days or years) which is the lowest common multiple of the two component cycles. Moiré fringes are caused by a similar effect.

Right ascension The right ascension of a star is a measure of its position eastwards along the celestial equator from the first point of Aries.

Sanculotide The name given to the epagomenal days (q.v.) in the French Republican calendar. Note that modern French dictionaries spell it 'Sansculottide'.

Sankrânti A term from Hindu astronomy that denotes the instant that the sun enters a sign of the zodiac.

Saros A period of about 6585.32 days (a little over 18 years) after which the sun, moon, and nodes of the moon's orbit return to almost the same relative position. It contains 223 lunations. If there is an eclipse on a certain day, there will be another after the elapse of the saros—an observation known to the ancients.

Season A somewhat ill-defined period of the year marked by its meteorological characteristics or the sort of agricultural activity which must be conducted within it. In Europe and North America, the four seasons are punctuated by the two equinoxes and the two solstices.

Second Until the adoption of the SI system of units, the second was defined as the 1/86400 part of a mean solar day. When it was discovered, by comparison with an atomic clock, that it was not constant, the second was redefined as:

The duration of 9 192 631 770 periods of the radiation corresponding to the transition between the two hyperfine levels of the ground state of the caesium-133 atom.

Secular In astronomical parlance, a secular change is one that is continuous, that is, neither periodic nor irregular.

Secular year In the context of the calendar, a secular year is a centurial year, or one divisible by 100; for example, 100, 1900, 5000.

Serpet The Egyptian name for Soth (q.v.).

Siddhântas An Indian astronomical treatise.

Sidereal day The period between successive culminations (q.v.) of the same fixed star (q.v.).

Sidereal year The period in which the sun moves through a complete circle round the ecliptic relative to the stars. It is slightly longer than the tropical year because of the effect of precession (q.v.).

Signs of the zodiac A series of equally spaced points round the ecliptic which mark the passage of the sun in the course of a year. On account of the effect of precession (q.v.), their location relative to the stars is changing slowly.

Small circle A circle on the surface of a sphere; the two with different diameters. Lines of latitude are small circles on the earth, and the two tropics are small circles on the celestial sphere.

Solar cycle The period of 28 (4 × 7) years after which the sequence of dominical letters repeats. Also known as the 'Sunday letter cycle'.

Solar equation The solar equation (using the term 'equation' in a sense now archaic) is the number of days by which the dates of the notional paschal moons must be increased to account for the decrease in the average length of the year.

Solar longitude The angle measured from the vernal equinox of the sun in its journey round the ecliptic.

Solar month The period during which the sun passes between two equally spaced points on the ecliptic whose solar longitudes differ by 30°. Due to the ellipticity of the earth's orbit, the solar months are not all of exactly the same length.

Solstice The times at which the sun is furthest from the celestial equator. At the summer solstice the day is longest and the night is shortest; at the winter solstice, the day is shortest and the night longest. Derived from the Latin for 'standstill of the sun'.

Soth The Greek name for the star Sirius.

Sothic cycle A cycle of about 1460 years in which the heliacal rising of Sirius returns to the same day of the wandering year (q.v.).

Sunday letter See *Dominical letter*.

Sunday letter cycle See *Solar cycle*.

Synodic period of moon See *Lunation*.

Tithi A term used by Hindu astronomers and others to denote the time taken for the angular separation of sun and moon to change through 12°.

Tropical year The period between successive equinoxes. The effect of nutation is to make this vary from year to year by 20 minutes or so. Other effects lead to a gradual decrease in the long term. See also *Sidereal year*.

Tropics The points on the ecliptic which are most distant from the equator. The sun reaches the tropic of Cancer at the summer solstice and the tropic of Capricorn at the winter solstice.

Tzolkin A 260-day period used in the Mayan calendar.

Universal time (UT) A time-scale based on the rotation of the earth; until 1928 it was known as Greenwich Mean Time. There are several versions in which various corrections are applied. See 'Co-ordinated Universal Time (UTC)' and 'Leap second'.

Vague year See *Wandering year*.

Victorian cycle See *Paschal cycle*. It is named after Victorius, one-time Deacon of Aquitaine.

Wandering year A year that makes little attempt to remain in synchrony with the seasons. Usually applied to the Egyptian year. The Islamic year is a more extreme example. Known in Latin as 'annus vagus' which translates to 'wandering year'. Sometimes misleadingly called a 'vague year'.

Watch **1.** A device by which the time is observed by means of a dial with moving hands; that is, it is watched. See 'Clock'. **2.** A device for telling the time and small enough to be carried on the person. **3.** A division of time used on board ship and in other situations.

Week A seven-day period now in use throughout the world.

Year A period of about 365 days. **1.** A period based upon the time taken by the sun to perform a complete circle round the ecliptic—or equivalently, the earth to move once round its orbit. It is necessary to distinguish between the sidereal year (q.v.) and the tropical year (q.v.). **2.** A period of an integral number of days more or less in synchrony with the sun's motion round the ecliptic—the astronomical year.

Zenith The point on the celestial sphere above you. See also *Nadir*.

Zodiac A series of constellations close to the ecliptic through which the sun passes. Distinguish between the constellations of the zodiac and the signs of the zodiac (q.v.). The sign called Aries is now located in the constellation of Pisces.

Further Reading

T H E number of books and papers in journals devoted to aspects of the calendar is huge and encompasses such disparate fields as anthropology, archaeology, astronomy, classical literature, history, and mathematics—to mention the most obvious. No attempt can be made in this book to provide more than a representative selection of those likely to be of interest to the non-specialist and which can be found with relative ease.

One may distinguish several genres of works which discuss calendars in one way or another. Among these, represented here, are:

- Accounts by classical authors and historians
- Accounts by medieval computists
- Early treatises on chronology
- Treatises by nineteenth-century scholars
- Recent reports of research by archaeologists, historians, and astronomers
- Propaganda for calendar reform
- Mathematical works on calendar conversion and the date of Easter (including computer programs)
- Recent popular works

With few exceptions, I have restricted the selection to those written in English. To become really erudite in the subject, you would need to be able to read Greek, Latin, Chinese, and Sanskrit as well as most of the European languages and languages such as Sumerian, Egyptian, and the like; it would probably take you several lifetimes to acquire the necessary expertise.

You may also be interested in several of the publications of Tarquin Publications, Stradboke, Diss, Norfolk, England. This enterprising publisher produces a number of accounts of various aspects of astronomy, sundials, nocturnals, and so on, which include cut-out models which are suitable for older children (up to the age of 90 or more!). They will even sell you a plastic clockwork orrery (recommended).

Those wishing to further explore the subject on the internet might try: http://sal.cs.uiuc.edu/~nachum/calendars.html. This gives access to a large number of internet sites which have information on calendars. Of course, these are ephemeral and their content is of unknown accuracy.

A *Classical authors*

The following works by classical Roman authors are among the most important sources of our knowledge of the Roman calendar and others. Some are available in translation.

A1   Censorinus: *De die natali liber*

A2   Dio Cassius: *Roman history*

A3   Plutarch: *Lives*

A4   Macrobius: *The Saturnalia*

A5   Varro: *De lingua Latina*

B *Original sources in Latin and Arabic*

Several works stand out as important sources of our knowledge of the history of calendars and of chronology. These include:

B1   The venerable Bede: *De temporibus* and *De tempora ratione* in *The collected works of the venerable Bede* by J. A. Giles. Medieval Academy of America, Cambridge Mass., 1943.

Bede's master works on the calendar and Easter. I know of no translation into English.

B2   C. Clavius: *Explicatio Romani calendarii a Gregorio XIII P.M. restituti*, Rome 1603; reprinted in Opera, volume 5, Mainz 1612. It is available on microfilm.

Clavius' original explanation of the Gregorian reform in Latin. It contains the compendium of Lilio's proposal.

B3   al Biruni (1000): *Vestiges of the past*, translated as *The chronology of ancient nations* by E. Sachau. London, 1879.

An inexhaustible source of information on Oriental and medieval calendars and chronology by a medieval Arabic scholar.

B4   J. J. Scaliger (1583): *De emendatione temporum*. The first comprehensive work on chronology. There is a two-volume biography of Scaliger by Anthony Grafton.

C *Important works*

The nineteenth century showed a great blossoming of interest in the calendar; many scholarly books were written, particularly by German scholars. Many have not been translated into English. Among the most important are:

C1   L. Ideler (1825): *Handbuch der chronologie*. Berlin.

A classic work on chronology; some of it is now out of date.

C2   N. V. de St Allais and others (1814–44): *L'art de verifier les dates et les faites historiques*; 44 volumes in French. Paris.

A compendious but sometimes inaccurate compilation of chronological information.

C3   F. K. Ginzel (1906–14): *Handbuch der mathematischen und technischen chronologie*; 3 volumes in German. Leipzig. Reprinted 1960.

Another classic work which is the most comprehensive to be written. Although recent research has meant that some of it is now dated, it remains the most important modern treatise on calendars and chronology.

c4   M. P. Nilsson (1920): *Primitive time reckoning*. Oxford University Press.

A classic in the old style of ethnology describing time-keeping practices of numerous tribes and nations.

c5   J. J. Bond (1969): *Bond's handybook for verifying dates*. Bell and Dalby, London.

A classic with extensive tables for the conversion of dates and details of different eras, and such like. Widely known in the past, but now dated.

D *Encyclopaedias*

There are articles on the calendar and chronology in the major encyclopaedias. Particularly recommended are those in the *Encyclopaedia Britannica*.

D1   *Encyclopedia Britannica*, various editions.

The articles in the various editions have different emphases. The 11th edition (1910) has a good mathematical account.

D2   Ed. J. Hostings (1921): *The encyclopaedia of religion and ethics*, Clark, Edinburgh.

Authoritative articles on calendars from all over the world, many concentrating on their role in religious affairs.

D3   *Collier's encyclopedia* (1988).

A short article with a focus on calendar reform.

D4   Abraham Rees (1819): *The cyclopaedia or universal directory of arts, sciences, and literature* (with the assistance of eminent professional gentlemen). Longman, Hurst, Rees, Orme, and Brown, London.

Volume 5 of this 39-volume work contains useful articles on the calendar. Mainly of historic interest.

D5   *The dictionary of scientific biography*, volume 15, supplement 1. C. Scribner & Sons, New York, 1978.

This has several articles on various calendars in the 'Topical essays'.

D6   *The Encyclopedia of Science and Technology*. McGraw Hill, 1987.

The article on 'Time' defines various modern aspects of time.

E *Almanacs*

Various almanacs, including the *Nautical almanac*, contain useful general articles.

E1   *Whitaker's almanac*. J. Whitaker and Sons, published annually.

As well as a wealth of other information concerning the United Kingdom, Whitaker has a short but useful introduction to the various calendars of the world and on the measurement of time.

E2　J. K. Fotheringham (1935): 'The calendar' in *The nautical almanac and astronomical ephemeris*. HMSO, London.

An authoritative article by one of the first archeoastronomers.

E3　Anon. (1961): 'The calendar' in *The explanatory supplement to the astronomical ephemeris and the American ephemeris and nautical almanac*. HMSO, London.

This contains a useful article which lists the dates on which various countries adopted the Gregorian calendar.

E5　L. E. Doggett (1992): 'Calendars' in *Explanatory supplement to the astronomical ephemeris and the American ephemeris and nautical almanac*. University Science Books, Mill Valley CA, USA.

This contains a useful article and several algorithms.

F　*Semi-popular and general accounts*

There are a number of books, or books with significant chapters, devoted to the calendar; several were written by protagonists of calendar reform. Among those written in English and available in the larger libraries are:

F1　A. Philip (1921): *The calendar, its history, structure and improvement*. Cambridge University Press.

One of the few comprehensive books in English with an emphasis on calendar reform.

F2　P. W. Wilson (1937): *The romance of the calendar*. George Allen and Unwin, London.

A discursive with much out-of-the-ordinary information, some of which may be fanciful.

F3　E. Achelis (1943): *The calendar for everybody*. Putnam, New York.

A book by a prominent calendar reformer.

F4　R. D. Panth (1944): *Consider the calendar*. Columbia University, New York.

Another book by a reformer.

F5　S. A. Ionedes and M. L. Ionedes (1946): *One day telleth another*. Edward Arnold, London.

A history of astronomy with a long chapter on the calendar.

F6　H. Watkins (1954): *Time counts*. Neville and Shearman.

Has some interesting historical information.

F7　E. Achelis (1955): *Of time and the calendar*. Heritage, New York.

Another book by a reformer.

F8 B. Richmond (1956): *Time measurement and calendar construction*. E. J. Brill, Leiden.

Another book by a calendar reformer. It contains information difficult to find elsewhere.

F9 K. G. Irwin (1963): *The 365 days*. Thomas Y. Crowell & Co., New York.

A somewhat eccentric book.

F10 I. Adler and R. Adler (1967): *The calendar*. John Day, New York.

F11 W. M. O'Neil (1972): *Time and the calendar*. Manchester University Press.

Interesting and useful. O'Neil taught the history of astronomy.

F12 M. Bear (1989): *Days, months and years*. Tarquin Publications, Stradbroke, Norfolk, England.

A simple account of the calendar.

F13 M. Westheim (1993): *Calendars of the world*. Oneworld Publications, Rockport, Massachusetts.

A simple account emphasizing the role of calendars in religious life.

G *Mathematics of calendars*

There are many published methods for converting dates from one calendar to another, usually to the Julian or Gregorian calendars. Here is only a selection of modern methods.

G1 J. V. Uspensky and M. A. Heaslet (1939): *Elementary number theory*. McGraw-Hill, New York and London.

An introduction to the theory of numbers which discusses its application to calendrical problems including the calculation of the date of Easter.

G2 W. A. Schocken (1976): *The calculated confusion of calendars*. Vantage Press, New York.

A somewhat obscure discussion of several calendar conversions including the calculation of the date of Easter.

G3 O. L. Harvey (1983): *Calendar conversions by way of Julian day number*. American Philosophical Society, Philadelphia.

A description of a dozen or more calendars with methods for their interconversion.

G4 D. A. Hatcher (1984): Simple formulae for Julian day numbers and calendar dates. *Journal of the Royal Astronomical Society*, Volume 25, p. 53.

G5 D. A. Hatcher (1985): Generalized equations for Julian day numbers and calendar dates. *Journal of the Royal Astronomical Society*, Volume 26, p. 151.

The basis of the method for converting regular calendars described in Part III.

G6 J. P. Parisot (1986): Additif to the paper of D. A. Hatcher. *Journal of the Royal Astronomical Society*, Volume 27, p. 506.

G7   N. Dershowitz and E. M. Reingold (1997): *Calendrical calculations.* Cambridge University Press.

A comprehenive set of algorithms for calendar conversions written in LISP. It also contains historical notes.

Books on popular mathematics sometimes include chapters on the calendar. Among them are reference H13 and:

G8   M. Kraitchik (1949): *Mathematical recreations.* Allen and Unwin, London.

Kraitchik presents nomograms for obtaining the day of the week for Julian and Gregorian dates, but it is not clear if they always work.

The conversion of Jewish and Islamic dates is exhaustively covered in reference S31. There are a large number of books and articles which purport to give methods of converting Islamic dates to and from the Gregorian calendar. Most use fractional numbers and many are wrong, according to Burnaby.

H   *The calculation of the date of Easter*

I make no claim that this bibliography of methods of calculating the date of Easter even approaches being complete; it contains little that was not published in English.

Most modern methods for calculating the date of Easter, according to the Gregorian canon, derive from the works of Gauss and Delambre.

H1   C. F. Gauss (1800): in *Monatliche correspondenz vom freyherrn von zach*, volume ii, p. 129. See also his 'Berechnung des osterfestes' in *Werke*, volume 6, p. 79. Handschriftliche Bemerkung, Gottingham, (K.Ges.Wiss), 1874.

H2   A. Delambre (1817): *Connaissance des temps.* Paris.

The clearest and most comprehensive work in English is that of Butcher. It is the only work in English I have found which discusses Easter in sufficient detail as to be readily comprehensible.

H3   S. Butcher (1877): *The ecclesiastical calendar: its theory and construction.* Hodges, Foster, and Figgis, Macmillan, London.

A work of 269 pages; it is somewhat repetitive and has a few errors. It discusses both the Dionysian and Gregorian canon and contains proofs of the algorithms of Gauss, Delambre, and *Nature.*

See also by the same author: *General proof of Gauss' rule for finding Easter Day.* Hodges, Foster, and Figgis, Dublin.

H4   'A New York correspondant' (1876): To find Easter for ever. *Nature*, volume 13, p. 338, London.

A succinct and usable algorithm based on Delambre's method. The algorithm is quoted without explanation in the next four references.

H5   Sir Harold Spencer-Jones (1961): *General astronomy*, 4th edn. Edward Arnold.

H6   *Journal of the British Astronomical Association*, volume 88, p. 81, 1977.

H7   J. Meeus (1988): *Astronomical formulae for calculators*. Willmann-Bell, Richmond, Virginia, USA.

Meeus also gives an algorithm for calculating Easter according to the Dionysian canon.

H8   D. Willey (1985): *Mathematical Gazette*, **volume 69, no. 45**, p. 291, 1985.

The following have also presented algorithms:

H9   A. De Morgan (1845): *Companion to the almanac for 1845*. Society for the Diffusion of Useful Knowledge, London.

H10   W. S. B. Woolhouse (1910): 'Calendar' in *Encyclopaedia Britannica*, 11th edn. Cambridge University Press.

The algebraic treatment is not complete.

H11   M. J-M. Oudin (1940): Étude sur la date de Pâques (in French). *Bulletin Astronomique*, **second series, volume XII**, pp. 391–410.

Discusses errors in the formulae given by Gauss and Delambre. This paper was published after the German troops had entered Paris; this issue of the journal was printed on poor acidic paper; read it soon, it will not last much longer.

H12   B. D. Yallop (1964): *Algorithms for calculating the dates of Easter*. NAO technical note no. 64, H. M. Nautical Almanac Office.

This contains algorithms and reference to several other papers.

H13   T. H. O. Beirne (1965): *Puzzles and paradoxes*. Oxford University Press.

A book of popular mathematics containing several algorithms for Easter.

H14   D. E. Knuth (1969): *The art of computer programming*. Addison Wesley.

An adaption of Clavius' method.

H15   G. W. Walker (1945) in *Popular astronomy*, volume 53, pp. 162 and 218.

Walker investigates the frequency of occurrence of the various possible dates of Easter.

Whitrow in reference M4 also discusses the algorithms of Gauss and Delambe as does Schocken in reference G2.

I *The day of the week*

Many formulae for calculating the day of the week of any date have been published. Here is one by the author of *Alice in wonderland*:

I1   Lewis Carroll (1887): To find the day of the week for any given date. *Nature*, **March 31**, p. 517.

and another from *Nature*:

12   W. E. Johnson (1906): The day of the week for any date. *Nature*, volume 74, no. **1916**, p. 271.

A simpler algorithm for the day of the week by W. E. Johnson of King's College, Cambridge.

J *Conversion tables*

Many people prefer to convert dates with the aid of tables rather then using an arithmetical algorithm. For some calendars with an empirical element, this is the only way. This is a selection of some of the tables available.

J1   F. Parise (1982): *The book of calendars*. Facts on File, New York.

This contains brief descriptions of a large number of calendars and extensive tables for their conversion. It contains no references or sources of information.

J2   Hsüeh Chung San (1956): A Sino-western calendar for 2000 years: 1–2000 AD. Beijing.

J3   W. E. van Wijk (1938): *Decimal tables for the reduction of Hindu dates from the data of the Surya-Siddhanta*. Martinus Nijhoff, The Hague.

An excellent account of Hindu calendars, with worked examples of conversions which precede the table. Has a short bibliography.

J4   S. S. Gahlot (1979): *Historians' calendar (1544 AD to 1643 AD)*. Hindi Sahitya Mandir, Ratanada, Jodhpur.

J5   S. S. Gahlot and G. L. Devra (1980) *Indian calendars (AD 1444 to AD 1543)*. Hindi Sahitya Mandir, Sojati Gate, Jodhpur.

Reference S31 contains extensive tabulations for Jewish and Islamic dates, while references S9 and S10 contain tables for Hindu dates. The book by Bond (C5) also contains conversion tables. Reference N2 contains tabulations of the dates of Easter spanning several centuries.

K *History of number and arithmetic*

There are innumerable books on the history of counting and number systems; I list only a few which I have found useful.

K1   L. Hogben (1936): *Mathematics for the million*. Unwin, London.

An old popular classic but not to every taste.

K2   O. Neugebauer (1969): *The exact sciences in antiquity*. Dover.

A useful standard book on ancient numbering, mathematics, and astronomy.

K3   G. Flegg (1983): *Numbers, their history and meaning*. Andrew Deutsch, London.

A comprehensive history of number and arithmetic.

K4   O. A. W. Dilke (1987): *Reading the past: mathematics and measurement*. British Museum Press.

A brief history of numbering, mathematics, and measurement.

K5   F. J. Swetz (1987): *Capitalism and arithmetic*. Open Court.
     Despite the dull sounding title, this is a highly readable and recommended
     account of early Renaissance arithmetic. It includes a translation of a fifteenth-
     century textbook of arithmetic and many insights into the life of the times.

K6   D. Wells (1987): *The Penguin dictionary of curious and interesting numbers*.
     Penguin, London.
     A curious description of numbers with a strong bias towards their purely
     mathematical properties.

K7   B. Hayes (1995): Waiting for 01-01-00. *The American Scientist*, volume 83,
     p. 12.
     An almost exciting account of what might well happen on 1 January 2000 as
     computers try to come to terms with the start of the new millennium.

L   *Languages and the history of writing*

L1   A. Gaur (1992): *A history of writing*. Cross River Press, London.
     An interesting and readable account of the origin of writing.

L2   A. Robinson (1995): *The story of writing*. Thames and Hudson, London.
     A more popular account.

L3   K. Katzner (1995): *The languages of the world*. Routledge, London.

L4   M. D. Coe (1994): *Breaking the Maya Code*. Penguin, London.
     An entertaining account of the decipherment of Mayan writing and of the
     personalities involved. It also gives full translations of a few statements from
     Mayan stela.

M   *Time and its measurement*

M1   Ed. H. Shapley (1962): *Time and it mysteries*. Collier Books, New York.
     Eight lectures by eight authors on various aspects of time, including the
     calendar.

M2   A. J. Turner (1985): *Water-clocks, sand-glasses, fire-clocks*. The time Museum,
     Rockford, Illinois.
     Includes a history of these devices.

M3   J. T. Fraser (1987): *Time the familiar stranger*. University of Massachusetts
     Press, Amherst.
     A comprehensive discussion of time in all its aspects by an author well known
     in the field.

M4   G. J. Whitrow (1988): *Time in history*. Oxford University Press.
     Views of time and its measurement from prehistory to the present day, includ-
     ing the calendar.

M5   A. Aveni (1990): *Empires of time*. I. B. Tauris, London.

A discursive book on several aspects of the calendar and of time keeping in general.

M6   H. Tait (1990): *Clocks and watches*. British Museum Publications, London.

A history of mechanical clocks and watches.

M7   A. Borst (1993): *The ordering of time*; translated from the German by A. Winnard. Polity Press, Cambridge, England.

Discusses various aspects of time and calendars in Western Europe.

M8   G. Dohrn-van Rossum (1996): *History of the hour*; translated from the German by T. Dunlap. University of Chicago Press, London and Chicago.

A history of Clocks and time keeping customs.

N *Chronology*

There is a large literature devoted to chronology; here are just two examples. There are also articles on chronology in the major encyclopaedias. Neugebauer provides an important discussion in reference P3.

N1   E. J. Bickerman (1980): *The chronology of the ancient world*. Thames and Hudson, London.

A standard treatise on chronology with tables of archons and consuls and ancient kings.

N2   Ed. C. R. Cheney (1981): *Handbook of dates for students of English history*. Royal Historical Society.

A useful and popular book dealing with more recent, mainly English, chronology. It contains a useful bibliography.

O *Astronomy*

Accessible introductions to astronomy include the following, which I have found clear and interesting; there are many more I have not consulted.

O1   L. Hogben (1936): *Science for the citizen*. Allen and Unwin, London.

One of the great popularizations of astronomy and other sciences.

O2   Sir Harold Spencer Jones (1961): *General astronomy*, 4th edn. Edward Arnold, London.

Contains a clear and lucid account of the basic facts about the solar system.

O3   Ed. I. Ridpath (1987): *Norton's 2000.0 star atlas and reference handbook*. Longman.

A useful set of definitions and explanations of astronomical terms.

P *History of astronomy and archeoastronomy*

Several books were written earlier this century; in more recent years there has been a remarkable increase in interest in archeoastronomy, a field devoted to astronomy in ancient civilizations and preliterate societies.

P1   J. L. E. Dreyer (1953): *A history of astronomy from Thales to kepler.* Dover.

      A classical history of astronomy first published in 1906. More concerned with
      theoretical aspects of the science.

P2   O. Neugebauer (1969): *The exact sciences in antiquity.* Dover.

      A classical work on Babylonian, Egyptian, and Greek science and mathemat-
      ics, with special emphasis on archaeological discoveries.

P3   O. Neugebauer (1975): *A history of ancient mathematical astronomy.* Springer-
      Verlag.

      This three-volume work contains useful chapters on the calendar and on
      chronology. It is required reading for any serious student of ancient astronomy.

P4   E. C. Krupp (1983): *Echoes of the ancient skies.* Oxford University Press.

      A modern book on archeoastronomy.

P5   J. Needham (1985): *Science and civilisation in China.* Cambridge University
      Press.

      Volume 3 contains 300 pages on the history of astronomy in China and related
      matters.

P6   W. M. O'Neil (1986): *Early astronomy from Babylon to Copernicus.* Sidney Uni-
      versity Press.

      An interesting chapter on early astronomical instruments; useful bibliography.

P7   H. Thurston (1994): *Early astronomy.* Springer-Verlag.

      A warmly recommended account of ancient, and not so ancient, astronomy up
      to the time of Kepler.

Q   *Megalithic monuments*

From the vast number of books on Stonehenge and other megalithic structures, here
is a small, modern selection.

Q1   A. Thom (1967): *Megalithic sites in Britain.* Clarendon Press, Oxford.

      Professor Thom did much to establish the astronomical function of Megaliths.
      Here, he summarizes his ideas, including those on the megalithic calendar.

Q2   J. E. Wood (1978): *Sun, moon, and standing stones.* Oxford University Press.

      A readable and enthusiastic account of attempts to provide an astronom-
      ical interpretation of megalithic monuments. Contains a comprehensive
      bibliography.

Q3   C. L. N. Ruggles (1984): *Megalithic astronomy.* Oxford British Archeological
      Reports, British Series 123.

      A critique of correlations between megalith alignments and astronomical
      events.

Q4   R. J. C. Atkinson (1990): *Stonehenge.* Penguin Books, London.

A very readable account of Stonehenge by a world authority. Written before the works of Thom became respectable.

Q5 A. Burl (1995): *A guide to the stone circles of Britain, Ireland and Brittany*. Yale University Press, New Haven.

Encyclopaedic; useful if you want to visit the circles yourself.

Q6 C. Chippindale (1994): *Stonehenge complete*. Thames and Hudson, London.

All about Stonehenge and how it has been interpreted down the centuries.

Q7 R. M. J. Cleal, R. Montague, and K. E. Walker (1995): *Stonehenge in its landscape*. English Heritage.

A definitive work on the archaeology.

Q8 J. North (1996): *Stonehenge: neolithic man and the cosmos*. Harper Collins, London.

A comprehensive analysis of the astronomical significance of Stonehenge and other neolithic constructions. Not very much on neolithic calendars.

R *Calendar reform*

Besides the books noted in section F, which were written by people wishing to advertise their proposal for reforming the calendar, a number of works devoted to reform have been written.

R1 Auguste Comte (1849): *Calendrier positiviste* (in French). Paris.

Comte's lofty proposals in his own turgid words.

R2 A. Philip (1914): *The reform of the calendar*. Kegan Paul, London.

Some English proposals.

R3 *The Journal of Calendar Reform*, **1931–55**. The World Calendar Association, New York.

The issues of this journal carry many scholarly, and not so scholarly, articles on various calendars and their reform.

R4 R. Poole (1988): *Time's alteration. Calendar reform in early modern England*. UCL Press London, England and Bristol, Pennsylvania.

An excellent and detailed historical account of calendar reform in England.

S *Calendars of the world*

There now follows works devoted to specific calendars. Most of the major encyclopaedias and books listed in section A contain articles on these; *The Journal of Calendar Reform* also carried articles on various calendars.

PALEOLITHIC AND NEOLITHIC CALENDARS

The speculative neolithic calendar is described by Thom in reference Q1.

s1 A. Marshack (1972): *The Roots of civilisation*. Weidenfeld and Nicholson, London.

Marshack's account of how he came to discover his paleolithic lunar calendars.

s2 A. Marshack (1986): 'North American Indian calendar sticks: the evidence for a widely distributed tradition' in *World archaeoastronomy* (ed. A. F. Aveni). Cambridge University Press.

Marshack describes more recent lunar calendars and their use.

BABYLONIAN AND NEAR EASTERN CALENDARS

s3 R. A. Parker and W. A. Dubberstein (1956): *Babylonian chronology, 626 BC–AD 75*. Brown University Press, Providence.

A technical account of the evidence.

s4 Y. Meimarus (in collaboration with K. Kritykakou and P. Bougia) (1992): *Chronological systems in Romano-Byzantine Palestine and Arabia*. The National Hellenic Research Foundation, Athens.

A detailed account of eras and calendars in the area.

EGYPTIAN CALENDARS

s5 R. A. Parker (1950): *The calendars of ancient Egypt*. Chicago University Press.

A detailed account including hieroglyphic quotations. Perhaps not for the casual reader.

s6 R. A. Parker (1978): 'Egyptian astronomy, astrology and calendrical reckoning' in *The dictionary of scientific biography*, volume 15, supplement 1. C. Scribner & Sons, New York.

A more readable account covering much the same ground.

s7 M. Clagett (1995): *Ancient Egyptian science. Volume II: calendars, clocks and astronomy*. American Philosophical Society, Philadelphia.

s8 B. E. Tumanian (1974): Measurement of time in Ancient Armenia. *Journal for the History of Astronomy*, **volume 5**, p. 91.

A short account of Armenian calendars.

INDIAN CALENDARS

There is an excellent article in the 1911 edition of the *Encyclopaedia Britannica*, under 'Hindu chronology'. There are several virtually incomprehensible accounts of the Indian calendars available; among the more digestible are:

s9 R. Sewell and S. B. Dikshit (1911): *The Indian calendar*. George Allen, London.

This ably describes the various calendars and provides tables with an accuracy of two days or better for converting Indian dates from AD 300 to 1900.

s10 R. Sewell (1912): *Indian chronology*. George Allen, London.

This is an extension and companion to the earlier book.

S11   Calendar Reform Committee (1955): *Report of the Calendar Reform Committee.* Council for Scientific and Industrial Research, New Delhi.

An account of the calendars of India commissioned by the Indian government.

S12   D. C. Sircar (1965): *Indian epigraphy.* Motilal Banarsidass, Delhi.

This contains an exhaustive account of the history and use of a multitude of Indian eras.

S13   S. K. Chaterjee (1985): 'Indian calendars' in *History of Oriental astronomy* (ed. G. Swarup, A. K. Bag, and K. S. Shukla). Proceedings of an International Union Colloquium No. 91. Cambridge University Press.

A brief but useful account of the different calendars of India.

S14   R. Sewell (1989): *The siddhantas and the Indian calendar.* Asian Educational Services. New Delhi.

Reprints of articles which extended Sewell's earlier books.

CHINESE AND JAPANESE CALENDARS

Later editions of the *Encyclopaedia Britannica* contain useful articles under 'Calendar'.

S15   W. Bramson (1880): *Japanese chronological tables.* Tokyo.

Description of Japanese eras, and so on.

S16   J. Needham (1959): *Science and civilisation in China.* Cambridge University Press.

Volume 3 contains a lengthy article on Chinese astronomy including a discussion of long calendrical cycles.

S17   Chen Jiujin (1983): 'Chinese calendars' in *'Ancient China's technology and science'.* China Knowledge Series, Foreign Languages Press, Beijing.

An excellent account of the lunar calendar and its improvement over the centuries.

S18   S. Nakayama (1969): *A history of Japanese astronomy.* Cambridge, Massachusetts.

Technical, but contains chapters on the history of the Japanese calendar and its reforms.

ATHENIAN CALENDARS

S19   W. K. Pritchett and O. Neugebauer (1947): *The calendars of Athens.* Harvard University Press, Cambridge, Mass.

A full account including chronology.

S20   J. Mikalson: *The Athenian year.* Princeton University Press.

CELTIC CALENDARS

There have been several attempts (mostly in French) to reconstruct and interpret the Coligny calendar, including one that is somewhat fanciful. The two most recent are:

s21   P. M. Duval and G. Pinault (1986): *Recueil des inscriptions Gauloises* in XLVth supplement to Gallia. Paris.

s22   G. Olmsted (1992): *The Gaulish calendar*. Rudolf Habelt, Bonn.

      Olmsted employed a computer to decipher all the elements.

ICELANDIC CALENDARS

s23   T. Vilhjalmsson (1993): 'Time-reckoning in Iceland before literacy' in *Archeoastronomy* (ed. C. L. N. Ruggles). Group D Publications, Loughborough.

ROMAN CALENDARS

s24   A. K. Michels (1978): *The calendar of the Roman republic*. Princeton University Press.

      A detailed work on the Roman calendar before Caesar's reform.

THE GREGORIAN REFORM

The Vatican held a conference in 1982 to celebrate the fourth centenary of the reform by Pope Gregory XIII. The proceedings were later published (in English).

s25   R. S. Bates (1952): 'Give us back our fortnight'. *Sky and Telescope*, **volume 11**, p. 267.

      An interesting account of the English reform.

s26   N. Swerdlow (1974): 'The origin of the Gregorian civil calendar'. *Journal for the History of Astronomy*, **volume 5**, p. 48.

      A discussion of the length of the year used by Lilio and Clavius.

s27   *The Gregorian reform of the calendar. Proceedings of the Vatican conference to commemorate its 400th anniversary: 1582–1982* (ed. C. Coyne). Pontifica Academia Scientiarum, Specola Vaticana, 1983.

      This contains several articles on the lead up to the reform and its aftermath:

s28   G. Moyer (1982): 'The Gregorian calendar' *Scientific American*, **May 1982**.

      A short but useful article.

s29   J. Dutka (1988): 'On the Gregorian revision of the Julian calendar'. *The Mathematical Intelligencer*, **volume 10**, p. 56.

      Discusses the history of the reform and the lengths of the year available.

s30   H. Dagnall (1991): *Give us back our eleven days*. Privately published, London.

      A short account of the reform and the author's search for evidence for the riots of 1752. It also contains much interesting information about the English reform.

THE ISLAMIC CALENDAR

S31   S. B. Burnaby (1901): *The elements of the Jewish and Muhamadan calendars.* George Bell & Sons, London.

A detailed acount of the history and mathematics of the Islamic calendar.

THE BAHÁ'I CALENDAR

S32   J. E. Esslemont (1970): 'Bahá'i calendar, festivals and dates of historic significance' in volume 13 of *The Bahá'i World.* The Universal House of Justice, Haifa.

S33   J. E. Esslemont (1980): *Bahá'u'lláh and the new era.* National Spiritual Assembly of the Baha'i.

An account of the history and beliefs of the Bahá'i Church with a section on the calendar by a member of the Church.

THE JEWISH CALENDAR

Reference S28 also contains a detailed account of the Jewish calendar with special emphasis on its arithmetical aspects. Reference T2 also contains useful information on the 364-day year.

S34   *Encyclopaedia Judaica.*

The article on the calendar must be taken as authoritative.

S35   A. Spier (1952): *The comprehensive Hebrew calendar.* New York.

THE MAYAN AND AZTEC CALENDARS

S36   F. G. Lounsbury (1978): 'Mayan numeration, computation and calendrical astronomy' in *Dictionary of scientific biography*, volume 15, supplement 1. C. Scribner & Sons, New York.

An excellent and clear account of the Mayan calendar.

There is also a good account in reference L4 and full quotations of dates from Mayan stelae. The solution to the problem of relating the Mayan and Julian calendars is discussed in V12.

THE FRENCH REPUBLICAN CALENDAR

Most of the literature on the French Republican calendar is, naturally enough, in French. Accounts in the English language encyclopaedias are meagre. The best accounts I have found in English are:

S37   G. Wilson (1988): 'The French Republican calendar' in *Antiquarian Horology: the Proceedings of the Antiquarian Horological Society*, **no. 3, volume 17**.

This splendid article is vividly illustrated with contemporary paintings and facsimiles.

S38   J. H. Stewart (1951): *A documentary survey of the French Revolution.* Macmillan, New York.

Contains an English translation of the decrees establishing the new calendar.

The following works are often quoted but I have been unable to locate a copy of either in the United Kingdom:

s39   G. Kessen (1937): *Le calendrier de la Republique Francaise*. Paris.

s40   P. Couderc (1937): *Le calendrier*. Paris

Several articles were published in French in 1989 to mark the bicentenary of the French Revolution. Notable amongst these was:

s41   Bronislav Baczko (1989): 'La Révolution mesure son temps' in *La Révolution dans la mesure du temps: calendrier Republicaine heure decimale 1795–1806*' sous la direction de Catherine Cardinale. Musee International de fonds, Suisse.

       This is a shortened version of a longer study by the same author entitled *Le calendrier révolutionaire*.

T *The week*

Reference C4 contains information about 'weeks' from many lands, but I have only discovered two books and an article in a journal in English devoted to the seven-day week; they are

T1   F. H. Colson (1926): *The week*. Cambridge University Press.

       This presents several of the arguments deployed in Chapter 21 and gives many pertinent references to the classical literature.

T2   E. Zerubavel (1989): *The seven day circle*. University of Chicago Press, Chicago and London.

       A very useful and compendious treatment of the history and sociology of the week with chapters on calendar reform.

T3   S. Gandz (1948): 'The origin of the planetary week. *The Proceedings of the American Academy for Jewish Research*, **volume 18**, p. 215.

       This article suggests a Jewish origin of the astrological or planetary week.

T4   Dio Cassius, translated by E. Cary (1984): *Roman history, books XXXVI-XL*. Loeb Classical Library, Heinemann, London.

       This contains Dio's classic description of the astrological week.

U *Easter*

In reference H3, besides describing the mathematics of the calculation of the date of Easter, Butcher frequently digresses and enlarges on historical matters.

u1   O. Pederson (1983): 'The ecclesiastical calendar and the life of the Church' in *Gregorian reform of the calendar*. Pontifica Academia Scientiarum, Vatican.

       An excellent history of Easter in a volume produced by the Vatican to celebrate the 400th anniversary of Gregory XIII's reform.

u2   Eusebius, translated by G. A. Williamson (1989): *The history of the Church*. Penguin, London.

This early history by Eusebius (260–339) is a classic and one of the prime sources of our knowledge of early Christianity.

V *Archaeology and ancient history*

Take your pick from the immense number of works on ancient history but the following general books are useful.

VI   Ed. G. Barraclough and N. Stone (1992): *The Times atlas of world history*, 3rd edn. Times Books, London.

> The cartographical approach is good for showing the ebb and flow of empires and cultures.

V2   J. M. Roberts (1991): *The Penguin history of the world*. Penguin Books, London.

> A comprehensive view starting at the end of the last Ice Age.

V3   B. M. Fagan (1993): *World prehistory*. Harper Collins, London.

> A synoptic and critical view of world prehistory.

Histories of some of the local regions which I have found useful include:

V4   E. M. Forster (1982): *Alexandria, a history and a guide*. Michael Haag.

> An short but excellent history of Alexandria.

V5   J. Reade (1991): *Mesopotamia*. British Museum Press.

> A brief account of the area.

V6   A. K. Basham (1954): *The wisdom that was India*. London.

> A popular account of Indian culture; it includes a short appendix on the calendars.

V7   R. Tharpar and P. Spear (1965): *The Penguin history of India*. Penguin, London.

V8   Yong Yap and A. Cotterell (1975): *The early civilisation of China*. Weidenfeld and Nicholson, London.

> A readable introduction to Chinese history with illustrations.

V9   S. Piggott (1975): *The Druids*. Thames and Hudson, London.

> A critical review of the evidence.

V10   S. Gruzinski (1987): *The Aztecs: rise and fall of an empire*. Thames and Hudson, London.

> An illustrated account of the Aztec Empire.

V11   F. Delanay (1989): *The Celts*. Grafton, London.

> A readable but somewhat romantic account of Celtic culture and history.

V12   M. D. Coe (1993): *The Maya*, 5th edn. Thames and Hudson, London.

> An excellent introduction to the Maya, with many illustrations. Discusses the calendar.

V13   N. Grimal, translated from the French by I. Shaw (1994): *A history of ancient Egypt*. Blackwell, Oxford.

Index

abacus 80
Abel, Leonardo 244
aberration 22
Abri Blanchard 131, 133
ab urbe condita 205
Achelis, Elisabeth 118
Actium, battle of 156, 218
agriculture, start of 5
d'Ailly, Pierre 240
Akkadians 147
Alaska 265
Alexander the Great 38, 95, 147, 222
d'Alexandre of Villediew 95
Alexandrian calendar 157, 159, 260, 290, 311, 321
al-Fazani 79
Alfonsine tables 38, 240, 250
Alfonso the Wise 38
algorithm 79, 290 i
al-Khwarizma 79, 290
Almagest 32, 38
al-Ma'mun 70, 147
Almanac of Shepherds 261
almanac maker, Chinese 162
altitude 23
Anastas 159
Anatolius 349
Andronicus of Kyrrhos 55
anniversaries 90
Anixamander of Miletus 76
Anno Domini 106, 217, 351
Anno Mundi 331
annus bissextile 100
antarctic circle 31
Antikythera 61
Anthony 156, 218
aphelion 25
apocalypse 10
Apollonius of Tyana 273
Archons of Athens 13, 106
Arctic circle 31, 204
Aries, first point of 23, 25, 28, 32
Aristarchus of Samos 33, 38, 106
Aristophanes 197
arithmetic 80,
Armelin, Gustav 117
Armenians 159
Armenian eras
    great 159
    little 159

armillary 32
Aryabhata 172, 182
Assyrians 147
astrolabe 50
astrological week 268–71
astronomy
    Arabic 33
    Babylonian 26, 28, 30, 146
    Chinese 31, 33, 41, 163
    European 249
    Greek 150
    history of 37–41
    Indian 36, 173
Athenian calendar 197
Athanasius, Saint 349
Ataturk, Kemel 235
Aubrey, John 136
Augustalis 349
Augustus 111–12, 156, 215, 218
azimuth 23
Aztec calendar 186

Bab 236–7
Babur 235
Babylon 147
Babylonian
    astonomy 26, 28, 30, 146
    calendar 147
    names of months 148, 222
    number system 146
    seasons 147
    writing 146
Bacon, Roger 239
Badi calendar 235–6
Baghdad 36, 67, 70, 147
Bahá'í calendar 15, 100, 234–6, 290, 311, 321
    start of year 237
Ballochroy 140
Barnard's star 22
Baron Samedi 281
Bartholomew, Saint 247
Bede, The venerable 203, 218, 239, 351
Beijing 169
Beltane 143, 202
Benedict, Saint 379
Bernoulli, Jacob 11
Berry, duc de 380
Bible, Moffat's translation 225
Biblical references

Exodus 267
Deuteronomy 267
Genesis 226
Isaiah 48
Matthew 218
Jewish calendar 221
Revelations 11
billion 73
bissextile year 214
black death 240
Black plan 119
blue moon 179
blue stones 138
Book of Common Prayer 299, 354, 374, 377
book of hours 379
Boyle, John, 4th Earl of Orrerey 61
Bradley, Dr 110, 253
Brahe, Tycho 33, 39, 41, 249
Bristol, riots in 255
bronze age year 143
Buddha 173
Buddhism 6
Buddhist eras 170
Burgess, Anthony 16
Burghley, Lord 252
Burnaby, S. B. 297
Bush Barrow lozenge 144
Butcher, S. 377
Byblos 67
Byzantium 219

Caesar, Julius 70, 81, 97, 111, 198, 208, 212, 215
Cai Lun 67
calculi 80, 182
calendar conversions 287–91
calendar, history of 129
calendar priests 5, 89, 90
calendars
    Alexandrian 290, 311, 321
    Aztec 186
    Babylonian 94
    Bahá'í calendar 15, 100, 234–6, 290, 311, 321
    Celtic 198–9
    Chinese 37, 96, 100–1, 165–70
    Coligny 199–201
    Coptic 100
    Egyptian 97, 100, 150–7, 290, 311, 321
    Eternal 159
    Ethiopian 100
    Fasli 159
    Gregorian 6, 98
    Icelandic
    Indian 37, 96, 100, 110–11, 119
    Islamic 93, 99, 290, 311, 321
    Jelali 235
    Jewish 96
    Julian 97–8, 100, 102, 212–17, 290, 311, 321
    Macedonian 290, 311, 321

Marti 235
Mayan 37, 100–1
Neolithic 141
Ottoman, financial 235
Persian 290, 311, 321
Positivist 113–14, 116
Republican (French) 100, 159, 257, 290, 311, 321
Roman 15, 97, 206–19, 290, 311, 321
Saka 119, 133, 184, 235, 290, 311, 322
Soor San 159
Taichu 165
Teutonic 203
Yazgererd 157, 235
calendars, classification
    arithmetic 90, 100, 287
    astronomical 90, 100, 287
    calculated 99
    empirical 99
    lunar 91, 99–100
    lunisolar 94, 98–9
    regular 287, 289, 310–25
calendar letter 301
calendar number 301
calendar round 86, 188, 192, 335, 339–40
calendar, uses of 6
Calendrier des Heros 112
calends, see Kalends
Calippic cycle 33, 96, 106, 355
Calippus of Cyzicus 96, 198
Cambodia 170
Candlemas 143, 202
Clency 112
Canopus 70
Canopus, decree of 111, 156–7
Canterbury, Archbishop of 252–3
carbon 14 dating 14, 134
Carroll, Lewis 299, 378
Cassius, Dion 269
celestial equator 18
celestial latitude 24
celestial longitude 24
Celtic calendar 198–9
    names of months 201
    seasons 201
Celtic
    church 351
    culture 135
    festivals 143, 202
    language 199
Censorinus 154, 203
Chacon, Pedro 244–5
chalakim 326
Champollion, Jean-Francois 67, 69
character of Jewish year 228, 332
Chester 256
Chesterfield, Lord 253–4
chestidnevki 278–9
China, history of 161

Chinese calendar 37, 96, 100–1, 165–70
Christian year, start of 218
Christianity 6
chronocracies 268
chronology 12
Chuquet, Nicholas 73
cicumpolar stars 18
cities, first 5
Citizen Decadi 276
civilization 37, 64, 127–8
Clavius, Christopher 38, 99, 244, 246, 249, 356,
    361, 366
Clavius' adjustment 371
Clements, William 58
Cleopatra 70, 156, 212, 215, 218
Cleostratus of Tenedos 84, 94
clocks
    accuracy of 59
    astronomical 60
    atomic 62
    caesium 63
    dials 60
    escapement 58
    first 44
    incense 52, 55
    mechanical 58
    mercury 52
    pendulum 58
    regulation 58
    shadow 47–8
    source of power 58
    water 51–2, 54–5
coins, dates on 6–7, 205
Coligny 199
Coligny calendar 199–201
colour days of Soviet week 278
Columbe, Jean 380
Committee of Public Instruction 257, 261,
    263
Communards 264
computational calendar 310
    start of 315
Comte, Auguste 113, 115
concurrents 306
Confucius 162
Congress of Chambers of Commerce 118
conguence relations 296, 302
conjunction 35
conjunction, grand 170, 183
Constantine 101, 219
Constantinople 219, 273
constellations 20, 22
consuls, Roman 13
continued fraction 337
Copernicus, Nicholas 17, 39, 249
Copts 158–9
Coptic Church 6, 158
Cotsworth, Moses Bruin 118, 122

Council of
    Antioch 348
    Arles 347
    Basel 240
    Chaldecon 158
    Constance 240
    Constantinople 348
    Laodicaea 348
    Lateran, 5th 241
    Trent 241
    Nicaea 102, 250, 347, 349, 351–2, 356, 365–6
Counting 72–80
    backwards, Roman 82
    binary 73
    Chinese 162
    decimal 72
    duodecimal 74
    finger counting 72
    Hawaian 75
    inclusive 81, 208
    Mayan 85
    multiplicative principle 73
    sexagesimal 77
Creation, date of 109, 224, 226, 326
Croze, Abb 117
crucifixion 345
Crusoe, Robinson 3, 266, 277
Ctesibus of Alexandria 52
cuckoos 30
culmination 20, 50
current time 81
cycles
    Calippic 33, 96, 106, 355
    Chinese 101–2, 169
    contrived 101
    Dionysian 88
    Hyppolytic 349
    Julian 102
    Mayan 102, 338
    Metonic cycle 85, 87, 95, 148, 165, 167, 223, 297,
        327, 348–9, 354–60
    Paschal 88, 350
    sexagesimal 163
    Sothic 156
    Sunday letter 303
    Victorian 218, 351
cycles, table of 105

Danti, Ignazio 241
dates, ways of writing 16, 104
days
    divisions of 44
    epagomenal 47, 153, 156, 188
    grouping of 7
    Julian 43, 288
    length of 43
    of March 361–3
    mean solar 25, 43, 387

number 287–8
period of time 7, 24, 43, 90–1
    siderial 43, 387–8
    solar 24
    start of day 43, 91, 288
    twenty-four hour 44, 46
daylight saving 62
day of week, calculation of 299–309
Dead sea scrolls 87, 230
decadis 276
decadic festivals 276
decans 46–7
Decemviri 207
decimal
    counting 72, 80
    hours 260
    watch 260
declination 23
Dee, Dr 252
Delambre, Jean Baptist Joseph 97, 263, 368
de Morgan 377
dendrochronology 14
diagonal star calendar 46
Diaspora 89, 221–2
Diego de Landa 67
dies fasti 208
Digges, Thomas 252
Dioceltian 159, 208, 351
Dionysian canon 351, 354
Dionysian era 217
Dionysius of Alexandria 348
Dionysius Exiguus 106, 210, 217, 350, 355, 359
Dioscurus 158
divine officium 379
dominical number 302
dominical letter 302
Dondi, Jacob 60
Dresden codex 186
Druids 135, 199
Davall, Peter 253
Dunstable Priory clock 58

Eanfleda 351
earthly branches 163–4
Easter, date of 119, 239, 249, 309, 345–53, 354–76
    rule for 349, 354
Easter act 352
Eastern Orthodox Church 247, 251
Eastman, George 118
Egyptian calendar 97, 100, 150–7, 290, 311, 321
    civil 153
    first lunar 152
    names of months 154
    reform 156
    second lunar 155
Einstein, Albert 41
eclipses 35
ecliptic 27, 35

obliquity of 27
elapsed time 81
Elizabeth I 252
English
    calendar reform 252–6, 352
    financial year 253
Enlightenment, the 112
Enoch, patriarch 226
Enoch book of 117, 229
epact 361–3, 370, 373
epagomenal days 47, 153, 156, 188
Epiphany, feast of 254
epoch 109, 288
eponymous dating 104
equation of time 25
equinox 27–8
    dates of 240
    mean 32
    position of 28
    precession of 29
    vernal 28
eras
    Christian 106
    Diocletian 159, 208
    Martyrs 159, 217
    Republican 106
    table of 107–8
Eratosthenes of Alexandria 38
Eternal calendar 159
Ethiopian church 6
Ethiopian calendar 100
Euctemon of Athens 95
Eumenes II 66
Exodus 151
extracalation 89, 179–80
extrahebdomadal days 116–17, 119

Fabrew d'Eglantine 110, 257, 259, 261, 282
Fasli calendar 159
Fawkes, Guy 202
feast
    Epiphany 254
    Terminalia 207
    St Gabriel of the Seven Dolours 101
    St Matthias 101
Feria Octava 116
Fête de Revolution 100
Fibonacci 79
Flammarion, Camille 117
Fogg, Phileus 62
fortnightly periods (Chinese) 166
Franciade 259
Franklin, Benjamin 62
French Republican calendar 159
Futhark 203

Galileo 41, 48, 58, 249
Gauss, Carl Frederick 296, 377

gematria 77
Genghiz-Khan 71
Gigli, Tommaso 243–4
Glastonbury thorn 254
gnomon, early use of 41
gnomon, length of shadow 24, 31, 167–8, 241
golden number 95, 102, 354–6
graffiti at Pompeii 271
Greaves, John 253
Gregorian calendar 6, 239–56
   intercalation 317
   reform 112, 247–52, 352–3
Gregorian canon 352, 354, 365
Gregory XIII 38, 97, 112, 120, 352, 365
Greek city states 196
Greenwich meantime 103
Greenwich, London 18, 61
Grosclaude, Professor 117–18
Grosseteste, Robert 239
Grundy, Solomon 284
Guo Shoujung 40, 167

haab 87, 188, 336, 338
Halafta, Jose ben 224, 226
Halloween 143
Hammurabi 148
Hanin, Emil 117
Harappa 172
Harriot, Thomas 376
Harrison, John 59
Harun el-Raschid 71
Hatcher, D. A. 310
heavenly stems 163–4
Hegira 233
Helgason, Oddi 204
heliacal rising 8, 26, 31, 46, 152
Henry VIII 241
Herod the Great 218
Herschel, Sir John 97, 251
Hesiod 7, 196
Hillel II 223
Hooke, Robert 11
Hipparchus of Nicaea 24, 29, 33, 38, 44, 82, 96, 106
historical tables
   Babylon 145
   China 161
   early history of man 127
   Egyptian 150–1
   India 172
   Jews 220
   Mesoamerica 187
   Rome 206
history of calendars 129
Hogarth, William 255
Homer 196
Hopi village 144
horae 379
horizon 18

Hoshanah Rabba 227
hour
   equinoctal 44
   temporal 44
   canonical 44
Hulaka 71
hunter gatherers 5
Husayn, Ali 236
Huygens, Christian 56
Hyppolytes 349, 374

ides 210, 323
Ignatius, patriarch 244, 250
Imbolg 143, 201
Incas 31, 140
inclusive counting 81, 208
incommensurateness of days, lunations and years
   8, 92
index number 369
Indian
   calendars 172–85
   calendrical cycles 183
   divisions of time 45, 174
   eras 184
   lunisolar calendar 177
   seasons 176
   solar calendar 174
   week 174
indiction 85, 101
intercalation 89, 93–4 100, 179, 207, 214, 232,
   297
intercalation cycles 289, 310, 318
integer arithmetic 291
integer division 292
Inter Gravissimus 245
interlocking cycles 102, 188
International Almanak Reform League 118
International Calendar Organisation 119
International Fixed Calendar League 118
International date line 62
Iranaeus 347
Isadore of Seville 351
Islamic calendar 231–5, 290, 311, 321
   names of months 232
Icelandic calendar 87, 204–5

Jain religion 6, 107
Januarius, Saint 255
Japan 170
Java 275
Jehovah's witnesses 10
Jelali calendar 235
Jerusalem, temple of 90, 222, 225
Jesuits 109
Jesus of Nazareth 218
   Birth of 14, 218
   crucifixion 345
   resurrection 345–6

Jewish calendar 89, 221–30, 290, 311, 321, 326–34
    names of months 222
Jewish
    captivity 147, 221, 267
    exodus 267
    history 220
    patriarchs 225
    people 6, 9
    temple 90, 222, 225
    time divisions 45, 223
    year 224, 228
Jimmu Tenno 170
John the Deacon 159
John of Wallingford 282
Josephus, Flavius 218, 345
Journal of Calendar Reform 118, 122
Jours complémentaires 258
Jours supplémentaires 263
Joyce, James 16
Joysprick 16
Jubilees, Book of 229
Julian calendar 212–17, 290, 311, 321
Jupiter 37, 84, 183, 188, 283
Justinian 70

kabisah 232
kalends 210, 323
Kali-yuga 104
Karthikadi year 178
Kassites 147
Kepler, Johannes 41, 249
keviah 228
Khayyam, Omar 234
Kiddamu 38
king lists 13, 148, 154
Kintraw 140
Koran 234

Lady Day 253
Lagrange, Louis 257
Lammas 143, 202
Laplace, Pierre Simon 264
last supper 345
Lauro, Giovanni 243
Lauri, Vincenzo 244
Lao Tse 162
Laos 170
League of Nations 122
leap year 100
Leonardo of Pisa 79
Lepidus, Marcos Aemilius 215
Les très riches heures 380
Liber Abaci 79
libraries
    Alexandria 38, 70, 158
    Baghdad 70
    Chinese 67, 70, 162, 165
Lilio, Aluise Baldasdsar 99, 243–4, 249–50, 361
Lilio, Antonio 243–4

long count 103, 188, 192, 335, 341
Lords of the Night 85, 193
Lughnasa 143
lunar almanac 356–7, 359, 369
lunar equation 366–7
lunar Zodiac 177
lunar tallies 131–3
lunation
    distribution in lunar almanac 359
    length of 91, 93, 388, 388
    notional 352
    start of 91
    time period 43, 91

Maccu Picu 40
Macedonian calendar 290, 311, 321
Macrobios 207, 214
Magellan 62
Mahivira 172, 184
Maimonedes 226
Manetho 154
Man Friday 265
March Hare 378
Marco Polo 165
Maréchal, Pierre Sylvan 112, 257–9
market weeks 275
market rings 275
Marshak, Alexander 130
Marti calendar 235
Martian calendar 122
Martinmas 202
Massa compoti 95
Massoretic text of Bible 225
Mastrofino, Marco 116
mathematical conventions 292
Matthias, Saint 101
Maya 186–7
    chronology 187
    counting 187
Mayan calendar 186–95, 335–42
    calendar round 86, 188, 192, 335, 339–40
    cycles 338
    day groups 192
    great cycle 193
    haab 336, 338
    long count 103, 188, 192, 335, 341
    trecena 336
    tzolkin 336
    uinal 336, 338
    vientena 336
Mecca 231
Medina 233
megaliths 8, 134
Melbourne war memorial 100, 142
Memphis 31
Mencius 162
Mercedonius 207
meridian 18, 20
Merlin de Douai 276

Mersenne, Gabriel 58
Meton 95, 148, 180, 198
Metonic cycle 84–5, 87, 95, 148, 165, 167, 223, 297, 327, 348–9, 354–60
Mill, John Stuart 113
millenarianism 10
millenium 9
millenium bug 10, 16
Millerites 10
Minerva, Temple of 95
minute 44
Mithraism 272–3
MOD function 292–4
Moffat's translation of Bible 225
Moghuls 235
Mohengo-Daro 172
molod 273
molod of Tishri 326
Monge, Gaspard 257
monophysism 158
Monterrey, rock carving 135
month
    anomalistic 388
    draconic 388
    mean synodic period 35, 388
    sidereal 358
    tropical 388
months, names of 15
    Arabic 154
    Armenian 154
    Babylonian 148, 222
    Celtic 201
    Egyptian 154
    Ethiopian 154
    Indian 178
    Islamic 232
    Jewish 222
    Persian 154
    Positivist 114
    Republican 261–2
    Roman 211–12, 323
months, days in 14, 212, 294–5, see also months, names of moon 7, 34
    blue 179
    mean synodic period 34–5
    moon rises 35
    nodes 35
    notional 348, 352, 354
    paschal 349
    phases of 34
    precession of nodes 35
    standstills 36
Moslems 6, 9, 157
MS DOS 16
Muhammad (the prophet) 231, 233, 236
Muhammad, Ali 236

Nabonassar 148
nadir 18

nakshatras 85, 177
Napoléon 262, 264, 277
Nasa'a 231
National Convention 258
Nativity 14, 218
Nature (journal) 119, 121–2, 309, 377
Nebuchanezzer 104, 221, 267
Needham, Joseph 70
Nehru 173
Nengo 170
neolithic calendar 141
nepreryvka 277
Newcomb, Simon 99
New Grange 140
Newton, Isaac 11, 41
Nicholas of Cusa 240
Nicobar Islands 132
Nile 8, 26, 152
nones 210, 323
north polar distance 23
Nostradamus 12
Numa Pompilus 207
numbers
    decimal 80
    Gobar 79
    Milesian 380
    sexagesimal 77
    zero 77–8
    see also written numbers, counting Attic 75
nundinae 208, 274–5

octaeteris 84, 94–5, 198
Ogam 199
O'Kelly 140
Olympiads 83, 106
Olympic era 198
Olympic games 95, 198, 247, 277
opposition 35
oracle bones 163
Oresme, Nicholas 62
Oswy of Northumbria 351
Ottoman financial calendar 235
Ovid 207

Pahlavi, Shah 184, 235
palaeolithic moon 130
panchangas 181
parallax 22
Parsees 157
paschal cycle 350
paschal moon 349, 360–1
paschal table 361–2
passover 345–8, 366
Paul of Middleberg 241
Paul, Saint 345–6
Peoples Republic of China 169
Pergamon 67
perihelion 25
perpetual calendar 116

Philip, Alexander 118–19
Philippines 265
Phillipus, Marcos 50
planets 36–7, 268
    and gods 272
    Jupiter 37, 84, 183, 188, 283
    orbital periods 37, 269
    Venus 87, 188
Pilate, Pontius 345
Pitatus, Petrus 249
Plato 197
Platonic solids 84
playing cards 83
Pliny 199
Plutarch 207, 269
Polycarp of Smyrna 346
Polycrates of Ephasus 347
Pompeii 271
Pompey 70, 147, 212
pontifices 207
Pontifex Maximus 112, 207, 215
popes 239–41
    Anicetus 346
    Gregory XIII 38, 97, 112, 120
    Hilary 350
    John I 350
    Victor 346
Positivist calendar 113–14, 116
    names of months 114
postponements 226–8, 233
Pozzuoli 255
Prayer, Book of Common 299, 354, 374
Prescelly mountains 138
precession of equinoxes 28–9, 163,
precession of nodes 35
primes 82–3
Priscus, Tarquin 207
proleptic dates 108, 233, 288
proper motion 22
Protestants 2, 247–8, 252, 352
Prutenic tables 39, 250
Prytany 198
Ptolemy, Claudius 17, 32, 38, 41, 43, 83, 148, 155
Ptolemy III, Eurgetes 111, 156
Ptolemy Sotor 70, 149–50
Pythagorean triangles 135
Pythagorus 136
Pythian 198

quarter days 202
quarter-remainder year 165
Quartodecimans 346–8
Quintadecimans 346–7
quipu 74–5

Rayleigh, Lord 109
reform of calendar 110–22
refraction 22

Regiomontanus 111, 241
regnal years 104
regulars 306
Reinhold, Erasmus 39
religion and the calendar 6
Republican calendar 15, 100, 159, 257–64, 290, 311,
    321
    names of days 398–9
    names of months 261–2
    start of year 258, 263
    sextile years 259, 263
Resurrection of Christ 345–6
Revelations, book of 11
revolutionary week 276–9
Ricci, Matteo 41
Richard I 226
Richmond, Broughton 119
right ascension 23
riots at Bristol 255
rites of passage 9
Robertson, John 119
Roman calendar 15, 97, 206–19, 290, 311, 321
    names of months 211–12
Robespierre 276
Roman Emperors 216
Roman Republic 206
Romulus 207
Romme, Gilbert 97, 113, 257–8, 276, 283, 321
Rowley, John 61
Rosetta stone 67–8
Rosh Hoshanah 224, 227–8
Russian revolution 277
Russian Olympic team 247

Sabbath 268, 272–3
Sacrobosco 239, 244
Saka calendar 119, 133, 184, 235, 290, 311, 321
Saladin 226
Salisbury Cathedral clock 58
saltus lunae 355
Samhain 143, 202
sanculotide 258, 260
Sanhedrin 110, 223
sankranti 175
Sanskrit 174
saros 36, 389
Sarsen stones 138
Sassenids 157
Sevile, Henry 252
Scaliger, Joseph Justus 12, 102
Scaliger, Julius Caesar 103
Searle, G. N. 119
seasons
    Babylonian 147
    Celtic 201
    Egyptian 83, 154
    Indian 176
    Les très riches heures 382

Sebokht, Severus 79
seconds 82–3
secular changes 25
Seleucid Empire 147, 149, 221
Seleucis Nicator 147, 149, 220
Serapis, temple of 70
sexagesimal cycle 163
sextile year 259, 263
siddhantas 174, 176
Shang dynasty 161, 163
Shih Huang-ti 70, 169
Siculus, Diodorus 136
sidereal day 24, 25
sidereal year 166–7, 389
Sileto, Guglielmo 244
Sirius 8, 31, 46 152, see also Sopdet
solar equation 365–7
Stevinus, Simon 80
Solon of Athens 94
solstices, determination of 8, 29
Soor San calendar 159
Sopdet 26, 46, 152
Sothis 26, 152, see also Sopdet
Sothic year 152
Sothic cycle 156
Sosigenes 51, 97, 212–15, 239
Soviet Union 99, 113, 159, 251
Star Oddi 204
stars
    Orion 46
    Polaris 19, 29
    Pole 19, 29
    Thubon 29
    Vega 29
Stonehenge 8, 31, 136–7
station stones 139
Stelae 193–4
Strabo 198
Stukely, William 136
Sumerians 147
Sun Yat Sen 169
sun, mean 25
Sunday letter 302
Sunday letter cycle 303
sundials 48–50
Su Sung 56, 58
Surt, Thorstein 204
Swerdlow, Noel 249
Synod of Whitby 351
Syntaxis, see Almagest

Tacitus 345
Taichu calendar 165
tally sticks 130, 132
Taoism 162
Terminaluia, feast of 207
Teutons 203
Teutonic calendar 203

Thailand 170
Thales of Miletus 76
Theodosius II 350
Theophilus 70, 350
thirds 82–3
Thom, Alexander 141
Thorgilsson, Ari Frode 204
time 42
    atomic 63
    current 81, 311
    elapsed 81, 311
    Greenwich meantime 63
    local time 61
    second, atomic 387
    units 45
    universal (UT) 63
time zones 61
tithi 36, 96–7, 181
Titus 222
Tower of the Winds 55–6, 241
transit 20, 24
trecana 191
Treviso arithmetic 80
trillion 73
tropical year 166, 168
tropics 28
Twelfth Night 254
tzolkin 87, 190, 191, 336

uinal 188, 336, 338
Ulugh Beg 41
Umar I 232–3
Uppsala 80
Ussher, Archbishop 11, 225

vague year 153
Varro 208
Vasselier, Joseph 112
Vedas 173
Veintana 190–1, 336
vernal equinox 28, 348, 352
Verne, Jules 62
Victorius 350
Victorian cycle 218
Vikings 204

Waku system 275
Wallis, John 253
Walsingham, Sir Francis 252
wandering year 153, 188
week 101, 265–84
    Assyrian origin 266
    astrological 271–3
    Babylonian 267
    end of 268, 274, 281
    interruption 265
    Islamic 233
    Jewish 266

market 265, 275
names of days of 266, 279, 391–7
planetary 210
revolutionary 276–9
seven day 149, 221
Wilson H. 253
Winkle, Rip van 3
Winnebago calendar stick 134
Wise men of the East 14
Woolhouse, W. S. B. 377
World Calendar Association 118
writing materials 6–7, 64, 66
Writing 65–6
written numbers 75–7, 82, *see also* numbers

Xing Yunho 168

Yazgererd III 157
Yazgererd calendar 157, 235
year 8, 30, 43
  anomalistic 390
  embolismic 100, 359
  length of 20, 32–3, 93, 168

lunar 390
lunisolar 98
mean solar 34, 390
measurement of length 30–1, 167
sidereal 31, 389
solar 98
start of 92, 128, 167
tropical 31, 389
Year of confusion 215
yoga 181
Yom Kippur 223
Yu Hsi 163
Yuletide 203

Zellor's congruence 302
zenith 18
zenith tube 31
Zhou dynasty 103, 165, 168
zodiac 26, 35–6
zodiac, signs of 26–7, 165, 172, 175
Zoroaster 157
Zoroastrians 6, 159

Printed in the USA/Agawam, MA
July 5, 2022

795208.001